# MULTIVARIATE
# STATISTICAL
# METHODS

## A First Course

# MULTIVARIATE STATISTICAL METHODS

## *A First Course*

**George A. Marcoulides**
*California State University, Fullerton*

**Scott L. Hershberger**
*University of Kansas*

**LEA** LAWRENCE ERLBAUM ASSOCIATES, PUBLISHERS
1997  Mahwah, New Jersey

Lawrence Erlbaum Associates, Inc., Publishers
10 Industrial Avenue
Mahwah, New Jersey 07430

Cover design by Kathryn Houghtaling

**Library of Congress Cataloging-in-Publication Data**
Marcoulides, George A.
   Multivariate statistical methods : a first course / George A.
Marcoulides, Scott L. Hershberger.
      p.    cm.
   Includes bibliographical references and index.
   ISBN 0-8058-2571-1 (c) — ISBN 0-8058-2572-X (p).
   1. Social sciences—Statistical methods.  2. SAS (Computer file).
3. Multivariate analysis—Data processing.  I. Hershberger, Scott
L.  II. Title.
HA29.M261233  1997
300'.1'5195—dc21                                    96-46287
                                                      CIP

Books published by Lawrence Erlbaum Associates are printed on acid-free paper,
and their bindings are chosen for strength and durability.

Printed in the United States of America
10  9  8  7  6  5  4

# Contents

# Preface

The purpose of this book is to introduce multivariate statistical methods to advanced undergraduate and graduate business students, although students of other disciplines will also find the book useful. The material presented is suitable for a one-semester introductory course and provides coverage of what we believe to be the most commonly used multivariate techniques. The book is not intended to be particularly comprehensive; rather, the intention is to keep mathematical details to a minimum while conveying the basic principles of multivariate statistical methods.

Like most academic authors, our choice of the material in the book and mode of presentation are a joint product of our research and our teaching. One way to articulate the rationale for the mode of presentation is to draw a distinction between mathematical statisticians who gave birth to the field of multivariate statistics, like Hotelling or Wilks, and those who focus on methods for data analysis and the interpretation of results. Possibly the distinction between Pythagoreans (mathematicians) and Archimedeans (scientists) is useful, as long as one does not assume that Pythagoreans are not interested in data analysis and Archimedeans are not interested in contributing to the mathematical foundations of their discipline. We certainly feel more comfortable as Archimedeans, although we occasionally indulge in Pythagorean thinking. Therefore, this book is primarily written for individuals concerned with data analysis, although true expertise requires familiarity with both approaches.

It is assumed that readers already have a working knowledge of introductory statistics, particularly tests of significance using the normal, $t$, $F$, and chi-square distributions, single and multiple factor analysis of variance, and simple and multiple linear regression. Basically, the material covered in a standard first semester statistics course is adequate, along with some ability in basic algebra. However, because coverage in most first semester statistics courses is rarely uniform, each of the elementary topics is introduced in relation to a multivariate statistical method. Although understanding and applying multivariate statistical methods requires some use of matrix algebra, the amount needed is relatively small as long as one is prepared to accept some of the mathematical details on faith.

One of the main reasons that multivariate statistical methods are gaining popularity these days is due to the availability of statistical packages to perform the laborious calculations. As such, access to statistical packages will be a necessity for using the multivariate statistical methods presented in this book. The most popular statistical packages currently on the market are: Statistical Analysis System (SAS), the Statistical Package for the Social Sciences (SPSS), and the Biomedical Computer Programs: P-Series (BMDP). Originally, the three packages had very distinctive users. SAS was perceived as most closely tied to statistics and was heavily used for agricultural and economic data analysis. SPSS was for social scientists, while BDMP was for biomedical applications. Nowadays, these three packages do not have distinct users, in part because of their similar capabilities and flavor. In fact, at times output from these packages is so similar that only seasoned users are able to distinguish between them. For these reasons, the three different statistical packages will not be stressed in this book. Only the SAS package will be used in this book.

It is obvious that statistical packages have all but eliminated the long hours spent in laborious hand computation. And, once a particular package is learned, it is easy to solve other related problems using the same conventions of the package's language or even other packages. However, the student of multivariate statistical methods must understand that there is a difference between knowledge of the syntax of a package to generate output and actual knowledge of a statistical method. It is extremely easy to generate results that are meaningless, especially because most packages can create truly enormous volumes of thoroughly confusing output by means of a few simple commands. With this in mind, the book tries to tie knowledge of the statistical package to underlying statistical theory.

Our book provides a basic theoretical grounding in the mathematics of multivariate statistical methods and teaches students to translate mathematics into and from the language of computer packages. Throughout the book, the computer is used as an adjunct to the presentation of a multivariate statistical method in an empirically oriented approach. Basically, the model adopted in this book is to first present the theory of a multivariate statistical method along with the basic mathematical computations necessary for the analysis of data. Subsequently, a real-world problem is discussed and an example data set is provided for analysis. Throughout the presentation and discussion of a method, many references are made to the computer, output is explained, and exercises and examples with real data are included.

It is our intention that each chapter in this book can be read independently. The first two chapters can be considered prerequisite reading. Chapter one sets the agenda for the rest of the book. Chapter two introduces the basic concepts of matrix algebra. Readers who are familiar with matrix algebra may skim or omit this chapter. In chapter three, we present the multivariate normal distribution as an extension of the univariate normal distribution. We also discuss the multivariate counterparts of the familiar univariate $t$ tests (Hotelling's $T$ test) and the $F$ test for the significance of overall differences among group means (e.g., Wilk's test). In chapter four, we discuss the multivariate analysis of variance (MANOVA) for factorial designs and extend the topics presented in the previous chapter. Chapter five is devoted to the topic of discriminant analysis and the process of creating linear combinations of a set of variables that best differentiate among several groups. In chapter six, the topic of canonical correlation analysis is discussed with emphasis on how to determine linear combinations of sets of variables that correlate most highly with each other. Chapter seven is devoted to presenting the topics of principal component analysis and factor analysis and showing how principal components fit into the broader context of exploratory factor analysis. Finally, chapter eight presents the confirmatory factor analytic model and its extensions to structural equation models (such as LISREL and EQS) are also discussed.

This book could not have been completed without the help and encouragement provided by many people. First and foremost, we would like to thank our former professors and mentors who provided both the seeds of knowledge and words of encouragement. We would also like to thank numerous colleagues and students who provided comments on initial drafts of the various chapters. We are particularly indebted to Professor Ronald H. Heck of the University of Hawaii for reviewing the book and providing valuable comments and insights, which substantially improved the book.

Thanks are also due to all the wonderful people at Lawrence Erlbaum Associates for their assistance and support in completing this book. Special thanks to Larry Erlbaum, Judith Amsel, and Joe Petrowski, who over the years have patiently worked with us on various projects. We are also thankful to Kathleen Dolan and Kathryn Scornavacca for their help during the various stages of the publication process. Finally, we would like to thank our families for their love and inspiration needed to complete such a project.

—*George A. Marcoulides*
—*Scott L. Hershberger*

# CHAPTER ONE

# Introduction

*Multivariate statistics* refers to an assortment of statistical methods that have been developed to handle situations in which multiple variables or measures are involved. Any analysis of more than two variables or measures can loosely be considered a multivariate statistical analysis. For example, researchers (e.g., Marcoulides & Heck, 1993) are interested in learning how organizational culture can play a key role in determining levels of organizational performance. A common hypothesis about this role suggests that if an organization possesses "strong culture" by exhibiting a well-integrated and effective set of specific values, beliefs, and behavior patterns, then it will perform at a higher level of productivity. Multivariate statistics are required to study the multiple relationships among these variables adequately and obtain a more complete understanding for decision making.

This admittedly rather loose introduction to the subject matter of this book is intentionally a very broad one. It would include, for example, multiple regression analysis, which involves the study of the relationship between one dependent variable and several independent variables. But these topics are almost always covered in an upper division first-semester introductory statistics course. And yet, multiple regression analysis will serve as an excellent introduction to the "true" multivariate technique of canonical correlation commonly used to examine relationships between several dependent and independent variables.

Pointing out the analogy between a common univariate statistical technique and the corresponding multivariate method is one of the principal didactic strategies used throughout this book. Besides the drawing of analogies, actual data analyses are presented using the computer. The book is not intended to be particularly comprehensive; rather, the intention is to keep mathematical details to a minimum while conveying the basic principles of multivariate statistical methods.

One way to articulate the rationale for the mode of presentation is to draw a distinction between mathematical statisticians, like Hotelling or Wilks, who gave birth to the field of multivariate statistics, and those who focus on methods for data analysis and the interpretation of results. Possibly the distinction between Pythagoreans (mathematicians) and Archimedeans (scientists) is useful, as long as one does not assume that Pythagoreans are not interested in data analysis and Archimedeans are not interested in contributing to the mathematical foundations of their discipline. Therefore, this book is primarily written for individuals concerned with data analysis, although true expertise requires familiarity with both approaches. But rigorous mathematical proofs and derivatives are eliminated. Only chapter 2 is devoted to providing the necessary background in matrix algebra for understanding the multivariate techniques. The reader who is interested in pursuing a more mathematical approach to multivariate statistics should consult such books as Morrison (1991) and Johnson and Wichern (1988).

## THE IMPACT OF COMPUTERS

One of the main reasons that multivariate statistical methods have gained popularity is due to the availability of statistical packages to perform the laborious calculations. The most popular statistical packages on the market are the Statistical Analysis System (SAS Institute, Inc., 1979, 1989a, 1989b), Statistical Packages for the Social Sciences (SPSS, Inc., 1990), and Biomedical Computer Programs: P Series[1] (Dixon, 1990a, 1990b). These three packages are available for mainframe computers and microcomputers (basically IBM-compatible and Macintosh computers). The microcomputer versions tend to have a more "user-friendly" environment feel in terms of setup, but the outputs match those of the mainframe versions of the programs.

Originally the three packages had very distinctive users. SAS was perceived as most closely tied to statistics, and was heavily used for agricultural and economic data analysis. SPSS was for social scientists, while BMDP was for biomedical applications. Nowadays, these three packages do not have distinct users, in part because of their similar capabilities and flavor. In fact, at times output from these packages is so similar that only seasoned users are able to distinguish between them. For these reasons the three different statistical packages are not stressed in this book. Only the SAS package is presented.

---

[1]BMDP Statistical Software, Inc. was recently purchased by SPSS, Inc.

## A SUMMARY OF SAS

The Statistical Analysis System (SAS) is a computer package that was originally developed at North Carolina State University for use in statistical research (SAS Institute, Inc., 1979). It has evolved over the years into a widely used and extremely flexible package. The syntax of SAS is relatively easy to learn. The information does not have to begin in any particular column but each statement in SAS begins with a one-word name (that tells SAS what to do), and ends with a semicolon (;). Any SAS program consists of at least two steps: (a) a DATA step for data and variable input, and (b) a PROC step to conduct a given analysis.

Consider the following age, height, and weight measurements obtained for five people:

| Age (years) | Weight (pounds) | Height (inches) |
|---|---|---|
| 30 | 130 | 60 |
| 25 | 185 | 70 |
| 20 | 190 | 65 |
| 32 | 140 | 75 |
| 22 | 160 | 62 |

The following lines would be used to input these data into SAS:

```
DATA MEASURES;
INPUT   AGE  1-2  WEIGHT  4-6  HEIGHT  8-9;
CARDS;
30  130  60
25  185  70
20  190  65
32  140  75
22  160  62
;
```

The DATA command provides the name for the data set that follows. The name chosen for this data set was MEASURES. The INPUT command names the variables to be studied and the columns in which the data are located (a free-format statement can also be used in which no column location is specified). The variable names used in this example represent the variables included in the data set. Any variable name can be used as long as it begins with a letter and does not exceed eight characters. If the response provided for a variable is alphabetic (e.g., the name of an employee, an employee's gender), a dollar sign ($) must follow that particular variable name. For example, if the data set MEASURES contained information on each person's gender, then a variable called GENDER $ could be used.

The CARDS command indicates to SAS that the data lines come next. It is important to note that a CARDS statement is the simplest way to tell the SAS system about the location of the data. This approach requires the data lines to follow immediately after the CARDS statement. Another way to tell the SAS system about data is to use the INFILE statement. For example, the following lines could be used with the INFILE statement:

```
DATA MEASURES;
INFILE 'MEASURES.DAT';
INPUT  AGE 1-2 WEIGHT 4-6 HEIGHT 8-9;
```

The INFILE statement indicates the file name from which the data are to be read, and the INPUT statement directs the SAS system to "retrieve the data from file MEASURES.DAT" according to the specified variables.

Once an SAS data set has been created, all types of statistical analyses can be performed using the PROC statement. PROC refers to the procedures to be performed with the data. Basically, SAS PROCs are computer programs that read a data set, perform various manipulations, and print the results of the analysis. For example, the PRINT procedure is generally used to read a data set, arrange the data values in easy-to-read form, and then display them for review. Table 1.1 shows an SAS session that consists of a DATA step and a PROC PRINT step using the MEASURES data set.

Another commonly used procedure in SAS is the PROC UNIVARIATE. This procedure is used very often because it provides easy-to-use variable values in tabular form (a common frequency table). Table 1.2 shows some output obtained by using the PROC UNIVARIATE for all the variables in the MEASURES data set.

The PROCedure step to perform a statistical analysis in SAS always begins with a PROC statement giving the name of the procedure to be run. Although only two simple procedures have been presented here, it is important to note that the SAS system contains hundreds of procedures that can be used for data analysis. Once the basic syntax of the PROC statement is understood

TABLE 1.1
Example SAS Output Using PROC PRINT

| | | SAS | |
|-----|-----|--------|--------|
| OBS | AGE | WEIGHT | HEIGHT |
| 1 | 30 | 130 | 60 |
| 2 | 25 | 185 | 70 |
| 3 | 20 | 190 | 65 |
| 4 | 32 | 140 | 75 |
| 5 | 22 | 160 | 62 |

TABLE 1.2
Example SAS Output Using PROC UNIVARIATE

SAS

| AGE | Frequency | Percent | Cumulative Frequency | Cumulative Percent |
|-----|-----------|---------|---------------------|--------------------|
| 20 | 1 | 20.0 | 1 | 20.0 |
| 22 | 1 | 20.0 | 2 | 40.0 |
| 25 | 1 | 20.0 | 3 | 60.0 |
| 30 | 1 | 20.0 | 4 | 80.0 |
| 32 | 1 | 20.0 | 5 | 100.0 |

| WEIGHT | Frequency | Percent | Cumulative Frequency | Cumulative Percent |
|--------|-----------|---------|---------------------|--------------------|
| 130 | 1 | 20.0 | 1 | 20.0 |
| 140 | 1 | 20.0 | 2 | 40.0 |
| 160 | 1 | 20.0 | 3 | 60.0 |
| 185 | 1 | 20.0 | 4 | 80.0 |
| 190 | 1 | 20.0 | 5 | 100.0 |

| HEIGHT | Frequency | Percent | Cumulative Frequency | Cumulative Percent |
|--------|-----------|---------|---------------------|--------------------|
| 60 | 1 | 20.0 | 1 | 20.0 |
| 62 | 1 | 20.0 | 2 | 40.0 |
| 65 | 1 | 20.0 | 3 | 60.0 |
| 70 | 1 | 20.0 | 4 | 80.0 |
| 75 | 1 | 20.0 | 5 | 100.0 |

one can easily use the other available procedures in the SAS system (for details see *SAS User's Guide*). Finding an SAS procedure that performs a statistical analysis and learning the statements to use a procedure is fairly straightforward. The difficulty is knowing which statistical analysis to use on a data set.

## THE SAS PACKAGE AND MULTIVARIATE ANALYSIS

It is extremely easy to generate results that are meaningless, especially because SAS can create truly enormous volumes of thoroughly confusing output by means of a few commands. There is clearly a need, therefore, to tie knowledge of a statistical package to underlying statistical theory. Each chapter in this book attempts to present both the basics of the statistical procedure and the necessary computer analysis. Throughout the book, the computer is used as an adjunct to the presentation of the various multivariate statistical methods in an empirically oriented approach. Basically, each chapter presents a real problem from the business field along with a data set for analysis. Subsequently, the theory of a multivariate statistical method is discussed along with the basic mathematical computations necessary for understanding the analysis of the data. Throughout the presentation and discussion of a method, many references are made to the appropriate SAS computer output.

Although understanding and applying multivariate statistical methods requires some use of matrix algebra, the amount needed is relatively small as long as one is prepared to accept some of the mathematical details on faith. As such, the next chapter provides the basics of matrix algebra needed to understand and apply the multivariate procedures presented in this book.

## EXERCISES

1. The real estate board in a wealthy suburb wants to examine the distribution of prices of single-family homes sold during the past year. The following sample of 25 homes was collected and their prices (in thousands) recorded:

$$254 \quad 249 \quad 219 \quad 415 \quad 270$$
$$171 \quad 272 \quad 413 \quad 253 \quad 201$$
$$225 \quad 315 \quad 298 \quad 560 \quad 401$$
$$275 \quad 316 \quad 256 \quad 340 \quad 265$$
$$219 \quad 328 \quad 214 \quad 279 \quad 299$$

Plot a histogram of the data and find the mean and standard deviation for this sample of homes.

2. The distributions of selected financial ratios for industries are regularly published in financial journals. The table that follows provides three quartile categories for the profit-to-net-worth ratio (%) for four industries.

| Industry | A | B | C |
|----------|-----|-----|-----|
| Soft drinks | 9.25 | 15.31 | 31.24 |
| Mobile homes | 7.29 | 17.33 | 39.28 |
| Automobile | 11.22 | 19.21 | 45.98 |
| Cement | 6.55 | 17.99 | 51.21 |

Find the mean and standard deviation for each quartile category and construct a histogram.

3. Data on sales (in units) were obtained for a random sample of 10 salespersons for a machine manufacturing company.

| Name | Region | Sales |
|---|---|---|
| John Makk | East | 1200 |
| Mary Smith | North | 2400 |
| Kay Sanders | South | 700 |
| Jack Adams | West | 3313 |
| Andrew Kats | West | 1198 |
| Sam Johnson | East | 1767 |
| Laura Gibson | West | 3999 |
| George Anthony | South | 1675 |
| Ted Long | North | 4252 |
| Gary Marks | South | 2766 |

Sort the data set by region and determine the average and total sales by region.

4. The data here represent the ages of a class of 20 MBAs from the International Business School at the time of their graduation from the program.

$$26 \quad 27 \quad 34 \quad 38 \quad 24$$
$$29 \quad 28 \quad 25 \quad 31 \quad 33$$
$$35 \quad 29 \quad 42 \quad 33 \quad 29$$
$$25 \quad 28 \quad 29 \quad 31 \quad 37$$

Construct a frequency distribution and determine what percentage of the class are age 30 or below. Find the measures of central tendency and variability. What can be said about the shape of this distribution?

5. A security guard monitoring the main entrance to a large office building decided to count the number of people entering the building between the hours of 7:45 a.m. and 8:05 a.m. The guard recorded the following data:

| Time | People | Time | People | Time | People |
|---|---|---|---|---|---|
| 7:45 | 3 | 7:46 | 5 | 7:47 | 8 |
| 7:48 | 2 | 7:49 | 10 | 7:50 | 8 |
| 7:51 | 7 | 7:51 | 12 | 7:52 | 16 |
| 7:53 | 18 | 7:54 | 18 | 7:55 | 19 |
| 7:56 | 22 | 7:57 | 25 | 7:58 | 29 |
| 7:59 | 35 | 8:00 | 15 | 8:01 | 14 |
| 8:02 | 10 | 8:03 | 10 | 8:04 | 8 |
| 8:05 | 7 | | | | |

Construct a frequency distribution for the arrival times. Find the average arrival time at this corporation. Assuming that everyone working in this building starts work at 8:00 a.m., what proportion of employees observed were late? What assumptions must be made in order to use this data to estimate the proportion of employees who arrive late on any given day?

# Basic Matrix Algebra

In very general terms, the reader probably already has an understanding of what a matrix is. For example, the reader might have already heard the term used to represent a correlation matrix or just as a general term for the tabular display of certain types of data. Matrix algebra simply goes beyond the mere tabular display of data and deals with the algebraic manipulation of matrices. For example, the common operations of addition, subtraction, multiplication, and division of numbers are simply extended to matrices. The information presented in this chapter provides the reader with the necessary background to understand the multivariate techniques presented in this book.

## MATRIX DEFINITIONS

A matrix is a rectangular array of numbers arranged in several rows and columns. For example, a matrix $X$ might be of the form

$$X = \begin{bmatrix} x_{11} & x_{12} & x_{13} & x_{14} & x_{15} \\ x_{21} & x_{22} & x_{23} & x_{24} & x_{25} \\ x_{31} & x_{32} & x_{33} & x_{34} & x_{35} \\ x_{41} & x_{42} & x_{43} & x_{44} & x_{45} \\ x_{51} & x_{52} & x_{53} & x_{54} & x_{55} \end{bmatrix}$$

Upper case letters are generally used to denote matrices. Lower case letters with subscripts denote the elements of the matrix. The first subscript denotes the row number and the second denotes the column number. The size of a matrix is specified by giving its dimension, that is, by specifying the number of rows and columns in the matrix. The matrix $X$ just presented is a 5 by 5 ($5 \times 5$) matrix because it has 5 rows and 5 columns. A familiar example of such a square matrix (one in which the number of rows is equal to the number of columns) is a correlation matrix displaying the correlation coefficients between observed variables.

A matrix of dimension $1 \times 5$ has only 1 row and 5 columns. Such a matrix is called a row vector. For example,

$$A = [\,1 \quad 2 \quad 3 \quad 4 \quad 5\,]$$

would be a row vector. A matrix of dimension $2 \times 1$ has 2 rows and 1 column and is called a column vector. For example,

$$K = \begin{bmatrix} 1 \\ 2 \end{bmatrix}$$

would be a column vector.

The *transpose* of a matrix, denoted by a prime (i.e., $X'$), is also an important concept to remember. The transpose of a matrix is simply a matrix or vector whose rows and columns have been interchanged. For example, if the matrix $B$ is given as:

$$B = \begin{bmatrix} 1 & 4 \\ 2 & 5 \\ 3 & 6 \end{bmatrix}$$

its transpose is

$$B' = \begin{bmatrix} 1 & 2 & 3 \\ 4 & 5 & 6 \end{bmatrix}$$

## MATRIX OPERATIONS

There are five important matrix operations: addition, subtraction, scalar multiplication, matrix multiplication, and matrix inversion. The first three matrix operations are basically defined on an element-by-element basis, whereas the last two are a little more complicated.

To make these matrix operations easy to follow, consider this example. Suppose that a small study was conducted and that each of five managers was evaluated on two performance ratings forms. Each manager was independently rated on the forms by a supervisor. The rating forms contained 10 items each (e.g., quality of work, observance of work hours, meeting deadlines, initiative, etc.). The ratings comprised a nine-point scale with ratings ranging from "not satisfactory" to "superior." The matrix $X$ presents the total score obtained by each manager on the two rating forms.

$$X = \begin{bmatrix} 5 & 5 \\ 4 & 6 \\ 3 & 2 \\ 4 & 4 \\ 4 & 3 \end{bmatrix}$$

Suppose that a second performance rating was reported after six months on the same five managers. These results are presented in the matrix $Y$.

$$Y = \begin{bmatrix} 7 & 7 \\ 6 & 6 \\ 5 & 2 \\ 4 & 4 \\ 4 & 7 \end{bmatrix}$$

## MATRIX ADDITION AND SUBTRACTION

To obtain the total score for the five managers in a matrix $T$, an element-by-element sum of the two matrices $X$ and $Y$ is performed. Thus, the total score matrix $T$ would be:

$$T = \begin{bmatrix} 5 & 5 \\ 4 & 6 \\ 3 & 2 \\ 4 & 4 \\ 4 & 3 \end{bmatrix} + \begin{bmatrix} 7 & 7 \\ 6 & 6 \\ 5 & 2 \\ 4 & 4 \\ 4 & 7 \end{bmatrix} = \begin{bmatrix} 12 & 12 \\ 10 & 12 \\ 8 & 4 \\ 8 & 8 \\ 8 & 10 \end{bmatrix}$$

To report the difference score for the five managers in a matrix $D$, an element-by-element subtraction of the two matrices $X$ and $Y$ is performed. Thus, an improvement in performance would be shown by positive values in the difference matrix ($D$) obtained by subtracting $X$ from $Y$:

$$D = \begin{bmatrix} 7 & 7 \\ 6 & 6 \\ 5 & 2 \\ 4 & 4 \\ 4 & 7 \end{bmatrix} - \begin{bmatrix} 5 & 5 \\ 4 & 6 \\ 3 & 2 \\ 4 & 4 \\ 4 & 3 \end{bmatrix} = \begin{bmatrix} 2 & 2 \\ 2 & 0 \\ 2 & 0 \\ 0 & 0 \\ 0 & 4 \end{bmatrix}$$

It is important to remember that in order to perform the matrix operations of addition and subtraction, the number of rows and the number of columns of the matrices involved must be the same. It should be clear to the reader that it would not be possible to add or subtract a $5 \times 2$ matrix with or from a $5 \times 3$ matrix.

## SCALAR MULTIPLICATION

Any matrix can be multiplied by a number called a scalar. To perform a multiplication of a scalar with a matrix, each (and every) element of the matrix is multiplied by that scalar. For example, one may wish to report the average performance for the five managers over the two rating periods. This would be obtained by dividing the elements of the matrix $T$ by 2 or by multiplying each element of $T$ by $\frac{1}{2}$. Thus, the average performance would be equal to:

$$A = \frac{1}{2}T = \frac{1}{2}\begin{bmatrix} 12 & 12 \\ 10 & 12 \\ 8 & 4 \\ 8 & 8 \\ 8 & 10 \end{bmatrix} = \begin{bmatrix} 6 & 6 \\ 5 & 6 \\ 4 & 2 \\ 4 & 4 \\ 4 & 5 \end{bmatrix}$$

Similarly, if one wanted to add (or subtract) a scalar to the matrix $T$, one simply adds (or subtracts) the scalar to every element in the matrix. For example, $T + 10$ would provide:

$$T + 10 = \begin{bmatrix} 22 & 22 \\ 20 & 22 \\ 18 & 14 \\ 18 & 18 \\ 18 & 20 \end{bmatrix}$$

This is the same as adding a $5 \times 2$ matrix with all elements in the matrix equal to 10.

## MATRIX MULTIPLICATION

Matrix multiplication is a little more complicated than the other matrix operations presented so far. To begin with, there is always a restriction on the type of matrices that can be multiplied. This restriction involves the dimensionality of the matrices. In matrix addition and subtraction, the number of rows and columns of matrices must be the same. For example, the matrices $X$ and $Y$ were both of dimension $5 \times 2$. In matrix multiplication, however, the number of columns of the first matrix must be equal to the number of rows of the second matrix. Thus, in order for the matrix product to be defined, the matrix $X$ ($5 \times 2$) can only be multiplied by $Y'$ (the transpose of $Y$), which is a $2 \times 5$ matrix.

When the number of columns of a matrix (the first) are the same as the number of rows of another matrix (the second), the two matrices are considered *conformable*. This indicates that a matrix product can be defined.

The reason that the number of columns in the first matrix must be the same as the number of rows in the second matrix (i.e., $X$ and $Y'$) is that during matrix multiplication the elements of each row of the first matrix are paired with the elements of each column in the second matrix.

An example should make things much clearer. Consider the following two matrices $A$ and $B$:

$$A = \begin{bmatrix} 1 & 4 \\ 2 & 3 \\ 3 & 5 \end{bmatrix}$$

$$B = \begin{bmatrix} 5 & 4 & 3 \\ 1 & 2 & 3 \end{bmatrix}$$

Notice that because $A$ is a $3 \times 2$ matrix and $B$ is a $2 \times 3$ matrix, the two are conformable (a matrix product can be defined). To determine the elements of the product matrix $C$, each entry in the first row of $A$ is multiplied by each column in $B$ and the resulting products are summed. This same process is repeated until all the elements of the matrix $C$ are determined. Thus, $c_{11}$ (the first element of $C$) is obtained as

$$c_{11} = a_{11}b_{11} + a_{12}b_{21}$$

or simply

$$c_{11} = (1)(5) + (4)(1) = 9$$

In a similar manner,

$$c_{12} = (1)(4) + (4)(2) = 12$$
$$c_{13} = (1)(3) + (4)(3) = 15$$
$$c_{21} = (2)(5) + (3)(1) = 13$$
$$c_{22} = (2)(4) + (3)(2) = 14$$
$$c_{23} = (2)(3) + (3)(3) = 15$$
$$c_{31} = (3)(5) + (5)(1) = 20$$
$$c_{32} = (3)(4) + (5)(2) = 22$$
$$c_{33} = (3)(3) + (5)(3) = 24$$

Arranging the elements in the matrix $C$, it is clear that the product of $A$ and $B$ is a $3 \times 3$ matrix:

$$C = \begin{bmatrix} 9 & 12 & 15 \\ 13 & 14 & 15 \\ 20 & 22 & 24 \end{bmatrix}$$

In the previous section the matrix $X$ represented the total score obtained by each of five managers on two rating forms. The transpose of the matrix $X'$ could be used to represent the scores of the five managers on the two ratings forms. Thus,

$$X' = \begin{matrix} \text{form 1} \\ \text{form 2} \end{matrix} \begin{bmatrix} 5 & 4 & 3 & 4 & 4 \\ 5 & 6 & 2 & 4 & 3 \end{bmatrix} \quad \text{and} \quad X = \begin{matrix} \text{manager 1} \\ \text{manager 2} \\ \text{manager 3} \\ \text{manager 4} \\ \text{manager 5} \end{matrix} \begin{bmatrix} 5 & 5 \\ 4 & 6 \\ 3 & 2 \\ 4 & 4 \\ 4 & 3 \end{bmatrix}$$

Following the rules of matrix multiplication already presented, the product of $X'X$ would be a $2 \times 2$ matrix:

$$P = X'X = \begin{bmatrix} 82 & 83 \\ 83 & 90 \end{bmatrix}$$

An examination of the elements of the product matrix $P$ reveals some very familiar statistical values. For example, consider the first element $p_{11}$, which is determined as

$$p_{11} = (5)(5) + (4)(4) + (3)(3) + (4)(4) + (4)(4) = 82$$

This element is precisely the sum of squares for the first rating form used in this example study. Similarly, the element $p_{22}$ is determined as

$$p_{22} = (5)(5) + (6)(6) + (2)(2) + (4)(4) + (3)(3) = 90$$

and is the sum of squares for the second rating form. The elements $p_{21}$ and $p_{12}$ are equal in value and can be conceptualized as the sum of cross-products. Because of their special importance in multivariate statistics, such matrices are generally referred to as sum of squares and cross-products matrices or simply SSCP matrices.

Another special case of matrix multiplication involves the multiplication of a row vector with a column vector of the same dimension. For example, consider

$$r = \begin{bmatrix} 1 & 2 & 3 \end{bmatrix} \quad \text{and} \quad c = \begin{bmatrix} 3 \\ 2 \\ 1 \end{bmatrix}$$

as row and column vectors. In order to determine the product of $r$ and $c$ (i.e., $p = rc$), the $1 \times 3$ row vector is multiplied by the $3 \times 1$ column vector. This results in the scalar value of 10 (a $1 \times 1$ matrix). However, the product of $c$ and $r$ requires that a $3 \times 1$ vector be multiplied by a $1 \times 3$ vector. This results in the following $3 \times 3$ matrix:

$$P = \begin{bmatrix} 3 & 6 & 9 \\ 2 & 4 & 6 \\ 1 & 2 & 3 \end{bmatrix}$$

It is important to note that matrices or vectors that are conformable to matrix multiplication in one order might not be conformable if the order is reversed. In addition, even in situations where they are conformable, there is no guarantee that the products will be the same. This is clearly evident in the preceding example.

## MATRIX INVERSION

The operation of matrix division (more commonly called inversion) is also very important in multivariate statistics. Many of the techniques that are presented in the next chapters rely on matrix inversion. In fact, as will become evident, matrix inversion can provide clues about the "quality" of an analysis. Although it is rarely the case that one will ever need to perform a matrix inversion by hand, an understanding of the process and the rules governing inverses will enable one to further understand multivariate data analysis.

Let us first begin with a definition of a special type of matrix known as a *diagonal matrix*. A diagonal matrix is a square matrix (row dimension equal to column dimension) that has elements that are all zeros (0s) except along the main diagonal (the diagonal from the upper left corner to the lower right corner of the matrix). For example, the matrix

$$D = \begin{bmatrix} 1 & 0 & 0 \\ 0 & 3 & 0 \\ 0 & 0 & 5 \end{bmatrix}$$

would be a diagonal matrix. A diagonal matrix whose diagonal elements are equal to unity is called an *identity matrix*, and is denoted by the letter $I$ (or with a subscript $I_n$ indicating its dimension). For example,

$$I = \begin{bmatrix} 1 & 0 & 0 \\ 0 & 1 & 0 \\ 0 & 0 & 1 \end{bmatrix}$$

would be an $I_3$ identity matrix.

In general terms, the process of matrix inversion is similar to performing a simple division by finding the reciprocal of a number that when multiplied by the original number will equal 1. For example, for the number 5 the reciprocal is $5^{-1}$ (i.e., $\frac{1}{5}$), and $5^{-1} \times 5 = 1$. Therefore, the reciprocal of the number is that number that, when multiplied by itself, equals 1. These concepts are used to determine matrix inverses, but they are slightly complicated by the fact that a matrix is an array of numbers.

The matrix analogue to the number 1 is simply an identity matrix of some dimension. And, given any matrix $A$, what we want to find is the inverse of this matrix (i.e., $A^{-1}$ if it exists) such that when multiplied by $A$ it will provide the identity matrix. That is, $A^{-1}A = I$ or $AA^{-1} = I$. For example, if the matrix $A$ is

$$A = \begin{bmatrix} 2 & 0 & 0 \\ 0 & 3 & 0 \\ 0 & 0 & 3 \end{bmatrix}$$

the matrix that when multiplied by $A$ will yield an identity matrix is

$$A^{-1} = \begin{bmatrix} \frac{1}{2} & 0 & 0 \\ 0 & \frac{1}{3} & 0 \\ 0 & 0 & \frac{1}{3} \end{bmatrix}$$

The reader should check and see that $A^{-1}A = I = AA^{-1} = I$.

Although there are some fairly elaborate computing rules for determining inverses, the reader should simply understand the basic principles for determining matrix inverses. If the reader is curious about the various rules, they can be found in any text on matrix or linear algebra. For the purposes of this book, the computer will generate all the inverses needed for the multivariate techniques. However, it is necessary that one at least understand the basic principles governing inverses.

To compute the inverse of a square matrix one must have a working familiarity of the concept of a *determinant*. The determinant of a matrix $A$ is denoted as $|A|$ and is a number that is commonly referred to as the *generalized variance* of the matrix. The determinant of a $2 \times 2$ matrix is quite easy to find. Simply multiply the elements along the main diagonal and then subtract the product of the elements along the secondary diagonal. For example, if the matrix $D$ is provided as

$$D = \begin{bmatrix} 1 & 1 \\ 3 & 4 \end{bmatrix}$$

then

$$|D| = (1)(4) - (3)(1) = 1$$

is the determinant of the matrix. Calculations of determinants for higher order matrices are much more complicated but basically involve the same principles.

Another important concept that relates to finding the inverse of a matrix is the *adjoint* of a matrix. The adjoint of the matrix $D$ is denoted by adj($D$). The adjoint of the matrix $D$ is equal to:

$$\text{adj}(D) = \begin{bmatrix} a_{22} & -a_{12} \\ -a_{21} & a_{11} \end{bmatrix} = \begin{bmatrix} 4 & -1 \\ -3 & 1 \end{bmatrix}$$

In order to find the inverse of the matrix $D$ (i.e., $D^{-1}$) each element of adj($D$) must be divided by $|D|$—the determinant. Thus, the inverse of the matrix $D$ is determined to be:

$$D^{-1} = \text{adj}(D)/|D| = \begin{bmatrix} 4/1 & -1/1 \\ -3/1 & 1/1 \end{bmatrix}$$

It can also be verified that $D^{-1}D$ provides an identity matrix.

It is possible that some matrices encountered in practice might not have an inverse. When the determinant of a matrix is zero, the inverse cannot be determined. The inverse cannot be determined because the adjoint of a matrix is divided by zero, and division by zero is forbidden [recall that $D^{-1} = \text{adj}(D)/|D|$]. If the inverse of a matrix does not exist, the matrix is called *noninvertible* or *singular*. The concept of singularity is very important for examining computer output from the various multivariate statistical analyses presented throughout this book. The concept of singularity is basically used as a way of alerting the user to a potential problem with the data. The presence of a singular matrix implies that the matrix has no inverse and therefore no meaningful statistical computations can be conducted. For example, in a correlation matrix singularity occurs when some of the observed variables are perfectly correlated and/or one of the variables is a linear combination of one or more of the other observed variables. In simple terms, the variables contain redundant information that is not really required to perform the statistical analysis. Unfortunately, a cursory examination of a correlation matrix would not necessarily detect a singularity problem. This potential problem is discussed in greater detail when each multivariate technique is presented.

If the determinant of a matrix is zero, then the matrix cannot be inverted and the matrix is singular. Unfortunately, there are some cases in multivariate analysis when the determinant of a matrix is not exactly zero (so an inverse can be found) but it is very close to zero (e.g., 0.00034). Because matrix inversion relies on a nonzero determinant, whenever a near-zero determinant is obtained it will produce large and quite unreliable numbers in the inverted matrix. Because multivariate techniques use inverted matrices, the solutions will also be unstable.

In common statistical terminology the presence of a near-zero determinant is a clue to the presence of the problem of *multicollinearity* (to be discussed more extensively in a subsequent chapter). The presence of multicollinearity is usually an indication that the observed variables are highly and linearly intercorrelated (i.e., correlation of .90 or greater), and in some multivariate techniques this produces major problems. In particular, in regression analysis the presence of multicollinearity implies that one is using redundant information in the regression model to determine a prediction. This prediction can be artificially inflated because of using the same variable twice. For example, if one is using a measure of intelligence along with some other set of variables (e.g., age, education) to predict future income, there is no need to use both the Stanford–Binet Intelligence Scales and the Wechsler Adult Intelligence Scales because they are both measuring the same thing.

## EIGENVALUES AND EIGENVECTORS OF A MATRIX

Most multivariate statistical techniques rely on determining the values of the characteristic roots of a matrix (commonly called *eigenvalues*) and their corresponding vectors (called *eigenvectors*). Fortunately, as in the case with matrix inverses, one is not often called on to compute by hand eigenvalues and eigenvectors. However, one does need to know what is meant by the terms and understand the important role they play in multivariate statistics.

By definition, the eigenvalue of any matrix (e.g., $A$) is a number (e.g., $\lambda$) such that

$$Ax = \lambda x$$

where the value of $x$ satisfying this equation is the corresponding eigenvector. Obviously, because this definition does not make the terms any more understandable, an example is in order.

Notice that if

$$Ax = \lambda x$$

one can determine that

$$Ax - \lambda x = 0$$

and from this (by collecting similar terms) that

$$(A - \lambda I)x = 0$$

which is a system of linear equations each having a right-hand side equal to zero.

Consider the following matrix $A$:

$$A = \begin{bmatrix} 2 & 2 \\ 2 & 2 \end{bmatrix}$$

Expanding the preceding system of linear equations using the matrix $A$ yields

$$\left( \begin{bmatrix} 2 & 2 \\ 2 & 2 \end{bmatrix} - \lambda \begin{bmatrix} 1 & 0 \\ 0 & 1 \end{bmatrix} \right) \begin{bmatrix} x_1 \\ x_2 \end{bmatrix} = 0$$

or

$$\left( \begin{bmatrix} 2 & 2 \\ 2 & 2 \end{bmatrix} - \begin{bmatrix} \lambda & 0 \\ 0 & \lambda \end{bmatrix} \right) \begin{bmatrix} x_1 \\ x_2 \end{bmatrix} = 0$$

This yields

$$\begin{bmatrix} 2 - \lambda & 2 \\ 2 & 2 - \lambda \end{bmatrix} \begin{bmatrix} x_1 \\ x_2 \end{bmatrix} = 0$$

It is obvious that a trivial solution exists when $\begin{bmatrix} x_1 \\ x_2 \end{bmatrix}$ is zero. A nontrivial solution, however, exists if and only if the determinant of the $\begin{bmatrix} 2 - \lambda & 2 \\ 2 & 2 - \lambda \end{bmatrix}$ matrix is zero. Thus, to obtain a nontrivial solution one must find the values of $\lambda$ for which the determinant of $\begin{bmatrix} 2 - \lambda & 2 \\ 2 & 2 - \lambda \end{bmatrix}$ is zero. Thus, the eigenvalues are found by solving the equation

$$\left| \begin{bmatrix} 2 - \lambda & 2 \\ 2 & 2 - \lambda \end{bmatrix} \right| = 0$$

where | | denotes the determinant of the matrix.

The concept of a determinant was presented in the previous section. For a $2 \times 2$ matrix it was shown to be equal to the difference between the products of the main and secondary diagonals. That is

$$(2 - \lambda)(2 - \lambda) - (2)(2) = 0$$

or

$$\lambda^2 - 4\lambda = 0$$

and solving for $\lambda$, the eigenvalues are $\lambda_1 = 0$ and $\lambda_2 = 4$.

To find the eigenvectors associated with these eigenvalues several steps must be followed. The first step is to write out the matrix $A - \lambda I$ for each eigenvalue obtained. This is accomplished by subtracting $\lambda$ from each diagonal element of $A$. The second step requires that the adjoint of the resulting matrix $A - \lambda I$ be determined [i.e., adj$(A - \lambda I)$]. Finally, divide the elements of any column of adj$(A - \lambda I)$ by the square root of the sum of squares of these elements. The resulting values are the elements of the eigenvector $(x_i)$. These are also the only values that can satisfy the condition that $x_i' x_i = 1$, which is commonly referred to as the *unit-norm condition*. The unit-norm condition is introduced in the determination of eigenvector because, in general, there are several vectors $x$ that can satisfy the systems of equations defined by $(A - \lambda I)x = 0$. It also follows, therefore, that two eigenvectors $x_i$ and $x_j$ (that satisfy the unit-norm condition) associated with two distinct eigenvalues of a symmetric matrix are orthogonal; that is, $x_i' x_j = 0$.

Thus, to find the eigenvectors associated with the eigenvalues of the matrix $A = \begin{bmatrix} 2 & 2 \\ 2 & 2 \end{bmatrix}$ first write out the matrix $A - \lambda I$ corresponding to $\lambda_1 = 0$. This provides the matrix

$$\begin{bmatrix} 2-\lambda & 2 \\ 2 & 2-\lambda \end{bmatrix} = \begin{bmatrix} 2 & 2 \\ 2 & 2 \end{bmatrix}$$

The adjoint of this matrix [i.e., adj$(A - \lambda I)$] is found to be $\begin{bmatrix} 2 & -2 \\ -2 & 2 \end{bmatrix}$. Finally, each element of the column $\begin{bmatrix} 2 \\ -2 \end{bmatrix}$ is divided by $\sqrt{2^2 + (-2)^2}$ . Thus,

$$x_1 = \begin{bmatrix} 0.707 \\ -0.707 \end{bmatrix}$$

is the eigenvector associated with $\lambda_1 = 0$. Using the same procedure, the eigenvector associated with $\lambda_2$ is found to be:

$$x_2 = \begin{bmatrix} -0.707 \\ -0.707 \end{bmatrix}$$

An examination of the first eigenvector reveals that it fulfills the unit-norm condition (i.e., $x_1' x_1 = 1$). Similarly, the two eigenvectors are orthogonal (i.e., $x_1' x_2 = 0$).

Although the computations of eigenvalues and eigenvectors are best left to computers, it is important that one understand the terms because they are used so frequently in multivariate statistics and because the properties of eigenvalues and eigenvectors can help in the characterization and assessment of a multivariate data analysis. This is basically the principle of the Cayley–Hamilton Theorem. The Cayley–Hamilton Theorem indicates that because a matrix behaves like its eigenvalues, knowledge of the eigenvalues is essential for data analysis. For example, it is well known that the product (which is abbreviated using the Greek letter $\Pi$—for readers unfamiliar with the $\Pi$ notation, this is just a shorthand notation for multiplication in exactly the same way $\Sigma$ is an abbreviation for addition) of the eigenvalues of a matrix $(A + I)$ is equal to the value of the determinant of the matrix (i.e., $\Pi\lambda_i = |A + I|$). It is also well known that the product of the eigenvalues of a matrix $A$ is equal to the determinant of the matrix (i.e., $\Pi\lambda_i = |A|$). This implies that if a matrix has one or more eigenvalues equal to zero then the matrix is singular (recall the discussion of singularity in the previous section), and matrix inversion cannot be conducted. Similarly, if the product of the eigenvalues is close to zero, there is a problem with multicollinearity and the matrix inverse is unstable.

There are some additional properties of eigenvalues that are commonly used in computer output as warning signs that will be discussed much more extensively in later chapters. The three most important are positive definiteness, positive semidefiniteness, and indefiniteness or being ill-conditioned. In summary, when all the eigenvalues of a matrix are positive ($\lambda > 0$), the matrix is called *positive definite*. When not all the eigenvalues of a matrix are positive ($\lambda \geq 0$), the matrix is called *positive semidefinite*. Finally, when the eigenvalues of a matrix are negative ($\lambda < 0$), the matrix is called *indefinite* or *ill-conditioned*. Later chapters demonstrate the cause for great concern regarding positive semidefinite and ill-conditioned matrices.

## EXERCISES

1. Let

$$X = \begin{bmatrix} 2 & 3 \\ 4 & -1 \end{bmatrix} \quad Y = \begin{bmatrix} 2 & 3 & 2 \\ 2 & -2 & 3 \end{bmatrix} \quad Z = \begin{bmatrix} 1 & 1 \\ -1 & 1 \\ 0 & 1 \end{bmatrix}$$

If possible, perform each of the following operations. If the operation cannot be performed, explain why.

(a) $X + Y$   (b) $Y + Z$   (c) $Y' + Z$   (d) $XY$
(e) $YZ$     (f) $YX$     (g) $X + (YZ)$  (h) $2X$

2. Consider the matrix $A = \begin{bmatrix} 2 & 0 \\ 1 & -1 \end{bmatrix}$. Find the eigenvalues associated with the matrix $A$.

3. Consider the matrices

$$A = \begin{bmatrix} 2 & 3 \\ 4 & -1 \end{bmatrix} \quad \text{and} \quad B = \begin{bmatrix} 2 & 3 \\ 4 & -1 \end{bmatrix}$$

Verify the following properties of the transpose:

(a) $(A')' = A$     (b) $(AB)' = B'A'$

4. Let

$$A = \begin{bmatrix} 2 & 2 \\ 2 & 2 \end{bmatrix} \quad \text{and} \quad B = \begin{bmatrix} 2 & 2 & 3 \\ 4 & 1 & -1 \end{bmatrix}$$

Are the matrices $A$ and $B$ positive definite, positive semidefinite, or neither?

5. Consider the matrix $K = \begin{bmatrix} 3 & 2 \\ 1 & 4 \end{bmatrix}$. Find the eigenvalues and eigenvectors associated with the matrix $K$.

# The Multivariate Normal Distribution and Tests of Significance

Most of the statistical techniques presented in this book are based on the assumption that the variables under study follow a multivariate normal distribution. Recall the importance of the normal distribution or bell-shaped curve for conducting hypothesis tests and constructing confidence intervals in univariate statistics. Although it is well known that observed data are never exactly normally distributed, the assumption of normality is vitally important for three main reasons. First, numerous continuous phenomena seem to follow or can be approximated by the bell-shaped curve. Second, the normal distribution can be used to approximate various discrete distributions. Third, and perhaps most important, the normal distribution provides the basis for statistical inference because of its relationship to the central limit theorem.

The central limit theorem essentially states that, regardless of the type of population distribution from which one draws a random sample, the sampling distribution of sample sums and means will be approximately normal under several conditions. If the population distribution is normal, the sampling distribution will be normal regardless of the sample size selected. If the population distribution is approximately normal, the sampling distribution will be approximately normal even for fairly small sample sizes. If the population distribution is not normal, the sampling distribution will be approximately normal if the sample size is large enough (at least 30).

The multivariate normal distribution is just a generalization of the quite familiar bell-shaped curve but extended to several dimensions. Moreover, the central limit theorem plays just as important a role in multivariate statistical inference as in the univariate case. Therefore, before we introduce the multivariate normal distribution, it is essential that the reader have a firm grasp of the concepts of the univariate normal distribution. The next section provides that foundation.

## THE UNIVARIATE STANDARD NORMAL DISTRIBUTION

The normal distribution does not really exist. Actually, it is a mathematical model that is used to represent observed data. The major value of the normal distribution lies in its ability to serve as a reasonably good model of many variables of interest. For example, the distribution of incomes of people in a population can be modeled reasonably well with the normal distribution. Most people are about average in income, but a few are quite poor and a few are quite rich.

The univariate normal distribution is a symmetrical bell-shaped curve with identical measures of central tendency (i.e., mean, median, mode). In addition, the observed outcomes of the one continuous random variable of interest have an infinite range (i.e., $-\infty < X < +\infty$). However, despite the theoretically infinite range, the practical range for observed outcomes is generally obtained within approximately three standard deviations above and below the mean.

The reader may not be aware that the mathematical expression representing the univariate normal distribution density function is:

$$f(x) = \frac{1}{\sqrt{2\pi\sigma^2}}\, e^{-(1/2)[(x-\mu)/\sigma]^2}$$

where $e$ is a mathematical constant approximated by 2.71828, $\pi$ is a mathematical constant approximated by 3.14159, $\mu$ is the population mean, $\sigma$ is the population standard deviation, $\sigma^2$ is the population variance, and $x$ is the continuous random variable of interest. A plot of this function for some observed variable of interest yields the familiar bell-shaped curve shown in Fig. 3.1.

It should be obvious that because the values of $e$ and $\pi$ are mathematical constants, a different normal distribution may be generated for any particular combination of the mean and variance (or standard deviation). Figure 3.2 shows two generated normal distributions for an observed variable $x$ that have the same mean but different variances. Although the probabilities of occurrences within particular ranges of the random variable $x$ depend on the values of the two parameters of the normal distribution, the mathematical expression used to generate these different distributions is computationally tedious.[1] For

---

[1] For readers familiar with calculus, the probability that $x$ takes a value between $a$ and $b$ is given by this integral:

$$P(a \leq x \leq b) = \int_{a}^{b} \frac{1}{\sqrt{2\pi\sigma^2}}\, e^{-(1/2)[(x-\mu)/\sigma]^2}\, dx$$

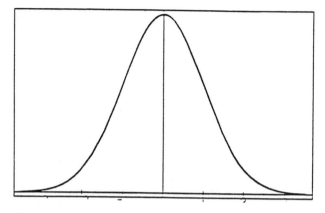

FIG. 3.1. The univariate normal distribution.

this reason a transformation formula is used to convert any normal random variable $x$ to a standard normal random variable $z$. The transformation formula is:

$$z = \frac{x - \mu}{\sigma}$$

This transformed variable $z$ represents a score from a standard normal distribution with $\mu = 0$ and $\sigma = 1$ no matter what the values of the mean and standard deviation in the observed data.

The real importance of the standard normal distribution derives from the fact that it is a theoretical distribution whose special characteristics allow one to determine probabilities very easily. Moreover, because any normal distribution can be transformed into the standard normal distribution by

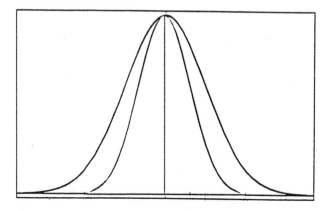

FIG. 3.2. Two different normal distributions.

employing the conversion formula just given, the standard normal distribution can be used as the single reference distribution for comparing a wide variety of otherwise not comparable statistics. Notice that when different normal distributions are converted to the standard normal, each of the transformed distributions has the same mean and standard deviation. There should be little concern about the fact that the original untransformed distributions have different means and standard deviations because these distributions have been converted into a single distribution, the standard normal distribution.

Using a table of the standard normal distribution (see Appendix A), various probability values of interest can be determined. Such a table represents the probability or the area under the standardized normal curve for the transformed $z$ value of interest. For example, assume that a production manager is investigating the amount of time it takes factory workers to assemble a particular machine part and determines that the data are normally distributed with $\mu = 10$ and $\sigma = 2$ minutes. To determine how likely it is for a factory worker to take more than 12 minutes to assemble the part, one simply determines the transformed $z$ value for this score and looks it up in the table of the standard normal distribution. This provides

$$z = \frac{x - \mu}{\sigma} = \frac{12 - 10}{2} = 1$$

which corresponds to a probability value of .8413 (obtained from the table of the standard normal distribution in the Appendix). Thus, it is 15.87% likely that a factory worker will take more than 12 minutes to assemble the part. In addition, if 100 workers are employed by the factory, one would only expect approximately 16 of them to take more than 12 minutes to assemble the machine part.

## UNIVARIATE SAMPLING DISTRIBUTIONS
## AND STATISTICAL INFERENCE

Two types of estimates can be constructed for any population parameter (e.g., mean or proportion): a point estimate or an interval estimate. Point estimates are single numbers used as estimates of the parameter of interest. For example, to estimate the population mean $\mu$, the sample mean $\bar{x}$ is typically used as the point estimate. An interval estimate is a range of values used as an estimate of the parameter. A confidence interval for the population mean $\mu$ is generally constructed as $(\bar{x} - z\sigma_{\bar{x}}, \bar{x} + z\sigma_{\bar{x}})$ where $\sigma_{\bar{x}}$ is the standard deviation of the sample mean. Although the purposes differ, establishing a confidence interval and conducting a significance test are formally identical in rationale. Basically the test statistic takes a value that leads to the rejection (at a given significance [$\alpha$] level) of a hypothesis about a parameter if the $100(1 - \alpha)\%$ confidence region fails to include the hypothesized parameter value.

The reader should be familiar with the fact that when random samples of size $n$ are drawn from a univariate normal distribution, the distribution of sample means will also be normally distributed with a mean equal to $\mu$ and standard deviation of $\sigma_{\bar{x}} = \dfrac{\sigma}{\sqrt{n}}$ (for a finite population [$N$] this value becomes $\sigma_{\bar{x}} = \dfrac{\sigma}{\sqrt{n}} \sqrt{\dfrac{N-n}{N-1}}$). The distribution is exactly normal if the population is normal, and approximately normal if $n$ is large enough. Thus, the transformation of $z$ can also be used to determine the probability of obtaining values of $\bar{x}$ that deviate from a given population mean by varying amounts. For example, suppose that a sample of 36 factory workers was observed by the production manager in the previous example and provided the mean value $\bar{x} = 12$. How likely is it that the population mean is equal to 10 if such a sample mean was observed? The test statistic using the $z$ formula has the value:

$$z = \frac{\bar{x} - \mu}{\dfrac{\sigma}{\sqrt{n}}} = \frac{12 - 10}{\dfrac{2}{\sqrt{36}}} = 6$$

which leads to a rejection of the hypothesis that $\mu = 10$ at the .05 level of significance against a two-sided alternative. Similarly, the 95% confidence interval for $\mu$ is $12 - 1.96(2/\sqrt{36})$, $12 + 1.96(2/\sqrt{36})$, or simply 11.35 and 12.65, which does not include the value of 10. This suggests that it is very unlikely that the observed sample mean ($\bar{x} = 12$) came from a population with $\mu = 10$. In fact, the probability of obtaining a sample mean of 12 or more when the population mean is really 10 is approximately zero [i.e., $p(z \geq 6) \cong 0$].

## OTHER SAMPLING DISTRIBUTIONS

The normal distribution is also related to several other sampling distributions. Of these, the most important are the $\chi^2$ (chi-square) distribution, the $t$ distribution, and the $F$ distribution. To examine the relationship between the normal and the $\chi^2$ distribution, suppose that a random sample of size 1 is repeatedly drawn from a normal distribution. Each observation can be expressed in standardized form as a $z$ score. The square of this score (i.e., $z^2$)

is simply called a $\chi^2$ with 1 degree of freedom. Similarly, consider two random samples of independent observations $x_1$ and $x_2$ drawn from a normal distribution expressed in standard form as $z_1$ and $z_2$ (two observations are said to be *independent* if the probability of one observation occurring is unaffected by the occurrence or nonoccurrence of the other). The sum over repeated independent sampling has a $\chi^2$ distribution with 2 degrees of freedom (i.e., $z_1^2 + z_2^2 = \chi^2$). Thus, if a random sample of size $n$ is drawn from a normal distribution with a mean $\mu$ and standard deviation $\sigma$, the observations will be distributed as a $\chi^2$ with $n$ degrees of freedom.

If the population standard deviation is unknown, the sample standard deviation $s$ is introduced as an estimate of $\sigma$. The appropriate test statistic in this case is the familiar $t$ statistic, defined as:

$$t = \frac{z}{\sqrt{\dfrac{\chi^2}{n-1}}}$$

which is generally expressed by replacing $z$ with $\dfrac{\bar{x}-\mu}{\sigma/\sqrt{n}}$ and $\chi^2$ with $(n-1)s^2/\sigma^2$ to provide the more common formula:

$$t = \frac{\bar{x}-\mu}{\dfrac{s}{\sqrt{n}}}$$

The $t$ statistic follows Student's $t$ distribution with $n - 1$ degrees of freedom, and it is well known that as the number of observations increases, the distribution gradually approaches the standard normal distribution. In fact, for very large samples the values are identical. Tables of critical values of the $t$ statistic are also provided in the Appendix.

## INFERENCES ABOUT DIFFERENCES BETWEEN GROUP MEANS

The $t$ statistic is also commonly used for comparing sample groups to determine whether observed differences between two independent group means arose by chance or represent true differences between populations. Of course, because this decision cannot be made with complete certainty, it is probabilistically determined. The problem, then, is to determine the probability of observing differences between the sample means of two groups under the assumption of no difference between the two groups (i.e., the null hypothesis of no difference is true). This involves determining both the distance between the difference of the means and estimating the standard error of the difference between the means. Because the sampling distribution of $t$ will vary from one sample to the next due to random fluctuation of both the mean and the standard deviation, the $t$ distribution used to model the sampling procedure is based on $n_1 - 1$ and $n_2 - 1$ degrees of freedom (or simply $n_1 + n_2 - 2$ degrees of freedom). Thus, the $t$ statistic for differences between means is defined as:

$$t = \frac{\bar{x}_1 - \bar{x}_2}{\sqrt{s^2(\frac{1}{n_1} + \frac{1}{n_2})}}$$

with

$$s^2 = \frac{(n_1 - 1)s_1^2 + (n_2 - 1)s_2^2}{n_1 + n_2 - 2}$$

When the sample sizes are equal (i.e., $n_1 = n_2 = n$), the cumbersome denominator simplifies and the $t$-statistic can be written as:

$$t = \frac{\bar{x}_1 - \bar{x}_2}{\sqrt{\frac{s_1^2 + s_2^2}{n}}}$$

For example, suppose that an advertising manager of a breakfast cereal company wants to determine whether a new package shape would improve sales of the product. In order to test the impact of the new package shape, a random sample of 40 equivalent stores is selected. Twenty (20) stores are randomly assigned as the test market of the new package (group 1), and the other 20 continue receiving the old package (group 2). The weekly sales during the time period studied can be summarized as follows:

$$\bar{x}_1 = 130 \quad s_1 = 10 \quad \text{and} \quad \bar{x}_2 = 117 \quad s_2 = 12$$

The $t$ test to determine whether there is a difference between the average sales of the two groups is:

$$t = \frac{130 - 117}{\sqrt{\frac{100 + 144}{20}}} = 3.72$$

which leads to the rejection of the hypothesis of no difference between the average sales at the .05 level of significance for a two-tailed test. Thus, the average sales at the stores receiving the product in the new package shape are higher than those of the stores receiving the old package.

## THE *F* STATISTIC AND ANALYSIS OF VARIANCE

It is also quite well known that the *t* distribution and the the $\chi^2$ distribution are related to the *F* distribution. The value of $t^2$ is identical to an *F* with 1 and $n - 1$ degrees of freedom (i.e., $t^2 = F_{1,n-1}$). The *F* distribution is basically the sampling distribution that is used for testing hypotheses about two or more population means (or variances). The *F* distribution provides a model of what happens when random samples of two or more means (or variances) are drawn from the same population and compared with one another. The *F* statistic is defined as the ratio of two independent chi-square variables, each divided by its degrees of freedom. Thus,

$$F = \frac{\chi^2/(n_1 - 1)}{\chi^2/(n_2 - 1)}$$

In general, the model assumes that randomly drawn sample means will probably differ from one another due to sampling error. This variation between sample means provides an estimate of error variability between groups. Likewise, the scores within each group will probably differ from one another according to the variation of scores in the population. In this case, the variation within groups provides another estimate of error variability. Because the variabilities between and within groups are independent sample estimates of sampling error, they will not be exactly the same. As such, the degree to which they fluctuate can be measured by dividing one by the other. The ratio of these variances forms an *F* distribution, and the shape of the *F* distribution depends on the degrees of freedom of the two variances. Tables of critical values of *F* for testing null hypotheses for group differences are also provided in the Appendix.

When two or more means are to be compared using the *F* distribution, the procedure is commonly referred to as an *analysis of variance* (ANOVA). ANOVA is one of the most frequently used procedures for comparing estimates of differences between groups. The variances of the groups are traditionally summarized in terms of the partition of sums of squares (SS)—that is, sums of squared differences between scores and their means. A sum of squares is simply the numerator of a variance [i.e., $s^2 = \dfrac{\Sigma(x - \bar{x})^2}{n - 1}$], and is defined as:

$$SS = \Sigma(x - \bar{x})^2$$

Suppose that the production manager of a large automobile assembly plant is investigating the average amount of time it takes workers in different shifts to assemble a particular automobile part. There are three different shifts at this plant: a morning, an afternoon, and an evening shift. To deter-

TABLE 3.1
Recorded Assembly Time for Shifts

| Morning | Afternoon | Evening |
|---------|-----------|---------|
| 1 | 3 | 4 |
| 3 | 6 | 7 |
| 2 | 5 | 7 |
| 4 | 4 | 5 |
| 2 | 5 | 4 |
| 1 | 4 | 4 |

mine whether there are any differences between the shifts, the production manager randomly selects workers from each shift and records the amount of time each worker takes to assemble the part. The results are presented in Table 3.1.

Assume that a mean is computed to represent the average scores over all the observed groups, called the grand mean, and symbolized as $\overline{X}_{GM}$. The variability between each score and this grand mean ($\overline{X}_{GM}$) can be examined, ignoring groups with which scores are associated. This is the total sum of squares ($SS_{total}$), which is equal to the sum of the variabilities that exists between each of the groups and within each of the groups: $SS_{total} = SS_{between} + SS_{within}$. This relationship between the various sum of squares is also called the partitioning of the total sums of squares. Such partitioning is derived from the following equation (whose validity can be demonstrated by using the rules of summation):

$$\sum_{j}\sum_{i}(x_{ij} - \overline{X}_{GM})^2 = n\sum_{j}(\overline{X}_j - \overline{X}_{GM})^2 + \sum_{j}\sum_{i}(x_{ij} - \overline{X}_j)^2$$

where $x_{ij}$ is the score for worker $i$ in group $j$, $\overline{X}_j$ is the mean for group $j$, and $\overline{X}_{GM}$ is the grand mean. The term on the left represents the total variability of all data. If the total variability were divided by the sample size minus 1 (i.e., $n - 1$), the sample variance would be computed.

The degrees of freedom in an ANOVA partition the same way as the sums of squares, namely:

$$df_{total} = df_{between} + df_{within}$$

where $n$ is the number of observed workers, $k$ the number of observed groups, $df_{total} = n - 1$, $df_{between} = k - 1$, and $df_{within} = n - k$.

Finally, by dividing the sums of squares by their corresponding degrees of freedom, one obtains the *variances*, which are more commonly referred

TABLE 3.2
ANOVA Results

| Source of Variation | SS | df | MS | F |
|---|---|---|---|---|
| Between groups | 29.77 | 2 | 14.88 | 9.66* |
| Within groups | 23.17 | 15 | 1.54 | |
| Total | 52.94 | 17 | | |

*$p < .05$.

to as mean squares. The ratio of the mean squares for between groups and within groups provides the $F$ test in an analysis of variance.

The results of an ANOVA are traditionally displayed in a table called an ANOVA source table. Table 3.2 provides the results for this example. Clearly, at the .05 level of significance, the results lead to the rejection of the hypothesis of no difference between the average amount of time it takes workers in different shifts to assemble the part (using the tables for the $F$ distribution in the Appendix, for $\alpha = .05$ with 2 and 15 degrees of freedom, $F = 3.68$). To determine precisely which shift takes the longest average amount of time, separate post hoc comparisons (i.e., $t$ tests) could be examined between the various shifts.

## THE MULTIVARIATE NORMAL DISTRIBUTION

The multivariate normal distribution is just a generalization of the univariate bell-shaped curve extended to several dimensions. As in the univariate case, a mathematical expression can be used to represent the multivariate normal distribution. For example, if two variables $x_1$ and $x_2$ are observed, the bivariate normal distribution density function is:

$$f(x_1, x_2) =$$

$$\frac{1}{2\pi\sigma_1\sigma_2\sqrt{1-\rho^2}} \exp\left\{\frac{-1}{2(1-\rho^2)}\left[\frac{(x_1-\mu_1)^2}{\sigma_1^2} + \frac{(x_2-\mu_2)^2}{\sigma_2^2} - 2\rho\frac{(x_1-\mu_1)(x_2-\mu_2)}{\sigma_1\sigma_2}\right]\right\}$$

where $e$ is a mathematical constant approximated by 2.71828, $\pi$ is a mathematical constant approximated by 3.14159, $x_1$, $x_2$ are the random variables of interest, $\mu_1$, $\mu_2$ are the population means of the $x_1$ and $x_2$ variables, $\sigma_1$, $\sigma_2$ are the population variances of $x_1$ and $x_2$, and $\rho$ is the correlation between the $x_1$ and $x_2$ variables.

An example plot of this function resembles the bell-shaped mound as shown in Fig. 3.3. Because the drawing of such a mound is relatively difficult to construct, it is customary to represent a bivariate normal distribution by drawing cross sections of the surface called isodensity contours (i.e., taking

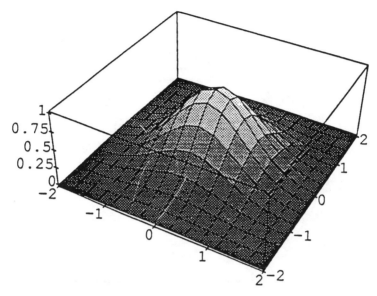

FIG. 3.3. Bivariate normal distribution.

cross sections of the density surface at various elevations generates a family of concentric ellipses). To determine the contour lines, the expression

$$\frac{(x_1 - \mu_1)^2}{\sigma_1^2} + \frac{(x_2 - \mu_2)^2}{\sigma_2^2} - 2\rho \frac{(x_1 - \mu_1)(x_2 - \mu_2)}{\sigma_1\sigma_2}$$

is set equal to a constant. The smaller the value of the constant is, the higher is the altitude of the contour. Figure 3.4 represents a bivariate distribution with several selected contour lines with a center at the point ($\mu_1 = 10$, $\mu_2 = 20$). This center is commonly referred to as the *centroid* of the bivariate population. In other words, it is the vector containing the means of the two observed variables.

As in the univariate case, a simple tranformation formula can be used to convert any number of random variables (e.g., $X_1$, $X_2$) to standardized $z$ variables. Recall that the $z$ score was defined as $z = \dfrac{x - \mu}{\sigma}$, and that its square is equal to $z^2 = \chi^2 = (\dfrac{x - \mu}{\sigma})^2$ or simply $\chi^2 = (x - \mu)(\sigma^2)^{-1}(x - \mu)$ (i.e., the product of the deviation of the scores to the means with the inverse of the variance). In the multivariate case the transformation formula is slightly more complicated, but it is still directly related to the univariate formula.

For the multivariate standardization, the transformation formula is represented as a $\chi^2$ (chi-square) and is written as an extension of $(x - \mu)(\sigma^2)^{-1}(x - \mu)$, except that because several variables are involved, matrix notation

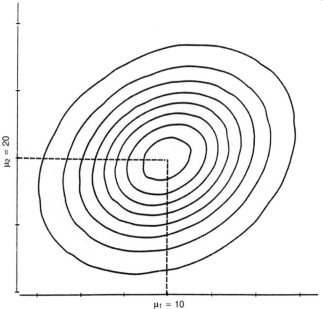

FIG. 3.4. Bivariate normal distribution with selected contour lines with a center $\mu_1 = 10$, $\mu_2 = 20$.

must be used. Thus, for a two-variable standardization, the $\chi^2$ is represented as:

$$\chi^2 = [x_1 - \mu_1,\ x_2 - \mu_2]'\Sigma^{-1}[x_1 - \mu_1,\ x_2 - \mu_2]$$

or simply $\chi^2 = x'\Sigma^{-1}x$, where $x'$ and $x$ are vectors representing the deviation of the scores to the means, and $\Sigma^{-1}$ is the inverse of the variance–covariance matrix between the two variables. For the bivariate case this inverse is equal to:

$$(1/1 - \rho^2)\begin{bmatrix} 1/\sigma_1^2 & -\rho/\sigma_1\,\sigma_2 \\ -\rho/\sigma_2\,\sigma_1 & 1/\sigma_2^2 \end{bmatrix}$$

Consequently, the value of $\chi^2$ can also be written as being equal to:

$$\chi^2 = \frac{1}{1 - \rho^2}\left[\frac{(x_1 - \mu_1)^2}{\sigma_1^2} + \frac{(x_2 - \mu_2)^2}{\sigma_2^2} - 2\rho\frac{(x_1 - \mu_1)(x_2 - \mu_2)}{\sigma_1\sigma_2}\right]$$

and it is now more obvious that in the event that only one variable is under investigation, the $\chi^2$ formula reduces to the simple case of $(x - \mu)(\sigma^2)^{-1}(x - \mu)$ previously presented.

Now suppose that the production manager who investigated the amount of time it takes workers to assemble a particular automobile part was also concerned about the quality of each part assembled. In order to assess quality, the production manager used a measure to rate the accuracy of assembly conducted by each worker. Assume that the same workers were observed on both variables of interest, namely, time and accuracy. The production manager determined that the observed data were normally distributed with a mean centroid of (10, 8) and a variance–covariance matrix of $\begin{bmatrix} 25 & 15 \\ 15 & 25 \end{bmatrix}$ [i.e., $\mu_1$ (time) = 10, $\mu_2$ (accuracy) = 8, $\sigma_1 = \sigma_2 = 5$, and $\rho = 0.6$].

To determine how likely it is for a factory worker to take more than 25 minutes to assemble a part and for that part to receive a quality measure of 1, one simply uses the multivariate tranformation formula to determine the value of $\chi^2$. One then looks up that value in the tables to determine the appropriate probability level. Thus, the value of $\chi^2$ is:

$$\chi^2 = \frac{1}{1 - 0.6^2}\left[\frac{(25 - 10)^2}{25} + \frac{(1 - 8)^2}{25} - \frac{2(0.6)(25 - 10)(1 - 8)}{25}\right] = 9.25$$

which corresponds to a probability value of approximately .01. This indicates that it is very unlikely that a worker will take so long to assemble the part and receive such a low quality score.

However, what if a random sample of 16 plant workers is observed on both variables of interest? What is the likelihood that their average assembly time and quality centroid will be equal to (25, 1)? As in the univariate case, a hypothesis concerning the centroid of a multivariate normal distribution with a known variance–covariance matrix $\Sigma$ can be tested taking into account the observed sample size.

## MULTIVARIATE SAMPLING DISTRIBUTIONS AND STATISTICAL INFERENCE

Recall that in the univariate case when random samples of size $n$ are drawn from a normal distribution, the distribution of sample means will also be normal. In fact, the distribution is exactly normal if the population is normal, and even if the distribution is not normal the distribution of sample means will be approximately normal if $n$ is large enough. This same principle holds in the multivariate case, and thus the $\chi^2$ can be used to determine the probability of obtaining centroid values (i.e., mean values on observed variables) that deviate from a given population centroid by varying amounts.

Taking into account the sample size in this example provides the following value for the $\chi^2$ test:

$$\chi^2 = n(\bar{x} - \mu)'\Sigma^{-1}(\bar{x} - \mu)$$

which (using the same observed data previously presented) provides a test statistic value of $\chi^2 = 148$. This observed $\chi^2$ corresponds to a probability value less than .00001 (the critical value from the $\chi^2$ table based on a significance level of $\alpha = .05$ with 2 df is only 5.991). This indicates that it is almost impossible to observe such a sample centroid when 16 workers are randomly selected from this automobile plant.

## HOTELLING'S STATISTIC

If the variance–covariance matrix must be estimated from the sample (as in the univariate case where the standard deviation was introduced as a population estimate), then the appropriate test statistic in the multivariate case becomes an extension to the familiar $t$ test (except that it is squared). The equation is very similar to the multivariate $\chi^2$, but the variance–covariance matrix $\Sigma$ is replaced by the unbiased estimate $S$. This multivariate test statistic was denoted $T^2$ by Hotelling (1931), and is commonly referred to as Hotelling's $T^2$. The value of Hotelling's $T^2$ is equal to:

$$T^2 = n(\bar{x} - \mu)'S^{-1}(\bar{x} - \mu)$$

Hotelling also showed that the $T^2$ statistic is related to the $F$ distribution by the following relationship:

$$\frac{n - p}{(n - 1)p} T^2 = F_{p,n-p}$$

where $p$ is the number of observed variables, and $n$ is equal to the sample size. It is important to note that for $p = 1$ (i.e., the univariate case), this reduces to the familiar relation between the $t$ statistic and the $F$ distribution previously presented as $t^2 = F_{1,n-1}$.

## INFERENCES ABOUT DIFFERENCES
## BETWEEN GROUP MEANS

The Hotelling statistic can also be used to determine the significance of observed differences between two sampled groups on some set of observed variables. Hotelling's $T^2$ for comparing two groups is simply an extension of the univariate $t$ statistic presented in the previous section. The univariate $t$ statistic for differences between two independent groups was defined as:

$$t = \frac{\overline{x}_1 - \overline{x}_2}{\sqrt{s^2(\frac{1}{n_2} + \frac{1}{n_2})}}$$

with

$$s^2 = \frac{(n_1 - 1)s_1^2 + (n_2 - 1)s_2^2}{n_1 + n_2 - 2}$$

In order to make the extension of this statistic to its multivariate equivalent, this equation is rewritten as:

$$t^2 = (\overline{x}_1 - \overline{x}_2) \left( s^2 \frac{n_1 + n_2}{n_1 n_2} \right)^{-1} (\overline{x}_1 - \overline{x}_2)$$

or simply as:

$$t^2 = \frac{n_1 n_2}{n_1 + n_2} (\overline{x}_1 - \overline{x}_2)(s^2)^{-1}(\overline{x}_1 - \overline{x}_2)$$

This formulation can be used to represent Hotelling's $T^2$ in the multivariate case. However, because there are several variables involved, matrix notation must be used. Thus, Hotelling's multivariate statistic can be represented as:

$$T^2 = \frac{n_1 n_2}{n_1 + n_2} (\overline{x}_1 - \overline{x}_2)' S^{-1}(\overline{x}_1 - \overline{x}_2)$$

where $S$ is the sum of the variance–covariance matrices of the two groups.

Hotelling (1931) also showed that this statistic is related to the $F$ distribution by the following equivalency:

$$\frac{n_1 + n_2 - p - 1}{(n_1 + n_2 - 2)p} T^2 = F_{p, n_1 + n_2 - p - 1}$$

When the number of observed variables is one (i.e., $p = 1$), this again reduces to the simple univariate relationship previously presented, $t^2 = F_{1, n_1 + n_2 - 2}$.

An example will undoubtably clarify the use of Hotelling's $T^2$ for comparing sample means. Suppose that two samples of Southern California homes are observed. The first sample consists of 45 homes with installed air-conditioning units ($n_1 = 45$), and the second sample consists of 55 homes without air-conditioning units ($n_2 = 55$). Assume that two measurements of electrical consumption in kilowatt-hours are observed in this study. The first is a measure of off-peak consumption during the summer months of June through August (the hours from midnight till noon can be considered off-peak consumption). The second is a measure of peak consumption during the same summer months. The Southern California Electric Company would like to determine whether there are any differences in the average electrical consumption of those homes with air-conditioning compared to those without. Assume that the following results were observed for the two types of homes sampled:

$$\text{Group 1: } \bar{x}_1 = \begin{bmatrix} 8.3 \\ 4.1 \end{bmatrix} \quad S_1 = \begin{bmatrix} 2 & 1 \\ 1 & 6 \end{bmatrix}$$

$$\text{Group 2: } \bar{x}_2 = \begin{bmatrix} 10.2 \\ 3.9 \end{bmatrix} \quad S_2 = \begin{bmatrix} 2 & 1 \\ 1 & 4 \end{bmatrix}$$

If univariate $t$ statistics are computed for the two measurements, depending on which measurement is chosen, one will arrive at different conclusions about the two groups. For example, if the two groups are compared on peak consumption, the results indicate that there are no significant differences between the groups at the 5% level of significance. The $t$ statistic for peak consumption is:

$$t = \frac{4.1 - 3.9}{\sqrt{\left[\frac{44(36) + 54(16)}{98}\right]\left[\frac{1}{45} + \frac{1}{55}\right]}} = 0.20$$

However, if off-peak consumption is compared, the results indicate that there are significant differences (at the 5% level) between the homes in terms of consumption during the summer months of June through August. Now the value of $t$ becomes:

$$t = \frac{8.3 - 10.2}{\sqrt{\left[\frac{44(4) + 54(4)}{98}\right]\left[\frac{1}{45} + \frac{1}{55}\right]}} = -4.75$$

This result indicates that homes without air-conditioning units have a greater amount of electrical consumption during off-peak hours than homes with air-conditioning units. It appears, therefore, that in a univariate analysis the choice of the "correct" measurement to compare the two groups can have an effect on one's conclusions. Unfortunately, another problem with conducting separate univariate analyses of multiple dependent variables in the same study is the escalation of error rates. When several dependent variables are included in a study, about 37% of the nonsignificant effects in the population are erroneously declared significant. A multivariate analysis, however, takes both measurements into account and also controls for inflated error rates. This permits a more accurate assessment of group differences.

In order to conduct Hotelling's $T^2$ test, one would first need to compute the following intermediate results:

$$\bar{x}_1 - \bar{x}_2 = \begin{bmatrix} 8.3 \\ 4.1 \end{bmatrix} - \begin{bmatrix} 10.2 \\ 3.9 \end{bmatrix} = \begin{bmatrix} -1.9 \\ 0.2 \end{bmatrix}$$

$$S = \frac{(n_1 - 1)S_1 + (n_2 - 1)S_2}{n_1 + n_2 - 2} = \begin{bmatrix} 2.00 & 1.00 \\ 1.00 & 4.90 \end{bmatrix}$$

and

$$S^{-1} = \begin{bmatrix} 0.56 & -0.11 \\ -0.11 & 0.23 \end{bmatrix}$$

Substitution of these values in Hotelling's $T^2$ equation yields:

$$T^2 = \frac{(45)(55)}{45 + 55} \begin{bmatrix} -1.9 & 0.2 \end{bmatrix} \begin{bmatrix} 0.56 & -0.11 \\ -0.11 & 0.23 \end{bmatrix} \begin{bmatrix} -1.9 \\ 0.2 \end{bmatrix} = 93.55$$

The corresponding $F$ statistic is determined to be equal to:

$$F = [97/(98)(2)]\ 93.55 = 46.29$$

which exceeds the 99th percentile of the $F$ distribution with 2 and 97 degrees of freedom (according to the given relationship between the $T^2$ and $F$ distribution). Based on this result, there is evidence to conclude that there are significant differences between the two groups of homes in terms of energy consumption. Of course, once this result is obtained, follow-up univariate analyses could be conducted to determine which of the two variables is more important.

## MULTIVARIATE ANALYSIS OF VARIANCE

Often more than two groups that have been observed on several variables need to be compared. When two or more groups are compared, the procedure is commonly referred to as a one-way multivariate analysis of variance (MANOVA). The MANOVA procedure was originally developed by Wilks (1932) on the basis of the generalized likelihood-ratio principle. The theory of the likelihood-ratio principle is beyond the scope of this book (for a treatment of this topic see Bickel & Doksum, 1977; Morrison, 1990). In general, the likelihood-ratio principle provides several optimal properties for reasonably sized samples, and is convenient for hypotheses formulated in terms of multivariate normal parameters. In particular, the attractiveness of the likelihood ratio presented by Wilks is that it yields test statistics that reduce to the familiar univariate $F$ and $t$ statistics. For the purposes of this book, it suffices to know that the likelihood-ratio principle led to the test statistic called Wilks's lambda ($\Lambda$). In the one-way MANOVA Wilks's $\Lambda$ is equal to:

$$\Lambda = \frac{|W|}{|B+W|} = \frac{|W|}{|T|}$$

where $|W|$ is the determinant of the within-groups SSCP matrix, $|B+W|$ is the determinant of the between and within-groups SSCP matrix, and $|T|$ is the determinant of the total sample SSCP matrix.

It is well known that Wilks's $\Lambda$ reduces to the univariate $F$ test when only one variable is under study (i.e., $p = 1$). When only one variable is considered, $|W| = SS_{within}$ and $|B+W| = SS_{between} + SS_{within}$. Hence, the value of Wilks's $\Lambda$ is:

$$\Lambda = \frac{SS_{within}}{SS_{within} + SS_{between}}$$

Because the $F$ ratio in a univariate ANOVA is traditionally formulated as

$$F = \frac{SS_{between}/df_{between}}{SS_{within}/df_{within}}$$

it should be clear that Wilks's $\Lambda$ can also be written as:

$$\Lambda = \frac{1}{1 + [(k-1)/(n-k)]F}$$

This indicates the relationship between $\Lambda$ and $F$ is somewhat inverse. The larger the $F$ ratio is, the smaller the Wilks's $\Lambda$. It follows, therefore, that the greater is the disparity between observed groups, the smaller is the value of Wilks's $\Lambda$.

The formula presented for Wilks's $\Lambda$ can be considered definitional because the actual computation of $\Lambda$ is made routine by the relationship between the determinant and the eigenvalues of a matrix (see chapter 2 for a discussion on eigenvalues). To make this relationship clear, consider Wilks's $\Lambda$ rewritten as:

$$\frac{1}{\Lambda} = \frac{|B+W|}{|W|} = |W^{-1}(B+W)| = |W^{-1}B + I|$$

Recall from chapter 2 that for any matrix A there are $\lambda_i$ eigenvalues, and that for a matrix $(A + I)$ there are $(\lambda_i + 1)$ eigenvalues. It is also well known that the product of the eigenvalues of a matrix is always equal to the determinant of the matrix (i.e., $\Pi\lambda_i = |A|$). Hence, $\Pi(\lambda_i + 1) = |A + I|$). Based on this information, the value of Wilks's $\Lambda$ can also be written as the product of the eigenvalues of the matrix $W^{-1}B$:

$$\Lambda = \frac{1}{\Pi(\lambda_i + 1)}$$

or simply

$$\frac{1}{\Lambda} = \Pi(\lambda_i + 1)$$

where $\lambda_i$ is the nonzero eigenvalues of the matrix $W^{-1}B$.

Wilks's $\Lambda$ is the oldest and most widely used criterion for comparing groups, but there are several others available. The alternative criteria are all functions of the eigenvalues of $W^{-1}B$. As such, although computing the eigenvalues of a matrix is best left up to a computer, understanding the eigenvalue approach to Wilks's $\Lambda$ gives additional insight into the other available criteria.

There are three other popular criteria that are also displayed as standard computer output for a MANOVA analysis. These are Hotelling's trace, Pillai's trace, and Roy's greatest root. The criteria are defined as follows:

Hotelling's trace $\quad T = \Sigma\lambda_i$

Pillai's trace $\quad V = \Sigma \dfrac{\lambda_i}{1 + \lambda_i}$

Roy's greatest root $\quad \theta = \dfrac{\lambda_{max}}{1 + \lambda_{max}}$

It is important to note that Pillai's and Hotelling's criteria are functions of the summation of the eigenvalues of $W^{-1}B$, whereas Roy's uses only the largest eigenvalue. It is for this reason that these test statistics often lead to different statistical results. Only with one eigenvalue (i.e., $p = 1$, the univariate case) will all four criteria provide identical results.

Although tables of $\Lambda$, $V$, $T$, and $\theta$ for assessing statistical significance exist, they are not widely available. Several authors have successfully investigated the notion that these tabled values can be reasonably approximated using an $F$ distribution (Rao, 1952, 1965; Schatzoff, 1964, 1966). Thus, it is common practice to assess the significance of a MANOVA by using the $F$ distribution approximation to these tests rather than looking up the exact value of the test statistic. Because the computational intricacies of these approximations can be quite tedious, the SAS and most other packages provide this information as standard output.

There has been considerable disagreement in the literature concerning which of the four multivariate criteria is "best," and the answer appears to be that none of them is uniformly superior. Schatzoff (1966) conducted a Monte Carlo study to examine the sensitivities of the test statistics. His results indicated that the best were Wilks's and Hotelling's. More recently, Olson (1976) conducted a study and found that for small sample sizes Pillai's was most appropriate. However, when the sample sizes are large, the test statistics appear to be equivalent. The best strategy is to examine all four criteria. If the results are similar, choosing the criterion to use is just a matter of personal taste. If the results are different, look for some consensus between at least two test criteria.

There has also been a bewildering array of follow-up tests that are designed to assess the relative importance of the individual dependent variables. In general, the procedure most frequently used is the follow-up univariate $F$ test. This strategy is based on the logic of Fisher's least significant difference procedure. The least significant difference procedure entails the computation of separate univariate ANOVAs on each dependent variable when a significant MANOVA has been observed. Any significant univariate $F$ test indicates the importance of that dependent variable to the multivariate analysis.

Many researchers argue that examining only the statistical significance of an effect can be misleading. This is because virtually any nonzero effect can be shown to be statistically significant with a suffiently large sample. For this reason, a measure of the practical importance of results is also used. The most common measure used is known as the eta-square ($\eta^2$) measure of association. This measure can be interpreted rather loosely as the proportion of variance in the dependent variables that is accounted for by the independent variables. Using Wilks's criterion this measure is defined as:

$$\eta^2 = 1 - \Lambda$$

and in the univariate case this value is $\eta^2 = 1 - (SS_{wit}/SS_{tot})$.

## AN EXAMPLE OF MULTIVARIATE ANALYSIS
## OF VARIANCE USING SAS

Marcoulides (1989b) conducted a study to compare the effectiveness of two types of instructional aids for improving student performance in an introductory statistics course. The two types of instructional aids used were an expert system program (ES), and a computer-assisted instructional program (CAI). Students who participated in the study were randomly assigned to three treatment groups (for the purposes of this example the original study is somewhat modified): (a) an ES as an adjunct to lectures, (b) a CAI program as an adjunct to lectures, and (c) a control group with lectures (LE) alone. At the conclusion of the study students were tested using a Statistics Achievement Test (SAT). The SAT test measured two cognitive abilities. It measured the ability to select and the ability to apply the appropriate statistical procedure to analyze a set of data. Thus, the SAT consisted of two separate parts. The first part, considered the knowledge part, consisted of 50 forced-choice questions that examined the students' knowledge of selecting the appropriate statistical procedure. The second part, the application part, consisted of 10 questions that examined students' ability to apply the appropriate statistical procedure. Scoring of the test was accomplished by assigning 1 point to each forced-choice item in the knowledge part having a correct response and 5 points to each application problem. Thus, each part was scored on a 50-point scale.

This study is an application of the classic experimental design method using a control group. The independent variable in this study was type of teaching aid used as an adjunct to classroom instruction. Although students were randomly assigned to one of three teaching conditions, all students attended the same lectures and received identical instruction throughout the study. The dependent variables in this study are the total scores on the two parts of the SAT. Table 3.3 presents the results of this study.

The deck setup to perform the statistical analysis in SAS on this example data set is presented in Table 3.4. Recall from chapter 1 that the DATA command provides the name for the data set that follows, and the INPUT command names the variables to be studied. Once an SAS data set has been created, all types of statistical analyses can be performed using the PROC statement. To perform a multivariate analysis on this data set two SAS procedures can be used: the PROC ANOVA or the PROC GLM. The main difference between the two procedures is that ANOVA is designed to handle balanced data (i.e., data with equal numbers of observations for every combination of factors in the study), whereas GLM can analyze both balanced and unbalanced data. There really is no preference for one procedure over the other when the data are balanced. In fact, when the data are balanced both procedures produce identical results. However, because PROC ANOVA does not check to see if the data are balanced, sometimes an unbalanced data set can incorrectly get analyzed as a balanced data set. Thus, if it is clear that the data are balanced use the PROC ANOVA, and otherwise use PROC GLM to conduct the analysis.

The data set presented in Table 3.3 is balanced, so either PROC ANOVA or PROC GLM can be used to generate the output presented in Table 3.5.

TABLE 3.3
Observed Student Scores on the SAT

| ES | | CAI | | LE | |
|------|------|------|------|------|------|
| K | A | K | A | K | A |
| 47 | 27 | 46 | 26 | 34 | 22 |
| 45 | 33 | 44 | 30 | 39 | 22 |
| 28 | 31 | 41 | 26 | 38 | 24 |
| 44 | 24 | 41 | 27 | 42 | 24 |
| 45 | 32 | 46 | 32 | 38 | 21 |
| 40 | 30 | 36 | 28 | 35 | 21 |
| 48 | 36 | 45 | 25 | 29 | 20 |
| 39 | 31 | 43 | 28 | 39 | 23 |
| 45 | 32 | 33 | 24 | 45 | 22 |
| 34 | 22 | 40 | 23 | 37 | 22 |

Several other statements are also needed in order to conduct this analysis. These are the CLASS, the MODEL, the MEANS, and the MANOVA statements (for complete details see the *SAS User's Guide*). The CLASS statement defines the independent variables used in the study. The MODEL statement names the dependent variables and the independent variables to be used in the analyses. The MEANS statement is used to produce mean values for the independent variables included in the CLASS statement and also provides post hoc comparison tests (in this case Fisher's least significant difference was selected but there are many others available in this procedure). Finally, the MANOVA statement provides the four commonly used multivariate tests of significance and provides univariate analysis of variance on each dependent variable.

As can be seen in Table 3.5, the four multivariate tests of significance indicate that there are significant differences between the three groups with respect to the scores from the two parts of the SAT test. As expected, the values of the tests and their $F$ approximation are somewhat different but they lead to the same statistical result. The eigenvalues used to produce these tests are also presented in Table 3.5. As can be seen, both eigenvalues are positive with the first value being the largest.

Table 3.6 presents the univariate analysis of variance generated by SAS for each dependent variable. The results indicate that there are significant differences between the groups only on the dependent variable SCORE2 (i.e., with $F_{2,27} = 16.64$, $p < .0001$). The $R$-square measure of association (notation used in SAS instead of the $\eta^2$ measure of association) of the practical importance of the results is also provided in Table 3.6. This value indicates that roughly 55% of the variance in SCORE2 can be accounted for by the type of instructional aid used. The multivariate equivalent of this value is

TABLE 3.4
SAS Setup for Example Data

```
DATA EXAMPLE1;
 INPUT GROUP SCORE1 SCORE2;
 CARDS;
1   47  27
1   45  33
1   28  31
1   44  24
1   45  32
1   40  30
1   48  36
1   39  31
1   45  32
1   34  22
2   46  26
2   44  30
2   41  26
2   41  27
2   46  32
2   36  28
2   45  25
2   43  28
2   33  24
2   40  23
3   34  22
3   39  22
3   38  24
3   42  24
3   38  21
3   35  21
3   29  20
3   39  23
3   45  22
3   37  22
;
PROC ANOVA;
 CLASS GROUP;
 MODEL SCORE1 SCORE2 = GROUP;
 MEANS GROUP/LSD;
MANOVA H=GROUP;
```

found by using Wilks's criterion (i.e., $\eta^2 = 1 - \Lambda = 0.56$), and indicates that about 56% of the variance of the dependent variables is accounted for by the type of instructional aid. Obviously, there is only an improvement of a little less than 1% when both dependent variables are considered. This would seem to indicate that the observed differences between the three groups are mainly on the application part of the SAT test.

## TABLE 3.5
### SAS Analysis of Variance Procedure, Multivariate Analysis of Variance

Characteristic Roots and Vectors of: E Inverse * H, where
H = Anova SS&CP Matrix for Group   E = Error SS&CP Matrix

| Characteristic Root | Percent | Characteristic Vector V'EV=1 | |
| --- | --- | --- | --- |
| | | SCORE1 | SCORE2 |
| 1.2362907859 | 98.16 | -0.00225688 | 0.06514313 |
| 0.0231867836 | 1.84 | 0.04055500 | -0.02172142 |

Manova Test Criteria and F Approximations for
the Hypothesis of no Overall GROUP Effect
H = Anova SS&CP Matrix for GROUP   E = Error SS&CP Matrix

S=2   M=-0.5   N=12

| Statistic | Value | F | Num DF | Den DF | Pr > F |
| --- | --- | --- | --- | --- | --- |
| Wilks' Lambda | 0.43703559 | 6.6646 | 4 | 52 | 0.0002 |
| Pillai's Trace | 0.57549230 | 5.4539 | 4 | 54 | 0.0009 |
| Hotelling-Lawley Trace | 1.25947757 | 7.8717 | 4 | 50 | 0.0001 |
| Roy's Greatest Root | 1.23629079 | 16.6899 | 2 | 27 | 0.0001 |

*Note.* Characteristic roots are eigenvalues; vectors referred to are eigenvectors. *F* Statistic for Roy's greatest root is an upper bound. *F* Statistic for Wilks' lambda is exact.

## TABLE 3.6
### SAS Analysis of Variance Procedure

Dependent Variable: SCORE1

| Source | DF | Sum of Squares | Mean Square | F Value | Pr > F |
| --- | --- | --- | --- | --- | --- |
| Model | 2 | 101.4000000 | 50.7000000 | 1.95 | 0.1616 |
| Error | 27 | 701.4000000 | 25.9777778 | | |
| Corrected Total | 29 | 802.8000000 | | | |

| R-Square | C.V. | Root MSE | SCORE1 Mean |
| --- | --- | --- | --- |
| 0.126308 | 12.67871 | 5.096840 | 40.2000000 |

Dependent Variable: SCORE2

| Source | DF | Sum of Squares | Mean Square | F Value | Pr > F |
| --- | --- | --- | --- | --- | --- |
| Model | 2 | 302.4666667 | 151.2333333 | 16.64 | 0.0001 |
| Error | 27 | 245.4000000 | 9.0888889 | | |
| Corrected Total | 29 | 547.8666667 | | | |

| R-Square | C.V. | Root MSE | SCORE2 Mean |
| --- | --- | --- | --- |
| 0.552081 | 11.47758 | 3.014778 | 26.2666667 |

TABLE 3.7
Analysis of Variance Procedure

```
T tests (LSD) for variable: SCORE1
```

NOTE: This test controls the type I comparisonwise error rate not the experimentwise error rate.

```
    Alpha= 0.05  df= 27  MSE= 25.97778
    Critical Value of T= 2.05
    Least Significant Difference= 4.6769
```

Means with the same letter are not significantly different.

| T Grouping | Mean | N | GROUP |
|---|---|---|---|
| A | 41.500 | 10 | 1 |
| A | | | |
| A | 41.500 | 10 | 2 |
| A | | | |
| A | 37.600 | 10 | 3 |

```
    T tests (LSD) for variable: SCORE2
```

NOTE: This test controls the type I comparisonwise error rate not the experimentwise error rate.

```
    Alpha= 0.05  df= 27  MSE= 9.088889
    Critical Value of T= 2.05
    Least Significant Difference= 2.7664
```

Means with the same letter are not significantly different.

| T Grouping | Mean | N | GROUP |
|---|---|---|---|
| A | 29.800 | 10 | 1 |
| B | 26.900 | 10 | 2 |
| C | 22.100 | 10 | 3 |

Table 3.7 provides the post hoc comparisons (i.e., *t* test) for the three groups on each dependent variable. As can be seen in Table 3.7, there are no differences between the three groups on the knowledge part of the SAT. There are significant differences between the three groups, however, on the application part of the SAT. In fact, based on the results provided, it appears that the highest performance on the application part of the SAT is from the students that used the expert system (ES) as an adjunct to lectures. This is followed by those that used the CAI as an adjunct to lectures.

The results of this study indicate that both types of instructional aids benefit students more than just receiving lectures. Of course, the ES approach seems to provide the best results. These differences were only observed across the cognitive category of application. The method of presentation of material has been a topic of continued interest in the literature. The use of computerized learning programs, as demonstrated in this study, can help students apply statistical procedures. It is generally expected that students taking a statistics course will eventually be involved in some form of decision making. As decision makers, they are expected to gather information and select and apply statistical procedures in such a way that alternative choices can be developed. Using the MANOVA procedure in this study, it was determined that computer-based learning programs can help students improve performance.

Although the style for reporting results in MANOVA might vary from one researcher to the next, it is essential that at a minimum one report the test statistic used, the $F$ test approximation, the univariate analyses of variance, the post hoc comparisons, and the measures of association. In this way other perusers can evaluate the results and arrive at similar conclusions.

## EVALUATING MULTIVARIATE NORMALITY

Multivariate normality is the assumption that each variable and all linear combinations of the variables observed are normally distributed. Unfortunately, to date the literature concerning the sensitivity of multivariate tests of significance to violations of this assumption is mixed and inconclusive. What does appear clear, however, is that the larger the sample size, the less effect nonnormality has on these tests. Thus, the central limit theorem plays just as important a role in multivariate statistical inference as in the univariate case. Nevertheless, the safest approach is to always check for potential violations of the normality assumption.

The assumption of normality should be checked for all multivariate distributions. However, for practical reasons it is usually sufficient to investigate the univariate and bivariate distributions of each variable observed. The simplest way to examine univariate normality is to examine two statistical indices: skewness and kurtosis. *Skewness* is an index that represents the symmetry of a univariate distribution. A normally distributed variable will have a skewness value of zero. Any deviation from this value indicates the direction of the skew in the distribution. If there is a positive skew, most of the observations are on the left of the distribution (i.e., the mean is not in the center). A negative skew implies that the observations are piled on the right of the distribution. *Kurtosis* has to do with the shape of the distribution in terms of its peakedness. The distribution can either be too thin or too flat. Actually, the terms *leptokurtic* and *platykurtic* are used to describe these types of distributions. A normally distributed variable will have a kurtosis of 3. However, in order to simplify things, the SAS computer package (and all others) uses 0 (by first subtracting 3) as the measure of kurtosis. As such, positive values indicate a leptokurtic distribution and negative values

TABLE 3.8
SAS Univariate Procedure

```
Variable=SCORE1
                                    Moments
      N              30           Sum Wgts           30
      Mean           40.2         Sum                1206
      Std Dev        5.261441     Variance           27.68276
      Skewness       -0.62753     Kurtosis           -0.23871
      USS            49284        CSS                802.8
      CV             13.08816     Std Mean           0.960603
      T:Mean=0       41.8487      Prob>|T|           0.0001
      Sgn Rank       232.5        Prob>|S|           0.0001
      Num ^= 0       30

Variable=SCORE2
                                    Moments
      N              30           Sum Wgts           30
      Mean           26.26667     Sum                788
      Std Dev        4.346488     Variance           18.89195
      Skewness       0.483742     Kurtosis           -0.87024
      USS            21246        CSS                547.8667
      CV             16.54754     Std Mean           0.793556
      T:Mean=0       33.09994     Prob>|T|           0.0001
      Sgn Rank       232.5        Prob>|S|           0.0001
      Num ^= 0       30
```

indicate a platykurtic distribution. In order to obtain the values of skewness and kurtosis in the SAS package, one can use the UNIVARIATE or the MEANS procedures. Table 3.8 presents a sample output of the UNIVARIATE procedure with the values of skewness and kurtosis for the two parts of the SAT test used in the previous example. As can be seen, both values are close to zero and satisfy the normality assumption (of course, the decision is somewhat subjective, but with small samples values in the interval ±1 are considered acceptable).

Although it is quite likely that the assumption of multivariate normality is met if all the observed variables are individually normally distributed, it is preferable to also examine bivariate normality. If the observations were generated from a multivariate normal distribution, each bivariate distribution should also be normal. This implies that the contours of a bivariate distribution would be ellipses, and any plot should conform to this structure by exhibiting a pattern that is nearly elliptical. Although generating plots of bivariate distributions is relatively straight forward, assessing their elliptical pattern is quite subjective and prone to error. A somewhat more formal method for judging bivariate normality is based on a plot of chi-square ($\chi^2$) percentiles and the distance measures of observations. The $\chi^2$ percentile for each observation is the $100(\text{observation number} - \frac{1}{2})/n$ percentile of the chi-square distribution with 2 degrees of freedom (because the number of variables is 2). The distance measure (called the generalized distance) is defined for each observation as:

$$D^2 = (x - \bar{x})'S^{-1}(x - \bar{x})$$

where $x$ is the observation vector for $x_1$ and $x_2$, and $\bar{x}$ is a mean vector for the two variables.

If the bivariate distribution is normal, the plot of the chi-square percentile and the generalized distance should resemble a straight line. A systematic curved pattern suggests a lack of normality. Any points far from the line indicate outlying observations that merit further attention (for a complete review of methods see Gnanadesikian, 1977).

Table 3.9 presents an example data set used to construct the plot displayed in Fig. 3.5. The data represent the assets ($X_1$) and net incomes ($X_2$) in millions of dollars for a randomly selected group of Greek corporations. The values for the generalized distance measure and the $\chi^2$ percentile for each observation are also presented in Table 3.9. For example, the value for the generalized distance measure for the first observation using

$$\bar{x} = \begin{bmatrix} 22.63 \\ 1.77 \end{bmatrix} \quad S = \begin{bmatrix} 96.56 & 8.04 \\ 8.04 & 1.33 \end{bmatrix} \quad S^{-1} = \begin{bmatrix} 0.02 & -0.12 \\ -0.12 & 1.51 \end{bmatrix}$$

was found to be:

$$\begin{bmatrix} 22.49 & -22.63 \\ 1.99 & -1.77 \end{bmatrix}, \begin{bmatrix} 0.02 & -0.12 \\ -0.12 & 1.51 \end{bmatrix} \begin{bmatrix} 22.49 & -22.63 \\ 1.99 & -1.77 \end{bmatrix} = 0.08$$

As can be seen in Fig. 3.5, the plot resembles a straight line, indicating that the bivariate distribution is normal.

TABLE 3.9
Example Data Set

| Corporation | $(X_1)$ | $(X_2)$ | $D^2$ | $\chi^2$ |
|-------------|---------|---------|-------|----------|
| Cyprus Cement | 22.49 | 1.99 | 0.08 | .10 |
| Hanseatic Shipping | 17.33 | 1.17 | 0.33 | .33 |
| Amathus Navigation | 16.05 | 1.29 | 0.45 | .58 |
| Hellenic Chemical | 16.63 | 0.94 | 0.54 | .86 |
| Elma Holdings | 22.14 | 1.05 | 0.69 | 1.20 |
| Cyprus Pipes | 24.13 | 1.17 | 0.82 | 1.60 |
| F. W. Woolworth | 9.02 | 0.23 | 2.16 | 2.10 |
| Zako Ltd. | 22.25 | 3.16 | 3.07 | 2.77 |
| Vassiliko Cement | 31.27 | 3.86 | 3.64 | 3.79 |
| Keo Ltd. | 44.98 | 2.81 | 6.18 | 5.99 |

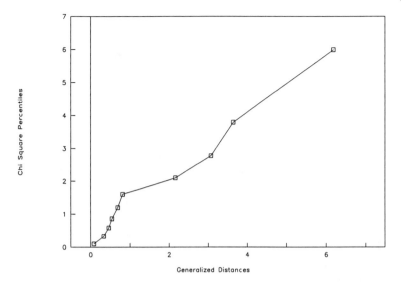

FIG. 3.5.  Chi-square plot of generalized distances in Table 3.9.

Two other assumptions are also considered important in multivariate statistical inference: linearity and homoskedasticity. The *assumption of linearity* implies that there is a straight-line relationship between any two observed variables or linear combinations of variables. A simple way of examining linearity is to generate scatterplots between pairs of observed variables and look for any systematic curve patterns. However, because the assumption of multivariate normality implies that there is linearity between all pairs of variables, once the assumption of normality is examined it is redundant to conduct a test of linearity.

The *assumption of homoskedasticity* implies that the variabilities of observed variables are similar to each other. This assumption is also referred to as the homogeneity of variance assumption. Although there are tests of this assumption (e.g., Box's $M$ test), an examination of normality is usually sufficient to assure homoskedasticity. When the assumption of multivariate normality is met, the observed variables have similarly shaped distributions and, as such, the variability in the distributions is similar.

If the assumption of normality is not a viable assumption, one alternative is to try to make the nonnormal data more "normal looking" by applying a tranformation to the data. Once the data have been transformed, normal theory analysis can then be carried out. In general terms, transformations are nothing more than a reexpression of the data in different units of magnitude. Many methods of transformation have been proposed in the literature. The most popular transformations included in the statistical packages are power tranformations (e.g., squaring an observation), square-root transformations, reciprocal transformations, and logarithm and arcsine transformations (see the *SAS User's Guide* for further details). Unfortunately, there is no one best method available that will always correct a nonnormality problem. At present, letting the data suggest a transformation method seems to be the best option available.

## EXERCISES

1. The imposition of a residential property tax is always a topic of debate in the United Kingdom. Currently, property tax is based on an old assessment system that dates back several hundred years. It is argued that this system produces inequity because newer homes tend to be assessed higher taxes than older homes. Assume that a new system is proposed in Parliament that is based on the market value of the house. Opponents of this new system argue that at least in the city of London some boroughs would have to pay considerably more tax than others. In order to conduct a study to address this concern, a random sample of 10 homes from each borough was selected and their property tax was assessed under both plans. The results are displayed in the accompanying table in thousands of pounds. Conduct an analysis and determine whether there are any differences between the boroughs with respect to the two tax systems.

| Borough | | | | | | | | | |
|---|---|---|---|---|---|---|---|---|---|
| A | | B | | C | | D | | E | |
| Old | New | Old | New | Old | New | Old | New | Old | New |
| 5 | 6 | 7 | 11 | 2 | 2 | 3 | 4 | 3 | 5 |
| 13 | 15 | 20 | 30 | 13 | 14 | 8 | 10 | 3 | 7 |
| 16 | 18 | 14 | 35 | 9 | 11 | 3 | 9 | 2 | 6 |
| 10 | 21 | 11 | 27 | 8 | 9 | 3 | 6 | 4 | 5 |
| 21 | 26 | 17 | 33 | 7 | 7 | 7 | 9 | 2 | 3 |
| 31 | 32 | 36 | 43 | 6 | 10 | 8 | 12 | 5 | 7 |
| 33 | 46 | 21 | 42 | 6 | 13 | 8 | 10 | 6 | 6 |
| 24 | 27 | 12 | 16 | 7 | 10 | 9 | 14 | 7 | 9 |
| 15 | 17 | 9 | 17 | 8 | 11 | 7 | 9 | 1 | 2 |
| 16 | 19 | 3 | 9 | 2 | 4 | 5 | 5 | 2 | 3 |

2. Data were collected on a class of 20 business students who took an introductory computer information systems course that required extensive computer work. Students from four different business majors participated in this course. The four majors were accounting (AC), marketing (MK), management science (MS), and management (MG). At the beginning and at

the end of the course, each student completed the Computer Anxiety Scale (CAS). The CAS is a measure of the perception held by students of their anxiety in different situations related to computers. The CAS contains 20 items that are rated on a 5-point scale reflecting a level of anxiety from "not at all" to "very much." As students rate their anxiety on each item from a low of 1 to a high of 5, the range of possible total scores is from 20 to 100, with higher scores indicating more self-reported anxiety. The table that follows provides the results of the study. Conduct an analysis of variance on this data and determine whether there are any significant differences among the majors with respect to their recorded level of anxiety both at the beginning and at the end of the course.

CAS Results

| Student | Major | Before | After |
|---------|-------|--------|-------|
| 1 | AC | 75 | 65 |
| 2 | AC | 65 | 64 |
| 3 | AC | 72 | 67 |
| 4 | AC | 79 | 76 |
| 5 | MK | 66 | 76 |
| 6 | MK | 54 | 77 |
| 7 | MK | 43 | 56 |
| 8 | MK | 33 | 45 |
| 9 | MK | 57 | 54 |
| 10 | MS | 34 | 24 |
| 11 | MS | 33 | 23 |
| 12 | MS | 30 | 20 |
| 13 | MS | 29 | 23 |
| 14 | MS | 37 | 27 |
| 15 | MS | 35 | 25 |
| 16 | MS | 41 | 31 |
| 17 | MG | 76 | 74 |
| 18 | MG | 78 | 68 |
| 19 | MG | 89 | 91 |
| 20 | MG | 67 | 77 |

3. The admissions officer of a business school is trying to determine what criteria are being used to admit students to the school's graduate programs. Data were collected on recent applicants who were categorized into three groups: (a) admit, (b) do not admit, and (c) borderline. The data are presented in the table that follows. The variables include X1, the undergraduate grade point average; X2, the graduate management aptitude test; and X3, an "index" for reference letters. Conduct an analysis of variance using the three variables to determine what criteria are being used for admission.

Student Scores

| Admit | | | Not Admit | | | Borderline | | |
|---|---|---|---|---|---|---|---|---|
| X1 | X2 | X3 | X1 | X2 | X3 | X1 | X2 | X3 |
| 2.96 | 596 | 17.6 | 2.54 | 446 | 18.1 | 2.86 | 494 | 17.0 |
| 3.14 | 473 | 13.4 | 2.43 | 425 | 19.7 | 2.85 | 496 | 12.5 |
| 3.22 | 482 | 10.3 | 2.20 | 474 | 16.9 | 3.14 | 419 | 21.5 |
| 3.29 | 527 | 22.3 | 2.36 | 553 | 23.7 | 3.28 | 371 | 22.2 |
| 3.69 | 505 | 20.5 | 2.57 | 542 | 19.2 | 2.81 | 447 | 13.0 |

4. The Internal Revenue Service (IRS) is always looking for ways to improve the format and wording of their tax return forms. Two new formats have recently been developed. To determine the best form, 30 individuals were asked to participate in an experiment. Each form was filled out by 10 different people. The amount of time taken by each person to complete the forms and the accuracy level (counted as number of arithmetic errors) for each person are recorded in the table that follows. Determine which form is the best form for the IRS to adopt.

| Form A | | Form B | | Form C | |
|---|---|---|---|---|---|
| Time | Accuracy | Time | Accuracy | Time | Accuracy |
| 9.0 | 5 | 6.2 | 4 | 7.5 | 3 |
| 8.6 | 7 | 6.9 | 7 | 6.7 | 4 |
| 9.3 | 8 | 7.8 | 9 | 5.3 | 5 |
| 8.3 | 5 | 6.6 | 7 | 6.5 | 7 |
| 9.9 | 8 | 7.1 | 3 | 8.1 | 2 |
| 11.2 | 10 | 9.0 | 5 | 6.4 | 5 |
| 9.1 | 9 | 10.1 | 7 | 7.2 | 6 |
| 6.9 | 7 | 7.6 | 6 | 7.3 | 5 |
| 8.3 | 5 | 7.7 | 7 | 4.1 | 1 |
| 9.8 | 9 | 8.4 | 4 | 5.5 | 3 |

5. A business student who was about to graduate with an MBA degree decided to survey various marketing companies in the Los Angeles area to examine which offered the best chance for early promotion and economic advancement. The student randomly sampled 25 different marketing companies and asked each company to provide data on how much time elapsed (on the average) after the initial hiring before each employee with an MBA received a promotion. The student also asked each company to report what, if any, salary increases (also averaged) went along with each promotion. Suspecting that there might be differences in both the promotion (reported as number of weeks) and salary increases (reported in dollars per month)

received depending on the size of the company, the student grouped the data by the size of the company. The data appears in the table that follows. Do the data provide sufficient evidence to indicate that there are differences among the various companies?

| | | Company Size | | | |
|---|---|---|---|---|---|
| Small | | Medium | | Large | |
| Time | $ | Time | $ | Time | $ |
| 20 | 100 | 30 | 50 | 10 | 0 |
| 36 | 200 | 10 | 75 | 25 | 250 |
| 76 | 50 | 24 | 100 | 12 | 500 |
| 12 | 125 | 27 | 25 | 12 | 50 |
| 36 | 75 | 24 | 200 | 24 | 200 |
| 24 | 250 | 12 | 400 | 8 | 50 |
| 36 | 500 | 16 | 50 | 16 | 100 |
| 32 | 80 | | | 18 | 75 |
| 12 | 100 | | | 26 | 50 |

# Factorial Multivariate Analysis of Variance

The one-way multivariate analysis of variance procedure was presented in chapter 3 as a generalization of the one-way univariate analysis of variance. The univariate analysis of variance (ANOVA) is one of the most frequently used procedures for comparing estimates of differences between groups on a single variable. When more than two groups are observed on several variables, a multivariate analysis of variance (MANOVA) is used. For example, if a production manager at an assembly plant is investigating the average amount of time it takes workers in three different shifts to assemble a particular automobile part, then an ANOVA can be used to determine whether there are any differences between the shifts. However, if the production manager also measures the rate of accuracy of assembly conducted by each worker, then a MANOVA must be used to determine whether there are any differences between the shifts on the two measures.

The analysis of variance procedures presented in the previous chapter focused on determining differences between groups from a single classification on some variables of interest. In design terminology a classification scheme is generally called a factor, and each condition within it represents a level of that factor. For example, the morning, afternoon, and evening groups are the three levels of the shift factor. Notice that the levels of a factor do not necessarily represent different amounts of the factor, as is demonstrated by the levels of the shifts factor. Studies that involve only one factor are commonly referred to as one-factor or one-way designs, which is why the analysis of variance used is generally referred to as a "one way."

There are many situations, however, where more than one factor is examined. For example, if the production manager is also interested in examining any potential differences between male and female assembly workers in the three shifts, then gender would be included in the study as a second factor. In such a two-factor design, in order to examine differences between groups, a factorial analysis of variance would be used. Factorial analysis of variance is just an extension of the one-factor model presented in the previous chapter.

In a two-factor analysis three types of questions can be asked about the observed data. First, is there a significant difference between the levels of the shifts factor? In other words, on the average, is there a difference with respect to the amount of time it takes workers in the different shifts to assemble the automobile part? Second, is there a difference between the levels of the gender factor? In this case, is there a difference between male and female assembly workers in the three shifts? Third, is there an interaction between the shifts factor and the gender factor? Specifically, are there differences between males and females in one shift that do not show up in other shifts?

Although factorial designs with many factors can be analyzed, this chapter is restricted to univariate and multivariate factorial designs with two independent variables. The analysis of variance applied to factorial designs with two factors is also called a two-way analysis of variance. Restricting the focus to the two-way case is not a major limitation, because the material presented can readily be extended to factorial designs with any number of factors. A second restriction is that the presentation is limited to designs with equal numbers of subjects in each cell. This simplifies the presentation of the method and the interpretation of results.

The two-factor multivariate analysis of variance is just a generalization of the two-factor univariate ANOVA but extended to several variables. Therefore, before introducing the two-factor MANOVA procedure, it is essential that the reader have a firm grasp of the concepts of the univariate two-factor ANOVA. The next section provides that foundation.

## UNIVARIATE TWO-WAY ANALYSIS OF VARIANCE

The purpose of the two-way analysis of variance (ANOVA) is to compare the mean scores from several groups in a factorial design in order to decide whether the differences between the means are due to chance or to the effect of the first factor (called the *main effect for factor A*), the second factor (the *main effect for factor B*), or a combination of certain levels of the first factor with levels of the second factor (called the *interaction effect*). An example will undoubtably clarify the use of the two-way ANOVA for comparing group means. Suppose that the National Automobile Insurance Company is investigating the driving habits of different age groups. Three

58                                                                 CHAPTER 4

age groups are selected to participate in this study: those 18–30 years old, those 30–50 years old, and those older than 50 years. For many years automobile insurance companies have charged young men higher premiums, reflecting what is believed to be this group's relatively poor driving record. The basic assumption for charging higher premiums is that because young men drive on the average farther than any other group they are more prone to accidents. The president of the National Automotive Insurance Company (NAIC), however, believes that different premiums should be charged according to the age and the gender of drivers. This is because drivers in some age groups drive considerably more than others. The president of NAIC believes that male drivers in the younger age group drive considerably more than female drivers in the same age group. The president also believes that male drivers in the older age group drive considerably less than female drivers in the same age group. To determine whether there are any differences between the various drivers, 600 drivers in a random sample were questioned concerning the number of miles they drove in the previous 12 months. The selected sample included 200 young, 200 middle-aged, and 200 older drivers. An equal number of male and female drivers from each age group were asked to participate. The summary data from this study are presented in Table 4.1.

The logic of a two-way analysis of variance is a direct extension of the rationale underlying one-way ANOVA. To review, in the one-way ANOVA the total variability of observed scores was partitioned into two parts, a between- and a within-groups part. In the case of a two-factor design, scores might differ from one another because of group differences, but the group differences are more complicated because there are two classification schemes. For example, in a two-way analysis scores might differ from one another because of group differences due to factor A, factor B, and/or the interaction of the two factors.

In a one-way ANOVA the total variability is partitioned into two components, each of which provides an estimate of the variability in the population

TABLE 4.1
Summary Data for Driving Example
(in Hundreds of Miles)

| Gender Factor | Age Factor | | |
| | Young | Middle-Aged | Older | |
|---|---|---|---|---|
| Male | $\bar{x} = 14$ | $\bar{x} = 8.5$ | $\bar{x} = 4.5$ | $\bar{X} = 9$ |
| Female | $\bar{x} = 6$ | $\bar{x} = 7$ | $\bar{x} = 8$ | $\bar{X} = 7$ |
| | $\bar{X} = 10$ | $\bar{X} = 7.75$ | $\bar{X} = 6.25$ | $\bar{X}_{GM} = 8$ |

Note. $\bar{x}$ = group mean, $\bar{X}$ = total group mean, and $\bar{X}_{GM}$ = grand mean.

(i.e., $SS_{total} = SS_{between} + SS_{within}$). In two-factor ANOVA, the total variability is partitioned into four components: the variability for each factor, the variability for the interaction between factors, and the variability within each of the groups (i.e., $SS_{total} = SS_A + SS_B + SS_{AB} + SS_{within}$).

This is basically equivalent to:

$$\sum_i \sum_j \sum_k (x_{ijk} - \overline{X}_{GM})^2 = nq \sum_j (\overline{X}_j - \overline{X}_{GM})^2 + np \sum_k (\overline{X}_k - \overline{X}_{GM})^2$$

$$+ n \sum_j \sum_k (\overline{X}_{jk} - \overline{X}_j - \overline{X}_k + \overline{X}_{GM})^2 + \sum_i \sum_j \sum_k (x_{ijk} - \overline{X}_{jk})^2$$

where

$x_{ijk}$ = score for person $i$ on factor A and B

$\overline{X}_j$ = mean for level $j$ of factor A

$\overline{X}_k$ = mean for level $k$ of factor B

$\overline{X}_{jk}$ = mean for each factor A and B combination

$\overline{X}_{GM}$ = grand mean

The degrees of freedom for the two-way ANOVA partition the same way as the sum of squares, namely:

$$df_{total} = df_{factor\ A} + df_{factor\ B} + df_{interaction} + df_{within}$$

for

$$
\begin{aligned}
df_{total} &= npq - 1 \\
df_A &= (p - 1) \\
df_B &= (q - 1) \\
df_{AB} &= (p - 1)(q - 1) \\
df_{within} &= pq(n - 1)
\end{aligned}
$$

where $n$ is the number of subjects in each cell, $p$ the number of levels of factor A, and $q$ the number of levels of factor B.

Finally, by dividing the sum of squares for each effect by their corresponding degrees of freedom, one obtains the mean square (MS) for each effect. Because the two main effects and the interaction effect are independent of each other, separate $F$-test statistics can be conducted for each. Thus, the following $F$ tests would be conducted:

Effect of factor A: $F = \dfrac{MS_A}{MS_{within}}$

Effect of factor B: $F = \dfrac{MS_B}{MS_{within}}$

Effect of interaction: $F = \dfrac{MS_{AB}}{MS_{within}}$

The results of a two-way ANOVA are traditionally displayed in an ANOVA source table. For each of the four sources of variability, the corresponding sums of squares, degrees of freedom, and mean squares are presented. Separate $F$ tests are then conducted for each of the three effects. Table 4.2 provides the results of the two-way ANOVA on the example data. Clearly, at the .05 level of significance the results indicate that all three effects are statistically significant. This would lead to the rejection of the hypothesis of no difference between the number of miles driven by the different groups examined.

As indicated, the two-way analysis of variance permits the testing of three hypotheses separately. One hypothesis addresses the main effect for the first factor, the second the main effect of the second factor, and the third hypothesis tests the effect of certain levels of one factor paired with certain levels of the other factor. The first two hypotheses are quite straightforward. The results presented in Table 4.2 indicate that there are significant differences among the three age groups and between male and female drivers with respect to their driving habits. The third hypothesis, the interaction hypothesis, is not as obvious. The interaction effect is viewed as a unique effect that cannot be predicted from knowledge of the effects of the other two factors. As such, the interaction is the effect that should always be examined first.

The best strategy to use in order to study a significant interaction effect is to present the data on an interaction graph. To accomplish this the mean values on the dependent variable are identified on the ordinate of the graph, and one of the two independent variables is identified on the abscissa. Either variable can be chosen to be identified on the abscissa. The other independent variable is then represented as two or more lines on the graph (depending on the number of levels of that variable). Figure 4.1 presents a

TABLE 4.2
Two-Way ANOVA Results for Driving Example

| Source of Variation | SS | df | MS | F |
|---|---|---|---|---|
| Gender effect | 600 | 1 | 600 | 61.6* |
| Age effect | 1425 | 2 | 712.5 | 73.2* |
| Interaction effect | 3325 | 2 | 1662.5 | 170.7* |
| Within groups | 5784.7 | 594 | 9.7 | |
| Total | 11,134.7 | 599 | | |

*$p < .05$.

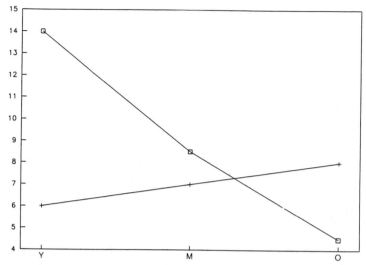

FIG. 4.1. Gender and age interaction—driver study. Male group, □; female group, +.

graph of the data from Table 4.1. The means are plotted with one line representing the male drivers and the other line representing the female drivers.

The interaction graph presented in Fig. 4.1 reflects the significant differences that were observed between gender and the three age groups. However, the graph also indicates that these differences depend on the combination of the two factors. For example, it is clear that for the younger group there are some major differences between male and female drivers. Males drive significantly more than females. Nevertheless, when the older age group is examined, it appears that these differences are reversed. Older females drive significantly further than males (at least with respect to these fictitious data). For middle-aged drivers there do not seem to be any significant differences between male and female drivers.

In general, interaction effects can take on several visual representations when graphed. Some common examples are shown in Fig. 4.2. There are essentially three different types of interaction effects that can occur: a disordinal interaction, an ordinal interaction, and no interaction. A disordinal interaction occurs when the lines in the graph cross. For example, Fig. 4.2a represents a disordinal interaction. Such an interaction indicates that the effects of the levels of each factor reverse themselves as the levels of the other factor change. A disordinal interaction was also encountered in the previous example study. An ordinal interaction occurs when the lines of the graph are not parallel and they do not cross. Figure 4.2b represents such an interaction. This pattern suggests a greater difference between groups at one level of a factor than at other levels of the same factor. For example, if

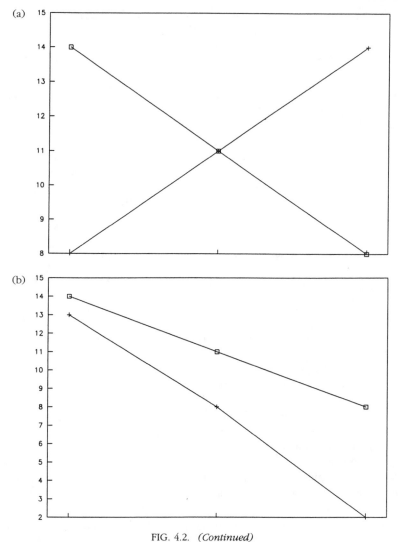

FIG. 4.2. *(Continued)*

differences between male and female drivers became more pronounced as age increased, then an ordinal interaction would be observed. Finally, parallel lines on a graph indicate the absence of an interaction. For example, Fig. 4.2c suggests that there is a similar pattern of means on one factor at each level of the other factor.

The presence of a significant interaction can certainly complicate the interpretation of a factorial analysis. A significant interaction is a clear signal that certain combinations of factor A and B have different effects than others. Because this implies that the effect of factor A depends on the level of factor

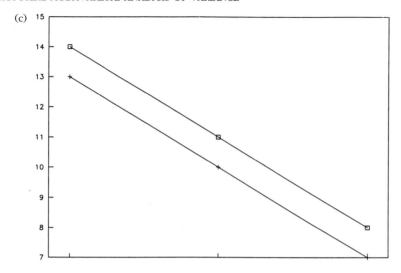

FIG. 4.2. (a) Example of disordinal interaction graph. (b) Example of ordinal interaction graph. (c) Example of no-interaction graph.

B with which it is combined, an interaction graph is used to facilitate interpretation of the interaction. Of course, in order to ensure that the interpretation of the effects of the interaction in terms of differences between means is accurate, one should also conduct post hoc statistical comparisons (i.e., *t* tests, one-way ANOVAs) between the various levels of each factor. Several statistical comparison procedures were described in chapter 3. These statistical comparisons, when used to assist in the interpretation of factorial designs, are commonly referred to as hypotheses of simple main effects (for further details see Kirk, 1982).

## FIXED, RANDOM, AND MIXED FACTORIAL DESIGNS

The two-way ANOVA presented in the previous section considered levels of factors A and B that are fixed. Clearly, the three levels of the age factor used in the driving study were not randomly sampled. Instead, the particular levels were included because they were the only levels of interest in the study. The National Automobile Insurance Company intends to draw inferences only for those levels observed. Such designs are considered fixed-effects designs. In contrast, if the study included levels of factors that were randomly selected (e.g., if ages could be randomly selected from among many age categories) then the design would be considered a random-effects design. Suppose, however, as is the case with the drivers study, that only the levels of one factor can be randomly selected. Obviously, the levels of the gender factor are fixed. In this situation, age can be a random factor

TABLE 4.3
$F$ Ratios for Various Factorial Designs

| $F$ | Fixed | Random | Mixed<br>(A = Random, B = Fixed) | Mixed<br>(A = Fixed, B = Random) |
|---|---|---|---|---|
| Factor A | $\dfrac{MS_A}{MS_W}$ | $\dfrac{MS_A}{MS_{AB}}$ | $\dfrac{MS_A}{MS_W}$ | $\dfrac{MS_A}{MS_{AB}}$ |
| Factor B | $\dfrac{MS_B}{MS_W}$ | $\dfrac{MS_B}{MS_{AB}}$ | $\dfrac{MS_B}{MS_{AB}}$ | $\dfrac{MS_B}{MS_W}$ |
| Interaction | $\dfrac{MS_{AB}}{MS_W}$ | $\dfrac{MS_{AB}}{MS_W}$ | $\dfrac{MS_{AB}}{MS_W}$ | $\dfrac{MS_{AB}}{MS_W}$ |

but gender is a fixed factor. Factorial designs with one or more random factors and one fixed factor are called mixed-effect designs.

The importance of knowing whether the design is fixed, random, or mixed is due to the differences in the types of hypotheses that can be tested in each design and the inferences drawn from these hypotheses. Computationally, the fixed-effects factorial ANOVA is the same as the random- and mixed-effects factorial ANOVA. The sum of squares, the degrees of freedom, and the mean squares are all computed the same way. However, for testing the three effect hypotheses (i.e., factor A, factor B, and interaction), the $F$ ratios are calculated differently. A comparison of the $F$ ratios for the fixed-, random-, and mixed-effect designs is given in Table 4.3. The major differences between the $F$ ratios for the various effects are with respect to the appropriate mean squares term (i.e., the correct error term) to use in the denominator (for further discussion see Kirk, 1982; Shavelson, 1988). For example, to test the significance of factor A in a fixed-effects design, the $F$ ratio would use the mean square for within groups ($MS_W$) as the correct error term. However, if the design is random the mean square for interaction ($MS_{AB}$) becomes the appropriate error term.

## TWO-WAY MULTIVARIATE ANALYSIS OF VARIANCE

The two-way multivariate analysis of variance (MANOVA) is a generalization of the univariate two-factor ANOVA but extended to several dependent variables. As discussed in chapter 3, the MANOVA procedure was originally developed by Wilks (1932) on the basis of the generalized likelihood-ratio principle. Recall from chapter 3 that the attractiveness of the likelihood ratio is that it yields test statistics that reduce to the familiar univariate $F$ and $t$ statistics. In the one-way MANOVA, Wilks's $\Lambda$ is equal to:

$$\Lambda = \frac{|W|}{|B+W|} = \frac{|W|}{|T|}$$

where $|W|$ is the determinant of the within-groups SSCP matrix, and $|T|$ is the determinant of the total sample SSCP matrix.

In any one-factor analysis, whether a univariate or a multivariate analysis, only one type of question can be asked about the observed data. Is there a difference between the groups on the one or more observed measures? In a two-factor analysis, however, three types of questions can be asked about the observed data. First, is there a difference between the levels of factor A? Second, is there a difference between the levels of factor B? Third, is there an interaction between the two factors? The procedures for examining these questions in a two-way ANOVA were discussed in the previous section but they are also directly applicable to the multivariate case.

The logic of a two-way MANOVA is an extension of the rationale underlying the univariate two-way ANOVA. The only difference is that in the MANOVA Wilks's $\Lambda$ is used as the criterion for comparing groups. However, because there are three hypotheses that must be tested, Wilks's $\Lambda$ for a one-way MANOVA must be slightly modified. The following general formula for Wilks's $\Lambda$ can be used for testing the three different hypotheses in the two-way MANOVA (this formula can also be used for examining higher order designs as long as the different hypotheses are identified):

$$\Lambda = \frac{|S_{\text{error}}|}{|S_{\text{effect}} + S_{\text{error}}|}$$

where $|S_{\text{error}}|$ is the determinant of the error SSCP matrix appropriate for the model under study, and $|S_{\text{effect}} + S_{\text{error}}|$ is the determinant of the effect SSCP matrix plus the appropriate error term.

It is important to note that in the one-way MANOVA the effect under study is the between-group difference, and thus the appropriate error in the model is the within-groups one. Hence, for a one-way MANOVA (using the general formula for Wilks's $\Lambda$), $|S_{\text{error}}| = |W|$ and $|S_{\text{effect}} + S_{\text{error}}| = |B + W|$. This provides the same test statistic presented for Wilks's $\Lambda$ in the previous chapter.

The general formula presented for Wilks's $\Lambda$ is considered definitional because the actual computation of $\Lambda$ is made routine by using the relationship between the determinant and the eigenvalues of a matrix (see chapter 3 for a complete discussion). In chapter 3, the value of Wilks's $\Lambda$ for the one-way MANOVA was determined by finding the product of the eigenvalues of the matrix $W^{-1}B$ (i.e., $\Lambda = 1/\Pi(\lambda_i + 1)$, where $\lambda_i$ are the nonzero eigenvalues of the matrix $W^{-1}B$). For the two-factor MANOVA, the value of Wilks's $\Lambda$ is determined by finding the product of the eigenvalues of the matrix $S_{\text{error}}^{-1}S_{\text{effect}}$ (which for a one-way MANOVA would be the same as $W^{-1}B$).

To make this relationship clear, consider the general formula for Wilks's $\Lambda$ rewritten as:

$$\frac{1}{\Lambda} = \frac{\mid S_{error} + S_{effect} \mid}{\mid S_{error} \mid} = \mid S_{error}^{-1} (S_{error} + S_{effect}) \mid$$

$$= \mid S_{error}^{-1} S_{effect} + I \mid$$

Based on this information, the value of Wilks's $\Lambda$ for each effect tested can be written as the product of the eigenvalues of the matrix $S_{error}^{-1} S_{effect}$:

$$\Lambda = \frac{1}{\Pi(\lambda_i + 1)}$$

or simply

$$\frac{1}{\Lambda} = \Pi(\lambda_i + 1)$$

where $\lambda_i$ are the nonzero eigenvalues of the matrix $S_{error}^{-1} S_{effect}$.

Recall from the discussion of the two-factor ANOVA that three hypotheses can be tested. As such, in determining Wilks's $\Lambda$ for the two-way MANOVA, the substitution of the effect SSCP matrix depends on the hypothesis being tested. To test for the effects of the first factor, the SSCP matrix for factor A would be used. To test for the interaction effect, the SSCP matrix for the interaction between the two factors would be used.

Another important issue that was discussed in the preceding section concerned knowing whether the design is fixed, random, or mixed. The importance of knowing the type of design examined is due to the fact that the three effect hypotheses and the $F$ ratios are calculated differently. The same concerns exist in the multivariate case. The only difference is that in the multivariate case it is the choice of the error SSCP matrix that depends on the type of model examined. A comparison of the SSCP matrices for determining Wilks's $\Lambda$ for a fixed-, random-, and mixed-effect design is given in Table 4.4. Once again, the major differences between Wilks's $\Lambda$ for the various effects are with respect to the appropriate error term to use. For example, to test the effect of factor B in a fixed-effects design requires that the within-groups SSCP matrix be used as the error term [i.e., $\Lambda = \mid S_W \mid / \mid S_B + S_W \mid$].

TABLE 4.4
Error SSCP Used to Test Effects in Two-Way MANOVA

| Effect | Fixed | Random | Mixed (A = Random, B = Fixed) | Mixed (A = Fixed, B = Random) |
|--------|-------|--------|-------------------------------|-------------------------------|
| Factor A | $S_W$ | $S_{AB}$ | $S_W$ | $S_{AB}$ |
| Factor B | $S_W$ | $S_{AB}$ | $S_{AB}$ | $S_W$ |
| Interaction | $S_W$ | $S_W$ | $S_W$ | $S_W$ |

For a random-effects design the appropriate error term is the SSCP matrix for interaction [i.e., $\Lambda = |S_{AB}|/|S_B + S_{AB}|$].

As discussed in chapter 3, Wilks's $\Lambda$ is the oldest and most widely used criterion for comparing groups. But there are several other criteria that can be used for testing effects in a factorial MANOVA. The three most popular criteria that are also displayed as standard computer output for any type of MANOVA analysis are (a) Hotelling's trace, (b) Pillai's trace, and (c) Roy's greatest root. These three criteria are defined as follows:

Hotelling's trace: $T = \Sigma \lambda_i$

Pillai's trace: $V = \Sigma \dfrac{\lambda_i}{1 + \lambda_i}$

Roy's greatest root: $\theta = \dfrac{\lambda_{max}}{1 + \lambda_{max}}$

where $\lambda_i$ are the nonzero eigenvalues of the matrix $S_{error}^{-1}S_{effect}$, and $\lambda_{max}$ is the largest eigenvalue of the matrix $S_{error}^{-1}S_{effect}$.

Although tables of the four criteria for assessing statistical significance exist, they are not widely available. However, several authors have successfully investigated the notion that these tabled values can be approximated using an $F$ distribution (Rao, 1952; Schatzoff, 1964, 1966). It is therefore common practice to assess the significance of any MANOVA results by using the $F$ distribution approximation to these tests. Fortunately, the SAS package (and most others) provides this information as standard output. As also discussed in chapter 3, there has been considerable disagreement in the literature concerning which of the four criteria is "best." Currently the best strategy is to examine all four criteria. If the results are similar, choosing the criterion to use is a matter of personal preference. If the results are different, look for some concensus between at least two test criteria before reporting the conclusions.

Another recommended strategy when conducting any multivariate analysis is to examine a measure of the practical importance of the results. The most common measure used is the eta-square ($\eta^2$) measure of association. This measure can be interpreted rather loosely as the proportion of variance of the dependent variables that is accounted for by the independent variables. Because there are three types of effects examined in a two-way MANOVA, there are also three measures of association that can be computed. The first represents the contribution of factor A. In other words, this measure represents the relative importance of the effect of factor A. The other two measures of association represent the contributions of factor B and the interaction effect. Using Wilks's $\Lambda$ criterion, a general measure of association can be defined as:

$$\eta^2_{\text{effect}} = 1 - \Lambda_{\text{effect}}$$

Thus, to determine the contribution of factor A, this value is set equal to $\eta^2_A = 1 - \Lambda_A$.

## AN EXAMPLE OF TWO-WAY MANOVA USING SAS

Gibson and Marcoulides (1995) conducted a study to compare the leadership styles of service industry managers from different countries. Three countries were investigated in this study: the United States, Australia, and Sweden (for the purposes of this example the original study is somewhat modified). Male and female managers from each country were compared in terms of three leadership styles: consensus, consultative, and autocratic. The countries were deliberately chosen because of certain important cultural similarities and differences that researchers have addressed over the past few years in the literature (for a complete discussion see Gibson & Marcoulides, 1995).

In order to conduct this study, each manager was asked to complete the Leadership Effectiveness Questionnaire (Flamholtz, 1986). The Leadership Effectiveness Questionnaire (LEQ) consists of questionnaire items that are designed to assess the dimension of leadership style exhibited by each manager. The LEQ was designed for use in corporate settings where managers desire feedback concerning their leadership style. According to several researchers, the questionnaire appears to stand-up well to common tests of validity and reliability (Baker, 1986; Scalberg & Doherty, 1983). The dependent variables in this study were the total scores received by each manager on the three leadership styles as measured by the LEQ.

This study is a classic two-factor multivariate design. The two factors examined are country and gender. There are three levels of the country factor (i.e., United States, Australia, and Sweden) and two levels of the gender factor (i.e, male and female). The levels of each factor included in this study were not randomly selected. Instead, the particular levels were included because they were the only levels of interest in the study. Thus, the design is a fixed-effects design. The dependent variables in this study were the total scores received by each manager on the leadership styles of consensus (SCORE1), consultative (SCORE2), and autocratic (SCORE3). Table 4.5 presents the results of this example study.

The deck setup to perform the statistical analysis in SAS on this example data set is presented in Table 4.6. Recall from chapter 1 that the DATA command provides the name for the data set that follows, and the INPUT command names the variables to be studied. Of special importance in the deck setup is the @@ symbol. This is a special type of pointer control statement used in SAS (for a complete discussion see the chapter on SAS Statements Used in the DATA Step in the *SAS User's Guide: Basics*). These

TABLE 4.5
Data from Study of Leadership Styles

| | LEQ Scores | | |
|---|---|---|---|
| | Consensus | Consultative | Autocratic |
| Male managers | | | |
| United States | 5 | 6 | 9 |
| | 5 | 7 | 7 |
| | 3 | 7 | 8 |
| | 4 | 5 | 7 |
| Australia | 3 | 2 | 8 |
| | 6 | 4 | 9 |
| | 7 | 5 | 9 |
| | 5 | 2 | 8 |
| Sweden | 4 | 3 | 3 |
| | 6 | 3 | 3 |
| | 5 | 3 | 4 |
| | 8 | 4 | 3 |
| Female managers | | | |
| United States | 5 | 4 | 9 |
| | 8 | 6 | 9 |
| | 6 | 2 | 8 |
| | 4 | 3 | 9 |
| Australia | 7 | 2 | 8 |
| | 5 | 2 | 8 |
| | 3 | 1 | 7 |
| | 7 | 4 | 8 |
| Sweden | 6 | 2 | 3 |
| | 5 | 2 | 4 |
| | 4 | 1 | 3 |
| | 4 | 2 | 3 |

pointers are used to help the SAS system keep track of its position as it reads values from the input lines. Essentially the @@ symbol (called a double trailing at-sign) is useful when each input line contains values for several observations. For example, say that the data have been input with each line containing the scores of managers from several country and gender variables. The SAS system will read first a gender value, then a country value, and then the three dependent scores (i.e., the scores from the three leadership styles) until all input values in the line have been read. The SAS system only releases a line held by a trailing @@ when the pointer moves past the end of the line. This approach is one of many that can be used to save considerable time and space in SAS deck setups. Alternatively, one could simply write out the data using a single line for each entry in the design.

To perform a factorial multivariate analysis of variance on this data set two SAS procedures can be used. The PROC ANOVA or the PROC GLM can be used. As discussed in chapter 3, the main difference between the two

TABLE 4.6
SAS Setup for Factorial MANOVA

```
DATA LEADERS;
 INPUT GENDER COUNTRY SCORE1 SCORE2 SCORE3 @@;
 CARDS;
 1 1 5 6 9 1 1 5 7 7 1 1 3 7 8 1 1 4 5 7
 1 2 3 2 8 1 2 6 4 9 1 2 7 5 9 1 2 5 2 8
 1 3 4 3 3 1 3 6 3 3 1 3 5 3 4 1 3 8 4 3
 2 1 5 4 9 2 1 8 6 9 2 1 6 2 8 2 1 4 3 9
 2 2 7 2 8 2 2 5 2 8 2 2 3 1 7 2 2 7 4 8
 2 3 6 2 3 2 3 5 2 4 2 3 4 1 3 2 3 4 2 3
 ;
PROC ANOVA;
 CLASS GENDER COUNTRY;
 MODEL SCORE1 SCORE2 SCORE3 = GENDER COUNTRY GENDER*COUNTRY;
 MEANS GENDER COUNTRY/LSD;
MANOVA H=GENDER COUNTRY GENDER*COUNTRY/PRINTH PRINTE;
```

procedures is that PROC ANOVA is designed to handle balanced data, whereas PROC GLM can analyze data with missing observations. When there are no missing observations both procedures produce identical results. However, because PROC ANOVA does not check to see if any data are missing, sometimes an unbalanced data set can incorrectly get analyzed as balanced. Thus, if it is clear that there are no missing observations use the PROC ANOVA, otherwise use PROC GLM to conduct the analysis. The data presented in Table 4.5 are balanced, so either PROC ANOVA of PROC GLM can be used to generate the output presented in Tables 4.7–4.9.

Several other statements are also needed in order to conduct this analysis. These are the CLASS, the MODEL, the MEANS, and the MANOVA statements (for complete details consult the *SAS User's Guide*). The CLASS statement defines the independent variables used in the study. In this example there are two: GENDER and COUNTRY. The MODEL statement names the dependent variables (i.e., SCORE1, SCORE2, and SCORE3) and the independent variables according to how they are to be used in the factorial design. Recall that in a two-way factorial design there are three important effects that must be examined. These include the main effects for each factor and the interaction effect. Accordingly, the MODEL statement defines these three effects. The MEANS statement is used to produce mean values for the independent variables included in the CLASS statement, and it also provides post hoc comparison tests (in this case Fisher's least significant difference, LSD, was selected, but there are many others available in this procedure). The post hoc comparisons are essentially separate analyses on each dependent variable. Finally, the MANOVA statement provides the four commonly used multivariate tests of significance and provides univariate ANOVAs on each dependent

TABLE 4.7
SAS Analysis of Variance Procedure, Multivariate Analysis of Variance

## (a) Gender

Characteristic Roots and Vectors of: E Inverse * H, where
H = Anova SS&CP Matrix for GENDER   E = Error SS&CP Matrix

| Characteristic Root | Percent | Characteristic Vector V'EV=1 | | |
|---|---|---|---|---|
| | | SCORE1 | SCORE2 | SCORE3 |
| 1.6150426965 | 100.00 | -0.12982657 | 0.28290585 | -0.17823853 |
| 0.0000000000 | 0.00 | 0.14057526 | 0.02163720 | 0.01101824 |
| 0.0000000000 | 0.00 | -0.05607142 | 0.01117652 | 0.39174461 |

Manova Test Criteria and Exact F Statistics for
the Hypothesis of no Overall GENDER Effect
H = Anova SS&CP Matrix for GENDER   E = Error SS&CP Matrix

S=1   M=0.5   N=7

| Statistic | Value | F | Num DF | Den DF | Pr > F |
|---|---|---|---|---|---|
| Wilks' Lambda | 0.38240293 | 8.6136 | 3 | 16 | 0.0012 |
| Pillai's Trace | 0.61759707 | 8.6136 | 3 | 16 | 0.0012 |
| Hotelling-Lawley Trace | 1.61504270 | 8.6136 | 3 | 16 | 0.0012 |
| Roy's Greatest Root | 1.61504270 | 8.6136 | 3 | 16 | 0.0012 |

## (b) Country

Characteristic Roots and Vectors of: E Inverse * H, where
H = Anova SS&CP Matrix for COUNTRY   E = Error SS&CP Matrix

| Characteristic Root | Percent | Characteristic Vector V'EV=1 | | |
|---|---|---|---|---|
| | | SCORE1 | SCORE2 | SCORE3 |
| 21.638433878 | 92.42 | -0.01673804 | -0.05869342 | 0.42388279 |
| 1.7747519585 | 7.58 | -0.13967844 | 0.27679669 | -0.07519392 |
| 0.0000000000 | 0.00 | 0.14131567 | 0.02382179 | -0.00484511 |

Manova Test Criteria and F Approximations for
the Hypothesis of no Overall COUNTRY Effect
H = Anova SS&CP Matrix for COUNTRY   E = Error SS&CP Matrix

S=2   M=0   N=7

| Statistic | Value | F | Num DF | Den DF | Pr > F |
|---|---|---|---|---|---|
| Wilks' Lambda | 0.01591950 | 36.9368 | 6 | 32 | 0.0001 |
| Pillai's Trace | 1.59543476 | 22.3469 | 6 | 34 | 0.0001 |
| Hotelling-Lawley Trace | 23.41318584 | 58.5330 | 6 | 30 | 0.0001 |
| Roy's Greatest Root | 21.63843388 | 122.6178 | 3 | 17 | 0.0001 |

*(Continued)*

TABLE 4.7
*(Continued)*

*(c) Gender × Country*

```
              Characteristic Roots and Vectors of: E Inverse * H, where
         H = Anova SS&CP Matrix for GENDER*COUNTRY   E = Error SS&CP Matrix

Characteristic      Percent                  Characteristic Vector V'EV=1
    Root                          SCORE1           SCORE2           SCORE3

 1.3117969264       91.25         0.11442466     -0.23089306       0.34098156
 0.1258423694        8.75        -0.16283627      0.06012216       0.23219857
 0.0000000000        0.00         0.1231650       0.15395624       0.12316499

               Manova Test Criteria and F Approximations for
               the Hypothesis of no Overall GENDER*COUNTRY Effect
         H = Anova SS&CP Matrix for GENDER*COUNTRY   E = Error SS&CP Matrix

                          S=2   M=0   N=7
```

| Statistic | Value | F | Num DF | Den DF | Pr > F |
|---|---|---|---|---|---|
| Wilks' Lambda | 0.38421360 | 3.2709 | 6 | 32 | 0.0127 |
| Pillai's Trace | 0.67921224 | 2.9141 | 6 | 34 | 0.0211 |
| Hotelling-Lawley Trace | 1.43763930 | 3.5941 | 6 | 30 | 0.0084 |
| Roy's Greatest Root | 1.31179693 | 7.4335 | 3 | 17 | 0.0022 |

*Note.* $F$ statistic for Roy's greatest root is an upper bound. $F$ statistic for Wilks' lambda is exact.

variable. It is important to note that three effects are defined in this statement in order to obtain separate significance test for each important effect.

As can be seen in Table 4.7, the four multivariate tests of significance indicate that all three effects are statistically significant. As expected, the values of the different tests and their $F$ approximations are somewhat different but they all lead to the same conclusion. The eigenvalues and eigenvectors (referred to in the printout as Characteristic Roots and Vectors) used

TABLE 4.8
Mean Values from Study of Leadership Styles

| | *LEQ Scores* | | |
|---|---|---|---|
| | *Consensus* | *Consultative* | *Autocratic* |
| Male managers | | | |
| United States | 4.25 | 6.25 | 7.75 |
| Australia | 5.25 | 3.25 | 8.50 |
| Sweden | 5.75 | 3.25 | 3.25 |
| Female managers | | | |
| United States | 5.75 | 3.75 | 8.75 |
| Australia | 5.50 | 2.25 | 7.75 |
| Sweden | 4.75 | 1.75 | 3.25 |

TABLE 4.9
SAS Analysis of Variance Procedure

```
                    Class Level Information
        Class                Levels              Values
        GENDER                 2                  1 2
        COUNTRY                3                  1 2 3
            Number of observations in data set = 24
```

Dependent Variable: SCORE1

| Source | DF | Sum of Squares | Mean Square | F Value | Pr > F |
|---|---|---|---|---|---|
| Model | 5 | 7.20833333 | 1.44166667 | 0.61 | 0.6956 |
| Error | 18 | 42.75000000 | 2.37500000 | | |
| Corrected Total | 23 | 49.95833333 | | | |

| R-Square | C.V. | Root MSE | SCORE1 Mean |
|---|---|---|---|
| 0.144287 | 29.58919 | 1.541104 | 5.20833333 |

Dependent Variable: SCORE1

| Source | DF | Anova SS | Mean Square | F Value | Pr > F |
|---|---|---|---|---|---|
| GENDER | 1 | 0.37500000 | 0.37500000 | 0.16 | 0.6958 |
| COUNTRY | 2 | 0.58333333 | 0.29166667 | 0.12 | 0.8852 |
| GENDER*COUNTRY | 2 | 6.25000000 | 3.12500000 | 1.32 | 0.2929 |

Dependent Variable: SCORE2

| Source | DF | Sum of Squares | Mean Square | F Value | Pr > F |
|---|---|---|---|---|---|
| Model | 5 | 49.33333333 | 9.86666667 | 7.25 | 0.0007 |
| Error | 18 | 24.50000000 | 1.36111111 | | |
| Corrected Total | 23 | 73.83333333 | | | |

| R-Square | C.V. | Root MSE | SCORE2 Mean |
|---|---|---|---|
| 0.668172 | 34.14634 | 1.166667 | 3.41666667 |

Dependent Variable: SCORE2

| Source | DF | Anova SS | Mean Square | F Value | Pr > F |
|---|---|---|---|---|---|
| GENDER | 1 | 16.66666667 | 16.66666667 | 12.24 | 0.0026 |
| COUNTRY | 2 | 30.33333333 | 15.16666667 | 11.14 | 0.0007 |
| GENDER*COUNTRY | 2 | 2.33333333 | 1.16666667 | 0.86 | 0.4410 |

Dependent Variable: SCORE3

| Source | DF | Sum of Squares | Mean Square | F Value | Pr > F |
|---|---|---|---|---|---|
| Model | 5 | 133.2083333 | 26.6416667 | 71.04 | 0.0001 |
| Error | 18 | 6.7500000 | 0.3750000 | | |
| Corrected Total | 23 | 139.9583333 | | | |

| R-Square | C.V. | Root MSE | SCORE3 Mean |
|---|---|---|---|
| 0.951771 | 9.361107 | 0.612372 | 5.54166667 |

*(Continued)*

TABLE 4.9
*(Continued)*

Dependent Variable: SCORE3

| Source | DF | Anova SS | Mean Square | F Value | Pr > F |
|---|---|---|---|---|---|
| GENDER | 1 | 0.0416667 | 0.0416667 | 0.11 | 0.7427 |
| COUNTRY | 2 | 130.0833333 | 65.0416667 | 173.44 | 0.0001 |
| GENDER*COUNTRY | 2 | 3.0833333 | 1.5416667 | 4.11 | 0.0338 |

to produce these results are also presented in the tables. As can be seen, all eigenvalues are positive or zero, with the first value being the largest.

The *R*-square measure of association (notation used in SAS instead of the $\eta^2$ measure of association) of the practical importance of the results is also presented in Table 4.9. This value indicates that roughly 14% of the variability in SCORE1 can be accounted for by knowing the gender and country of origin of a manager. For SCORE2 and SCORE3 the values are 67% and 95% respectively. It appears, therefore, that knowledge of the gender of a manager and the country of origin are important variables for determining the observed differences on each of the leadership styles. The multivariate measure of association for each effect is found by using Wilks's criterion for each effect. The multivariate measure of association indicates that about 62% (i.e., $\eta^2 = 1 - \Lambda = 1 - 0.38$) of the observed variance in the scores of the dependent variables can be accounted for by knowing the gender of a manager. Similarly, about 98% of the observed variance in the dependent variables can be accounted for by knowing the country of a manager.

The results presented in Table 4.7 indicate that there are significant differences between the three countries (country factor) and between the male and female managers (gender factor) with respect to their leadership styles. The results also indicate the presence of a significant interaction. Because this implies that the effect of the gender factor depends on the level of the country factor with which it is combined, an interaction graph is needed to facilitate the interpretation of the results.

Table 4.8 presents the mean values determined for the various levels. Figure 4.3 presents the graphs of this data. The mean values are plotted with one line representing the male managers in each country and the other line representing the female managers. An examination of the mean values for each group on SCORE1 (Fig. 4.3a) suggests the presence of a disordinal interaction. Such an interaction occurs when the lines cross and suggests that the differences between the male and female managers reverse themselves depending on their country of origin. The graph of the mean values for SCORE2 (Fig. 4.3b) suggests the presence of an ordinal interaction. This type of interaction suggests that there is a greater difference between the leadership styles of male and female managers across each country. Finally,

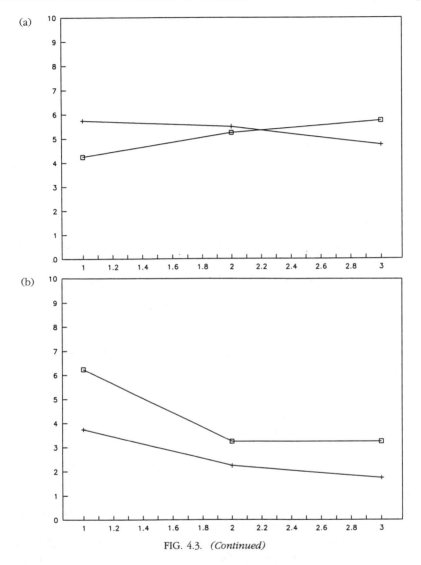

FIG. 4.3. *(Continued)*

the graph of the mean values for SCORE3 (Fig. 4.3c) also suggests that there is a disordinal interaction with respect to the differences between the leadership styles of male and female managers in each country. In particular, it seems that with an autocratic leadership style (i.e, SCORE3) female managers in the United States have higher scores than male managers.

Table 4.9 presents the univariate ANOVAs generated by SAS for each dependent variable. The results indicate that for SCORE1 (consensus style) there are no significant differences between any of the factors and no significant interaction between (e.g., for the gender factor with $F_{1,18} = 0.16$, $p$

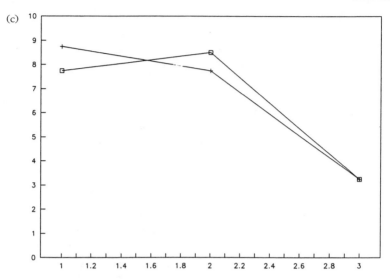

FIG. 4.3. (a) Country by gender interaction, SCORE 1. Males, □; females, +. (b) Country by gender interaction, SCORE 2. (c) Country by gender interaction, SCORE 3.

< .6958). An examination of the results for SCORE2 (consultative style) indicates that there are sigificant differences between the two factors but no significant interaction (i.e., gender factor with $F_{1,18} = 12.24$, $p < .0026$ and country factor with $F_{2,18} = 11.14$, $p < .0007$). Finally, an examination of the results for SCORE3 (autocratic style) indicate that there are significant differences between managers on the country factor (i.e., $F_{2,18} = 173.44$, $p < .0001$) and a significant interaction (i.e., $F_{2,18} = 4.11$, $p < .0338$). It appears, therefore, that although an examination of the interaction graphs is an essential first step toward exploring the results of a factorial analysis, a second step must include the examination of the univariate ANOVAs for each dependent variable. In addition, to disentangle entirely the differences between the various levels of the factors post hoc comparisons must be explored.

Table 4.10 provides the post hoc comparisons (more one-way ANOVAs) for the two levels of the gender factor on each dependent variable. For example, an examination of Tables 4.10a and 4.10f indicates that there are no differences between male and female managers from the various countries with respect to consensus leadership style. As can be seen in Table 4.10b, male managers from the United States exhibit a greater amount of consultative style than managers from the other countries. However, those differences in consultative leadership style are not observed in female managers. In addition, with respect to the autocratic leadership style, males from the United States and Australia are similar to each other but different from managers from Sweden. However, for female managers there are differences

TABLE 4.10

SAS Post Hoc Comparisons for Two Levels of Gender Factor

---

*(a) SCORE1, Male*

---

GENDER=1

Analysis of Variance Procedure

T tests (LSD) for variable: SCORE1

NOTE: This test controls the type I comparisonwise error rate not
the experimentwise error rate.

Alpha= 0.05  df= 9  MSE= 2.25

Critical Value of T= 2.26

Least Significant Difference= 2.3994

Means with the same letter are not significantly different.

GENDER=1

Analysis of Variance Procedure

| T Grouping | Mean | N | COUNTRY |
|---|---|---|---|
| A | 5.750 | 4 | 3 |
| A | | | |
| A | 5.250 | 4 | 2 |
| A | | | |
| A | 4.250 | 4 | 1 |

---

*(b) SCORE2, Male*

---

GENDER=1

Analysis of Variance Procedure

T tests (LSD) for variable: SCORE2

NOTE: This test controls the type I comparisonwise error rate not
the experimentwise error rate.

Alpha=0.05  df= 9  MSE= 1.138889

Critical Value of T= 2.26

Least Significant Difference= 1.7071

Means with the same letter are not significantly different.

GENDER=1

Analysis of Variance Procedure

| T Grouping | Mean | N | COUNTRY |
|---|---|---|---|
| A | 6.250 | 4 | 1 |
| B | 3.250 | 4 | 2 |
| B | | | |
| B | 3.250 | 4 | 3 |

---

*(c) SCORE3, Male*

---

GENDER=1

Analysis of Variance Procedure

T tests (LSD) for variable: SCORE3

NOTE: This test controls the type I comparisonwise error rate not
the experimentwise error rate.

Alpha= 0.05  df= 9  MSE= 0.5

Critical Value of T= 2.26

Least Significant Difference= 1.1311

Means with the same letter are not significantly different.

---

*(Continued)*

TABLE 4.10
*(Continued)*

---

GENDER=1

Analysis of Variance Procedure

| T Grouping | Mean | N | COUNTRY |
|---|---|---|---|
| A | 8.500 | 4 | 2 |
| A | | | |
| A | 7.750 | 4 | 1 |
| B | 3.250 | 4 | 3 |

---

*(d) SCORE1, Female*

---

GENDER=2

Analysis of Variance Procedure

T tests (LSD) for variable: SCORE1

NOTE: This test controls the type I comparisonwise error rate not
the experimentwise error rate.

Alpha= 0.05  df= 9  MSE= 2.5

Critical Value of T= 2.26

Least Significant Difference= 2.5292

Means with the same letter are not significantly different.

GENDER=2

Analysis of Variance Procedure

| T Grouping | Mean | N | COUNTRY |
|---|---|---|---|
| A | 5.750 | 4 | 1 |
| A | | | |
| A | 5.500 | 4 | 2 |
| A | | | |
| A | 4.750 | 4 | 3 |

---

*(e) SCORE2, Female*

---

GENDER=2

Analysis of Variance Procedure

T tests (LSD) for variable: SCORE2

NOTE: This test controls the type I comparisonwise error rate not
the experimentwise error rate.

Alpha= 0.05  df= 9  MSE= 1.583333

Critical Value of T= 2.26

Least Significant Difference= 2.0128

Means with the same letter are not significantly different.

GENDER=2

Analysis of Variance Procedure

| T Grouping | Mean | N | COUNTRY |
|---|---|---|---|
| A | 3.750 | 4 | 1 |
| A | | | |
| A | 2.250 | 4 | 2 |
| A | | | |
| A | 1.750 | 4 | 3 |

---

*(Continued)*

TABLE 4.10
*(Continued)*

*(f) SCORE3, Female*

GENDER=2
Analysis of Variance Procedure
T tests (LSD) for variable: SCORE3
NOTE: This test controls the type I comparisonwise error rate not
the experimentwise error rate.
Alpha= 0.05   df= 9   MSE= 0.25
Critical Value of T= 2.26
Least Significant Difference= 0.7998
Means with the same letter are not significantly different.

GENDER=2
Analysis of Variance Procedure

| T Grouping | Mean | N | COUNTRY |
|---|---|---|---|
| A | 8.750 | 4 | 1 |
| B | 7.750 | 4 | 2 |
| C | 3.250 | 4 | 3 |

between each country. Female managers from the United States exhibit the greatest amount of autocratic style, whereas female managers from Sweden exhibit the least.

The leadership style of managers has been a topic of continued interest in the management literature. The results of this example study indicate that the leadership styles of male and female managers from three observed countries are quite different. By using the factorial MANOVA procedure and all the fanfare that accompanies the procedure, some fairly complex relationships were disentangled.

Without doubt, there is an enormous amount of computer output that can be produced from analyzing the data from such a study. There is no cookbook recipe for selecting the appropriate output to present. And the style for reporting the results from a factorial MANOVA will often vary from one researcher to the next. Nevertheless, it is essential that one report all the test statistics used (e.g., Wilks's $\Lambda$), the $F$-test approximations, the graphs of the interaction effects, the univariate analyses of variance, the post hoc comparisons, and the univariate and multivariate measures of association. In this way others can evaluate the results and arrive at similar conclusions.

## MORE ADVANCED FACTORIAL MANOVAS

In this chapter the procedures for conducting two-way factorial ANOVAs and MANOVAs were discussed and some example cases were presented. There is no doubt that in real practice more advanced factorial designs can be encountered. However, the focus of this chapter was restricted to the two-way case. It would be impossible to discuss every type of design that one might encounter in real life without making the chapter inordinately long. As it turns out, the two-way MANOVA case can readily be extended to factorial designs with any number of factors. In fact, the extension is quite straightforward. Only three things need be remembered. First, in any type of design determine the SSCP matrix corresponding to the sums of squares in an ANOVA for each effect and their interactions. Second, select the appropriate error term (i.e., the correct SSCP matrix) following the same rules as presented in this chapter for an ANOVA. Finally, conduct a hypothesis test for each effect and each interaction using Wilks's $\Lambda$. Obviously, the actual computations for the various effects can be left to the computer.

## DESIGNS WITH UNEQUAL OBSERVATIONS

Another potential problem that can be encountered in practical situations is one of unequal observations in some level of a factor. The designs presented in this chapter only considered situations in which there were an equal number of subjects in each cell. Fortunately, for factorial designs with unequal observations the analyses are conducted in almost the same way. The only difference is that the significance tests from an analysis of variance are only approximate. There are several other estimation procedures that have emerged in an attempt to deal with this problem of unequal observations (many also deal with the problem of missing data). These procedures include unweighted means and maximum likelihood estimation (for a complete discussion see Marcoulides, 1990, 1996; Morrison, 1990).

## EXERCISES

1. A statistics professor is interested in comparing four statistical computer packages in terms of their ease of use. The four packages selected for this study were SAS, SPSS, BMDP, and SYSTAT. Forty graduate students enrolled in an advanced statistics course were selected to participate in this study. All students had taken the same prerequisite courses. The graduate students were asked to independently analyze two sets of data using a randomly assigned package. There were an equal number of Marketing and Management majors. The numbers of minutes of work required to analyze the data were recorded with the following results:

| | Computer Package | | | | | | | |
|---|---|---|---|---|---|---|---|---|
| | SAS | | SPSS | | BMDP | | SYSTAT | |
| Major | 1 | 2 | 1 | 2 | 1 | 2 | 1 | 2 |
| Marketing | 25 | 12 | 35 | 17 | 30 | 14 | 28 | 12 |
| | 26 | 10 | 27 | 19 | 33 | 15 | 35 | 16 |
| | 25 | 11 | 26 | 20 | 32 | 17 | 31 | 14 |
| | 28 | 12 | 32 | 20 | 37 | 11 | 33 | 11 |
| | 26 | 11 | 34 | 20 | 34 | 11 | 26 | 10 |
| Management | 35 | 13 | 37 | 16 | 33 | 10 | 31 | 14 |
| | 34 | 12 | 39 | 19 | 39 | 11 | 43 | 15 |
| | 33 | 11 | 40 | 18 | 41 | 13 | 47 | 15 |
| | 31 | 14 | 42 | 17 | 31 | 12 | 26 | 19 |
| | 30 | 14 | 36 | 15 | 38 | 10 | 36 | 16 |

Conduct an analysis and determine whether there are any differences between the packages with respect to ease of use. Does the major of a student contribute in any way to these differences?

2. The MARC SMOKE tobacco company is interested in studying the effect of nicotine in cigarettes on manual dexterity and steadiness. The company manufactures three brands of cigarettes: TARPS, LARPS, and BARPS. In addition, each brand is made with and without a filter. A psychologist conducts an experiment in which 30 randomly selected smokers are assigned to smoke a particular brand and type of cigarette and then complete a manual dexterity (M) and a steadiness (S) test (with scores ranging from a low of 0 to a high of 100). The results of the experiment are provided here. What conclusions can the psychologist draw about the effects of nicotine?

| | Cigarette Brand | | | | | |
|---|---|---|---|---|---|---|
| | TARPS | | LARPS | | BARPS | |
| Type | M | S | M | S | M | S |
| With filter | 38 | 27 | 30 | 38 | 40 | 43 |
| | 30 | 47 | 23 | 22 | 42 | 37 |
| | 44 | 39 | 37 | 31 | 38 | 36 |
| | 32 | 36 | 25 | 33 | 30 | 34 |
| | 42 | 35 | 35 | 27 | 50 | 41 |
| No filter | 24 | 28 | 40 | 46 | 45 | 52 |
| | 32 | 20 | 48 | 35 | 54 | 36 |
| | 16 | 22 | 32 | 45 | 34 | 50 |
| | 30 | 26 | 46 | 42 | 54 | 38 |
| | 18 | 25 | 32 | 39 | 34 | 44 |

3. The data provided next were collected on a class of 25 students enrolled in an introductory computer programming course. At the beginning of the course, each student provided data on the following variables: (a) major, Science (S), fine arts (F); (b) computer experience (reported as the number of computer courses taken prior to this course); (c) a computer anxiety score (53 to 265, with higher scores indicating more anxiety); and (d) a computer attitude score (26 to 130, with higher scores indicating more positive attitudes). Conduct an analysis to determine whether there are any differences between male and female students with respect to how they feel about computers. Does a student's major affect the results in any way?

| Student | Major | Gender | Number of Courses | Anxiety | Attitude |
|---------|-------|--------|-------------------|---------|----------|
| 1 | S | M | 5 | 75 | 120 |
| 2 | S | M | 4 | 65 | 109 |
| 3 | S | M | 3 | 72 | 99 |
| 4 | S | M | 7 | 89 | 97 |
| 5 | F | F | 0 | 109 | 52 |
| 6 | F | F | 1 | 123 | 48 |
| 7 | F | M | 0 | 155 | 56 |
| 8 | F | F | 1 | 111 | 68 |
| 9 | F | F | 0 | 123 | 56 |
| 10 | F | F | 0 | 127 | 67 |
| 11 | B | F | 1 | 121 | 80 |
| 12 | B | F | 2 | 88 | 81 |
| 13 | B | M | 3 | 77 | 107 |
| 14 | B | M | 2 | 99 | 37 |
| 15 | B | F | 1 | 122 | 78 |
| 16 | B | F | 3 | 106 | 69 |
| 17 | B | F | 2 | 57 | 102 |
| 18 | B | M | 5 | 67 | 55 |
| 19 | M | M | 6 | 45 | 76 |
| 20 | M | M | 8 | 108 | 44 |
| 21 | M | F | 5 | 45 | 120 |
| 22 | M | M | 4 | 56 | 121 |
| 23 | M | F | 3 | 67 | 111 |
| 24 | M | M | 6 | 66 | 110 |
| 25 | M | M | 2 | 78 | 111 |

4. The California Department of Education wishes to study differences in class sizes between private and public elementary, intermediate, and high schools in the cities of Fresno, Bakersfield, and Sacramento. A random sample of four private and public schools is selected at each school level from each city. The average class size for each school is recorded and

provided in the following table. Conduct an analysis and determine whether there are any differences in average class size between the three cities at each school level. Are there any differences between private and public schools with respect to class size?

| Educational Level | Cities | | | | | |
|---|---|---|---|---|---|---|
| | Fresno | | Sacramento | | Bakersfield | |
| | Pr. | Pub. | Pr. | Pub. | Pr. | Pub. |
| Elementary | 35 | 35 | 26 | 35 | 20 | 35 |
| | 36 | 40 | 28 | 38 | 19 | 38 |
| | 31 | 42 | 29 | 40 | 18 | 33 |
| | 30 | 48 | 25 | 45 | 16 | 37 |
| Intermediate | 28 | 33 | 25 | 29 | 20 | 28 |
| | 27 | 34 | 24 | 28 | 18 | 31 |
| | 26 | 35 | 25 | 26 | 19 | 31 |
| | 25 | 38 | 25 | 30 | 20 | 35 |
| High school | 35 | 45 | 33 | 40 | 34 | 43 |
| | 33 | 46 | 37 | 41 | 35 | 39 |
| | 32 | 48 | 35 | 42 | 30 | 36 |
| | 30 | 50 | 34 | 40 | 31 | 40 |

*Note.* Pr., private; Pub., public.

5. The Management and Training Department of the INXS Corporation is interested in studying the effects of various intervention strategies on the self-concept of male managers. Upper and lower level managers were randomly assigned to one of three intervention programs: (a) a naturalistic program, where managers together in wilderness areas developing climbing and other physical skills that depend on cooperation and trust; (b) a weight-lifting program, where the focus is on the development of physical ability; or (c) an in-class training program provided on the premises. Three measures of self-concept were taken: general self-concept, social self-concept, and physical self-concept. Self-concept theory leads to the predictions that intervention programs are more likely to be area specific than general (e.g., social self-concept is more likely to be affected in general), and that the area of self-concept affected is more likely to be directly related to the intervention program (e.g., social self-concept is more likely to be affected by the naturalistic program than physical self-concept). The data from the study are provided in the table that follows. Conduct an analysis to determine which of the self-concept measures is responsible for distinguishing among the various intervention programs. Are the groups distinguished along more than one dimension? (A picture in the form of a graph might help.)

| | Self-Concept Measures | | |
| --- | --- | --- | --- |
| | General | Social | Physical |
| Upper level managers | | | |
| Naturalistic program | 51 | 61 | 91 |
| | 52 | 72 | 71 |
| | 33 | 73 | 77 |
| | 45 | 55 | 71 |
| Weight-lifting | 33 | 22 | 88 |
| | 36 | 24 | 98 |
| | 37 | 25 | 89 |
| | 35 | 22 | 78 |
| In-class program | 44 | 33 | 33 |
| | 46 | 33 | 33 |
| | 45 | 33 | 34 |
| | 38 | 34 | 33 |
| Lower level managers | | | |
| Naturalistic program | 57 | 67 | 97 |
| | 56 | 77 | 77 |
| | 37 | 78 | 74 |
| | 45 | 57 | 72 |
| Weight-lifting | 38 | 28 | 88 |
| | 32 | 22 | 92 |
| | 33 | 23 | 83 |
| | 36 | 26 | 76 |
| In-class program | 41 | 31 | 31 |
| | 47 | 37 | 36 |
| | 43 | 36 | 35 |
| | 32 | 39 | 35 |

CHAPTER FIVE

# Discriminant Analysis

There are two types of problems that can be addressed using a discriminant analysis. The first problem, often referred to as a descriptive discriminant analysis, involves the process of describing group differences on the basis of information obtained from several observed variables. For example, a bank loan officer identifies customers according to their eligibility for a personal bank loan. The loan officer might use a descriptive discriminant analysis to determine that the primary differences between these two groups (loan eligible vs. loan ineligible) seem to lie in past credit history, current income, occupational status, amount of money in savings account, and age. Of course, it is important to note that in order for the loan officer to determine which variables most distinguish the two groups, individuals must first have been classified into one or the other of the two groups (e.g., in the preceding example this could have been accomplished by simply determining who had repaid a bank loan in the past and who had not). The second problem, often referred to as a predictive discriminant analysis, concerns the classification of an object (usually a person) to one of the groups. For example, if a new customer applies to the bank for a loan, the loan officer must determine into which of the two groups this new customer will be placed. In this chapter, both types of problems that can be addressed by a discriminant analysis will be discussed. The first part of the chapter discusses the problem of descriptive discriminant analysis and the second part the problem of predictive discriminant analysis.

## DESCRIPTIVE DISCRIMINANT ANALYSIS

A descriptive discriminant analysis is most frequently conducted following the detection of one or more significant main effects from a MANOVA (discussed extensively in both chapters 3 and 4). Based on this detection, one is then confronted with the problem of explaining why in fact these significant group differences exist. Recall from chapter 3 that a significant difference among the groups implies that two or more of the groups differ on one or more of the dependent variable means. In general, a descriptive discriminant analysis is concerned with identifying certain linear composites of the variables that produce the group differences.[1] In a way, these linear composites assist in identifying the underlying reasons that produce the group differences. Each linear composite may represent a different collection of underlying reasons.

One way to conceptualize the collection of underlying reasons is to consider them as a collection of latent variables (hypothetically existing variables, also referred to as constructs). For example, organizational culture is a latent variable that is used in the organizational theory literature as a way of describing the work environment of an company or firm. Intelligence is also a latent variable because although we use IQ tests to measure it, an IQ score itself is not intelligence but an inexact reflection of where an individual stands on intelligence. The term *inexact* is used because IQ tests presumably only measure a part of intelligence; a better approximation to the person's intelligence may be obtained if we also measure the person on creativity. Nevertheless, intelligence is a latent variable whose existence is implied by how well individuals do on tests believed to measure it. Other examples of latent variables used often are job satisfaction, locus of control, and quality of life. As such, latent variables are unobserved (unmeasured) variables whose existence is indicated by a set of observed (measured) variables. Any one of the observed variables is considered to measure only a part of the latent variable; therefore there is a need to combine the observed variables in a linear composite in order to obtain a more complete description of the latent variable.

As another example, consider the bank loan officer who determined that the primary differences between the two loan eligible versus loan ineligible groups seem to lie in past credit history, current income, occupational status, amount of money in savings account, and age. Assuming that a descriptive

---

[1]A bit of confusion surrounds the use of terminology for variables in discriminant analysis. Although we are still concerned with group differences (as in MANOVA), these group differences are predicted by a set of observed variables. This implies that group membership is the dependent or outcome variable. To simplify matters, whenever we refer to the *outcome variable* in a discriminant analysis, it is to the group membership variable we refer; whenever the word *variable* is used generically, the predictor variables are intended.

discriminant revealed a significant linear composite of these variables, the loan officer would attempt to interpret the meaning of the linear composite (latent variable) based on the degree to which each variable was associated with the linear composite. Thus, the loan officer may find that past credit history is very highly related to the linear composite, whereas the other variables have negligible relations with it. The loan officer may then interpret the linear composite as a "fidelity in paying debts" latent variable, choosing to ignore the other variables for the interpretation because of their negligible association with the composite. Unfortunately, not all exercises in interpreting a latent variable are this straightforward; in fact, the interpretation of a latent variable is sometimes considered an acquired skill. The topic of latent variables (also commonly referred to as components or factors), in particular their detection and interpretation, is discussed extensively in chapters 7 and 8.

## SELECTING THE DISCRIMINANT CRITERION

The main goal in a descriptive discriminant analysis is to construct a linear composite of the observed variables so as to maximize group differences. The linear composite to be constructed can be defined as:

$$Y = a_1X_1 + a_2X_2 + \cdots + a_pX_p$$

where there are $p$ variables used to separate the groups. The coefficients $a$ are chosen such that group differences on $Y$ are maximized. Recall from chapter 3 that one indicator of group differences was a significant $F$ ratio following a multivariate analysis of variance (MANOVA). Generally, the larger this $F$ ratio, the greater the group separation. Thus, if a linear composite is so constructed as to maximize this $F$ ratio, group separation will be maximized.

In order to define the coefficients $a$ that will accomplish this, the ratio of the between-groups SSCP matrix ($B$) to within-groups SSCP matrix ($W$) is written as a function of the coefficients $a$:

$$\frac{1}{[(k-1)/(n-k)]F} = \frac{B}{W} \propto \frac{a'Ba}{a'Wa} = \lambda$$

where

$$a = [a_1, a_2, \ldots, a_p]$$

and $B/W$ is proportional to ($\propto$) $a'Ba/a'Wa$.

In the preceding expression, $\lambda$ is referred to as the *discriminant criterion*, the criterion that maximizes group separation along the dimension specified by the coefficients in the $a$ vector. As indicated in chapter 4, $\lambda$ also happens to be the largest eigenvalue from the matrix product $(a' W^{-1} a)(a' Ba)$. Therefore, the coefficients $a$ are the eigenvectors associated with this eigenvalue. It was also pointed out in chapter 4 that for any matrix $A$ there are $\lambda_i$ eigenvalues. Thus, associated with each $\lambda_i$ is a $p$-dimensional eigenvector of coefficients. Of course, how many eigenvalues a matrix has depends on the rank of the matrix. The term rank refers to the number of nonzero determinants that can be extracted from the matrix. Generally, the rank of a matrix is equal to the number of rows $r$ or the number of columns $c$, whichever is smaller. In discriminant analysis, in which we are interested in the rank of the matrix product $(a' W^{-1} a)(a' Ba)$, the rank is equal to the smaller of either $p$ (the number of variables) or $k - 1$ (the number of groups minus one). Thus, the number of eigenvalues (discriminant criteria) is usually equal to $k - 1$, for it would be unusual to have more groups than variables.

## MAXIMIZING THE DISCRIMINANT CRITERION

Finding the coefficients $a$ that maximize the discriminant criterion ($\lambda$) involves finding the eigenvalues and eigenvectors of the matrix product $(a' W^{-1} a)(a' Ba)$. As explained in chapter 2, the easiest way to find eigenvalues is to solve the equation

$$(BW^{-1} - \lambda_1 I) = 0$$

which provides a solution to the largest eigenvalue ($\lambda_1$). After finding $\lambda_1$, the $p \times 1$ eigenvector $x_1$ associated with $\lambda_1$ is found by solving the set of $p$ equations:

$$(BW^{-1} - \lambda_1 I)x_1 = 0$$

The vector $x_1$ consists, within a constant of proportionality, of the coefficients $a$ that maximize the discriminant criterion.

One usually solves for eigenvalues and eigenvectors that are as many as the minimum of either the number of variables ($p$) or the number of groups less one ($k - 1$). The set of extracted eigenvalues and their associated eigenvectors possess two important properties. First, each successively extracted eigenvalue (discriminant criterion) separates the groups less well than the preceding eigenvalues. As such, the first eigenvalue ($\lambda_1$) provides the greatest group separation, the second eigenvalue ($\lambda_2$) provides the next, and so forth (i.e., $\lambda_1 > \lambda_2 > \lambda_3 > \cdots > \lambda_{k-1}$). The second property is that the eigenvalues ($\lambda$) are mutually uncorrelated with each other.

## DISCRIMINANT FUNCTIONS

Linear composites created by application of the coefficients $a$ to the variables are referred to as discriminant functions. For each discriminant function there is a different set of coefficients that provide maximal group difference. Thus,

$$Y_1 = a_1X_1 + a_2X_2 + \cdots a_pX_p$$

is the first discriminant function,

$$Y_2 = a_1X_1 + a_2X_2 + \cdots a_pX_p$$

is the second discriminant function, and

$$Y_{k-1} = a_1X_1 + a_2X_2 + \cdots a_pX_p$$

is the ($k - 1$)-th or last discriminant function. Once again, it is important to note that the first discriminant function will account for the greatest group differences (such that $Y_1 > Y_2 > \cdots > Y_{k-1}$) and that the functions are independent ($r_{Y_1Y_2} = r_{Y_1Y_{k-1}} = r_{Y_2Y_{k-1}} = 0$).

Figure 5.1a illustrates in what sense the discriminant functions provide for maximal group separation. In this figure, two groups are plotted on two observed ($x_1$ and $x_2$) variables. Assuming that there is group homogeneity of variance for the $W$ matrix and that the within-group correlation between the two variables $x_1$ and $x_2$ is zero, then the axis that connects the two groups' centroids is the discriminant function. In fact, there is no other axis that can be constructed to better separate the two groups. Of course, it is important to note that the discriminant function will only go through the group centroids when $x_1$ and $x_2$ are uncorrelated. When there is a correlation between the observed variables, the plot looks somewhat different. Figure 5.1b shows where the discriminant function is placed if the correlation between $x_1$ and $x_2$ is positive, and Fig. 5.1c shows where the discriminant function is placed if the correlation is negative.

## INTERPRETATION OF DISCRIMINANT FUNCTIONS

Earlier in the chapter it was noted that the discriminant functions can be considered like a latent variable subject to interpretation. Another way to conceptualize this point is to consider the scores $Y_1, Y_2, \ldots, Y_{k-1}$ generated by each discriminant function to represent a person's standing on the latent variable. Nevertheless, the problem of interpreting what these $k - 1$ dis-

(a)

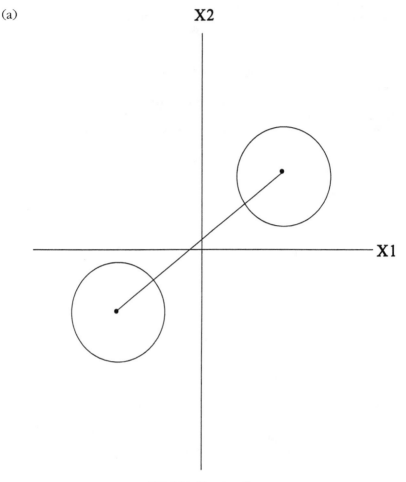

FIG. 5.1. *(Continued)*

criminant function scores represent remains. As it turns out, interpretation of the discriminant function is generally accomplished by identifying those variables most highly associated with the function. This task is made simpler by examining the magnitude of the discriminant function coefficients *a* for each variable on each function. Those variables that contribute most to the discriminant function (i.e., have the largest coefficients) will influence the interpretation of the discriminant function.

Consider an example in which a descriptive discriminant analysis was conducted for three loan applicant groups: (a) approved at preferred interest, (b) approved at usual interest rate, and (c) declined. In this example, two discriminant functions will be computed ($k - 1$) for the three loan applicant groups. These two discriminant functions are presented in Table 5.1a. Recall

(b)

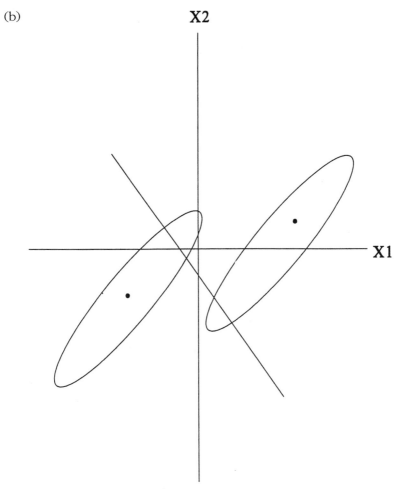

FIG. 5.1. *(Continued)*

from the previous section that the discriminant function coefficients presented are simply the two eigenvectors $(x_i)$ associated with the first two eigenvalues.

In most cases, however, the eigenvectors are routinely transformed (standardized) to yield what are referred to as raw coefficients. These transformations are determined using the following equations:

$$a_i = x_i \sqrt{N - k}$$

$$a_0 = -\sum_{i=1}^{p} a_i \overline{X}_{G_i}$$

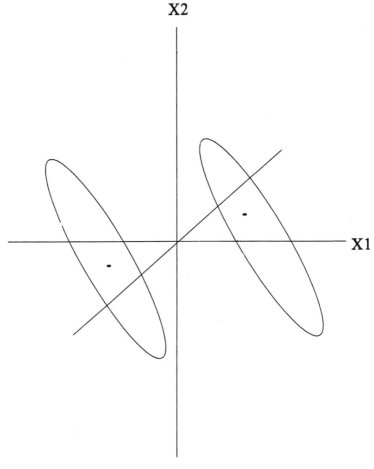

(c)

FIG. 5.1. (a) Plot of discriminant function when the observed variables X1 and X2 are uncorrelated. (b) Plot of discriminant function when the observed variables X1 and X2 are positively correlated. (c) Plot of discriminant function when the observed variables X1 and X2 are negatively correlated.

TABLE 5.1a
Discriminant Function Coefficients

|  | Eigenvector | |
| --- | --- | --- |
|  | 1 | 2 |
| Income | .01 | .06 |
| Savings | .01 | −.01 |
| Stocks | −.16 | −.17 |
| Rent/own | −.07 | −.29 |
| Occupation | .03 | .05 |
| Marital status | .06 | .16 |

TABLE 5.1b
Raw Discriminant Coefficients for Eigenvectors 1 and 2

| | Eigenvector | |
|---|---|---|
| | 1 | 2 |
| Income | .01 | .72 |
| Savings | .01 | −.01 |
| Stocks | −1.94 | −2.10 |
| Rent/own | −.88 | −3.55 |
| Occupation | .36 | .66 |
| Marital status | .75 | 1.97 |

TABLE 5.1c
Within-Class Standardized Discriminant
Coefficients for Eigenvectors 1 and 2

| | Eigenvector | |
|---|---|---|
| | 1 | 2 |
| Income | .04 | 2.78 |
| Savings | .50 | −.94 |
| Stocks | −.91 | −.98 |
| Rent/own | −.79 | −3.16 |
| Occupation | 1.21 | 2.21 |
| Marital status | .58 | 1.53 |

TABLE 5.1d
Pooled Within-Groups Discriminant Structure Coefficients

| | Eigenvector | |
|---|---|---|
| | 1 | 2 |
| Income | −.39 | .88 |
| Savings | −.33 | .86 |
| Stocks | −.43 | .82 |
| Rent/own | .84 | .23 |
| Occupation | .86 | .29 |
| Marital status | .73 | .16 |

where $a_i$ represents the raw coefficients, and $a_0$ represents the intercept (constant) of the discriminant function. The transformation of the eigenvector to raw coefficients ensures that the discriminant scores ($Y$) over all cases will have a mean of zero and a within-groups standard deviation of one. Thus, the discriminant scores here are comparable to $z$ scores, and represent in standard deviation units how far from the grand mean a particular case

TABLE 5.1e
Total-Groups Discriminant Structure

| | Eigenvector | |
|---|---|---|
| | 1 | 2 |
| Income | −.41 | .87 |
| Savings | −.36 | .85 |
| Stocks | −.46 | .81 |
| Rent/own | .86 | .22 |
| Occupation | .87 | .28 |
| Marital status | .75 | .16 |

TABLE 5.1f
Between-Groups Discriminant Structure

| | Group | |
|---|---|---|
| | 1 | 2 |
| Income | −.94 | .35 |
| Savings | −.92 | .39 |
| Stocks | −.95 | .30 |
| Rent/own | .99 | .04 |
| Occupation | .99 | .06 |
| Marital status | .99 | .04 |

TABLE 5.1g
Group Means on Discriminant Functions

| | Group | |
|---|---|---|
| | 1 | 2 |
| Reject | −.31 | .03 |
| Nonpreferred | .11 | −.14 |
| Preferred | .69 | .05 |

is. Table 5.1b presents the transformed values (i.e., the raw coefficients) of the eigenvectors for this example.

Although the values of the raw coeffiicients are useful for comparing individual scores on the discriminant function, these coefficients are not directly comparable across the observed variables. As such, comparison of the raw coefficients across observed variables is inappropriate because the standard deviations of the variables are not the same; thus a unit change in one variable is not the same as a unit change in another variable. In order to determine the relative importance of the observed variables (in a manner similar to how beta weights are used in multiple regression), standardized coefficients ($a_{i*}$) must be computed:

$$a_{i*} = a_i \sqrt{\frac{w_{ii}}{N-k}}$$

where $w_{ii}$ is the diagonal element for variable $i$ from the $W$ SSCP matrix. The standardized discriminant coefficients for the loan variables are presented in Table 5.1c.

Another way of describing the relationship between a variable and a discriminant function is through structure coefficients ($a_{i**}$). These structure coefficients express the correlation between the variable and discriminant function, and are computed using:

$$a_{ij**} = \sum_{k=1}^{p} r_{ik}a_{kj*} = \sum_{k=1}^{p} \frac{w_{ik}a_{kj*}}{\sqrt{w_{ii}w_{kk}}}$$

where $r_{ik}$ is the pooled within-groups correlation coefficient between variables $i$ and $k$. The computed structure coefficients for the loan variables are given in Table 5.1d. Obviously, at some point one should venture an interpretation of the two discriminant functions. In general, this interpretation is best made based on the values of the structure coefficients. For example, because the rent/own, occupation, and marital status variables have the highest correlations with the first function, one may interpret this function as a "life-style" latent variable. And because income, savings, and stocks correlate most highly with the second function, one may consider this a "monetary wealth" latent variable.

Three other issues need to be discussed with respect to the calculation and interpretation of discriminant coefficients. First, it should be noted that the formulas for the standardized and structure coefficients use elements from the within-group $W$ SSCP matrix rather than the between-group or total-group SSCP matrices (i.e., the $B$ or $T$ SSCP matrices). In practice, elements from these latter two matrices could be used instead of elements from the $W$ SSCP matrix to accomplish the coefficient transformations. However, it should be pointed out that elements from the $T$ matrix ignore intergroup mean differences, and elements from the $B$ SSCP matrix involve means and not individual/variate scores (rendering the elements not useful for interpreting the function). Thus, transformations using the $T$ and $B$ matrices are not likely to fully reveal the nature of the functions. Table 5.1e and Table 5.1f present the results of the transformation using the $T$ and $B$ SSCP matrices. Although the total structure matrix presented in Table 5.1e is somewhat similar to the within-structure matrix presented previously in Table 5.1d, there are still some slight differences. In addition, the results presented in Table 5.1f clearly demonstrate that the functions are essentially uninterpretable. It is important to note, therefore, that because the elements from the $W$ SSCP matrix take into account both group and individual differences, they are the most useful for conducting the transformations.

A second issue concerning the discriminant coefficients involves which of the two coefficients (standardized or structure) is better for interpretation. Unfortunately, the two types of discriminant coefficients can give different impressions concerning the meaning of the discriminant functions—which indeed they do in this example. To see this difference, simply compare the results presented in Table 5.1c with those presented in Table 5.1d. As can be seen from these two tables, there are some rather striking differences between the values on each of the discriminant functions. For example, the standardized coefficient for the INCOME variable on the first discriminant function is quite small (0.04) compared to that for the structure coefficient (−0.39). To date, there is some disagreement in the literature concerning which coefficients (standardized or structure) are better for interpretation. Each has its proponents, but the majority opinion appears to favor the structure coefficients. Huberty (1989) argued in favor of using the structure coefficients for two reasons:

1. The assumed greater stability of the correlations in small- or medium-sized samples, especially when there are high or fairly high intercorrelations among the variables.
2. The correlations give a direct indication of which variables are most closely associated with the latent variable the discriminant function represents.

In contrast, the standardized coefficients are partial coefficients (as are beta weights in multiple regression), with the effects of the other variables removed. One is rarely interested in interpreting the meaning of the regression equation; interest instead focuses on the unique contribution of each variable to the prediction of the criterion. In our opinion, an agnostic view to whether the structure or standardized coefficients are better seems more appropriate—after all, it all depends on one's purpose. Clearly, if interpreting the meaning of the discriminant function is the primary goal, then structure coefficients are preferable. On the other hand, if determining which variables are redundant for defining the discriminant function is of interest, then the standardized coefficients will be preferred.

The third issue we wish to discuss with respect to the calculation and interpretation of discriminant coefficients concerns how large a coefficient must be in order to use the variable to define the latent construct behind the discriminant function. Obviously, once again, this is a matter of opinion.

Usually, identifying a variable as aligned with a discriminant function depends on both the relative and absolute value of its coefficient. For instance, an examination of Table 5.1d reveals that rent/own, occupation, and marital status are clearly more similar to each other than the other three variables for the first function. Although the coefficients for the other three variable are also somewhat "high" (>.3) in an absolute sense, it is essential to recognize that both the relative and absolute value of each coefficient should be inspected. Thus, if three variables formed a cluster on a function with an average coefficient value of .2 and another three with an average coefficient value of .00, we would be hesitant to conclude that any of the variables defined the function.

## DISCRIMINANT FUNCTION PLOTS

A graphical device that can also be used for interpreting the discriminant functions is to graph each of the groups means on each of the discriminant functions. This graph is typically referred to as a discriminant function plot. A general equation for obtaining the coordinates of any set of $j$ groups on the discriminant functions is given by:

$$\overline{Y}_j = \overline{X}_j a_i$$

where $\overline{X}_j$ is the matrix of means for group $j$ on the variables. Note that group discriminant function means, as well as individual discriminant function scores, should be computed using the raw coefficients ($a_i$). A plot based on the three loan groups (whose means are provided in Table 5.1g) is shown in Fig. 5.2. Not surprisingly, the first function appears to contribute most to group separation.

## TESTS OF STATISTICAL SIGNIFICANCE
## IN DISCRIMINANT ANALYSIS

Although $p$ or $k - 1$ discriminant functions may be created, this by no means implies that significant group differences occur for each. For reasons of parsimony alone, an attempt should be made to retain as few discriminant functions as possible. In chapter 3 Wilks's $\Lambda$ was introduced for testing the significance of overall group differences in a MANOVA. Recall from chapter 3 that the significance of Wilks's $\Lambda$ is accomplished by using an approximation to the $F$ distribution. As it turns out, this same test may be used in discriminant analysis to examine which of the discriminant functions are significant. However, an important difference exists between the MANOVA and the discriminant analysis use of Wilks's $\Lambda$. In discriminant analysis a residualization approach is used instead. In the first step, all the discriminant functions

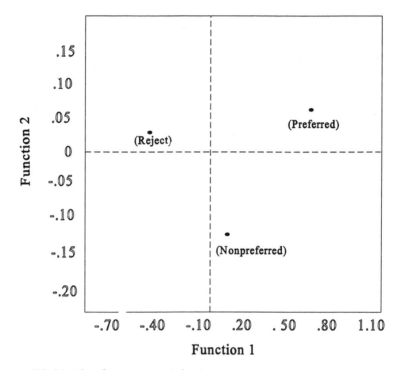

FIG. 5.2. Plot of group centroids for three groups in discriminant dimension.

are tested simultaneously; the null hypothesis is that all discriminant functions are equal to zero and the alternative is that at least one is significant. If the null hypothesis is rejected, then the largest (i.e., the first) discriminant function is removed and a test is made of the remaining functions (i.e., the residual) to determine if this is significant. At this stage, the null hypothesis is, only one (the largest) function differs from zero; the alternative hypothesis is, more than one function is significant. If the null hypothesis is retained, we stop and conclude that only one function is required to describe group differences. If the null hypothesis is rejected, a second residual is created by removing the first two functions. Similarly, the next null hypothesis is, only two functions are significant; the alternative hypothesis is, more than two functions are significant. The residualization/hypothesis testing continues until either the residual becomes nonsignificant or one runs out of functions to test.

An important point should be made concerning the appropriate interpretation of significant results using this residualization procedure: The significance of any individual discriminant function cannot be determined. The reason for this is rather complicated. Remember that when we test for the significance of a sample statistic, we really are not interested in the significance of that statistic in the sample but instead are interested in the significance of its population counterpart. However, due to sampling error, the largest discriminant function (or the first extracted) may not actually correspond to the largest in the population. A mathematical constraint exists for extracting the sample functions in decreasing order of magnitude; no such constraint exists on the corresponding population functions. It could well be the case, especially in small samples, that the smallest discriminant function in the population corresponds to the largest in the sample.

Of course, statistical significance isn't everything. Just because a significance test suggests that some number of functions is significant does not mean they will all be useful. For this reason we recommend that the practical significance of a discriminant function should be examined. At present two popular approaches exist for determining the practical significance of a discriminant function. The first approach requires one to determine how much of the total group difference is accounted for by increasing numbers of discriminant functions. This is a proportion of variance approach, in which one determines what proportion of the between groups variation is accounted for by retaining one, two, or more functions, as given by:

$$\frac{\lambda_1}{\sum_{j=1}^{r} \lambda_j}, \frac{\lambda_1 + \lambda_2}{\sum_{j=1}^{r} \lambda_j}, \ldots, \frac{\lambda_1 + \lambda_2 + \ldots + \lambda_r}{\sum_{j=1}^{r} \lambda_j} \times 100\%$$

where $\lambda$ is the eigenvalue, and $\sum \lambda$ the sum of the eigenvalues. One stops retaining functions as soon as a satisfactory amount of variance has been accounted for. As expected, what constitutes a "satisfactory" proportion of variance is a subjective matter (similar to the subjectivity involved in determining what constitutes a "large" function coefficient). In an absolute sense, it is desirable to account for a majority of the variance (>50%), and quite often, the first function will do that. On the other hand, in a relative sense, it hardly seems of value to retain a function that only increases the amount of variance accounted for by less than 10%.

The second approach to retaining functions of practical value goes hand-in-hand with the proportion of variance approach but takes into account the issue of interpretability. To illustrate this approach, we assume that two discriminant functions were retained (accounting for 80% of the variance cumulatively) and that these functions have a reasonable interpretation. Now, although retaining a third function would increase the variance accounted for to 90%, no reasonable interpretation can be made concerning the meaning of the third function (i.e., the variables most highly associated with it form a conceptually motley group). Thus, although the third discriminant function may account for 10% additional variance, it is difficult to justify retaining a function when the meaning of its contribution to group separation is ambiguous.

## AN EXAMPLE OF DESCRIPTIVE DISCRIMINANT
## ANALYSIS USING SAS

A simple example is used to illustrate the SAS program for conducting a descriptive descriminant analysis. This example involves four groups of individuals: Group 1, low IQ (70–90); Group 2, low–medium IQ (91–110); Group 3, medium–high IQ (111–130); and Group 4, high IQ. Each group was measured on 12 cognitive variables. We wish to perform a discriminant analysis to examine how well these 12 variables separate the groups, and what specific latent variables (discriminant functions) provide the best group discrimination. The deck setup to perform the descriptive discriminant analysis in SAS is presented in Table 5.2. In order to perform a descriptive analysis on the example data set, the PROC CANDISC should be chosen.

Table 5.3 displays the results for the residual test procedure for the number of significant discriminant functions. It is important to note the use of the word *canonical* along with discriminant analysis in the title, as well as the use of the word throughout. As explained in chapter 6 (the Canonical Correlation chapter), discriminant analysis may also be thought of as a canonical

TABLE 5.2
SAS Deck Setup for Discriminant Analysis Sample

```
DATA ONE;
INFILE DAT1;
INPUT
      GROUP VOCABLRY VFLUENCY READCOMP ARTHMTIC NFLUENCY
      WORDPROB FFLUENCY PERSPEED SPATIAL VMEMORY NUMSPANF
      NUMSPANB;
LABEL
      VOCABLRY='VOCABULARY TEST'
      VFLUENCY='VERBAL FLUENCY TEST'
      READCOMP='READING COMPREHENSION TEST'
      ARTHMTIC='ARITHMETIC TEST'
      NFLUENCY='NUMBER FLUENCY TEST'
      WORDPROB='ALGEBRAIC WORD PROBLEM TEST'
      FFLUENCY='FIGURAL FLUENCY TEST'
      PERSPEED='PERCEPTUAL SPEED TEST'
      SPATIAL='SPATIAL RELATIONS TEST'
      VMEMORY='VERBAL MEMORY TEST'
      NUMSPANF='MEMORY FOR NUMBERS-FORWARD'
      NUMSPANB='MEMORY FOR NUMBERS-BACKWARD';

RUN;
PROC CANDISC ALL;
CLASS GROUP;
VAR
      VOCABLRY VFLUENCY READCOMP ARTHMTIC NFLUENCY
      WORDPROB FFLUENCY PERSPEED SPATIAL VMEMORY NUMSPANF
      NUMSPANB;
```

TABLE 5.3
SAS System Canonical Discriminant Analysis,
Results of Residual Test Procedure

|   | Canonical Correlation | Adjusted Canonical Correlation | Approx Standard Error | Squared Canonical Correlation |
|---|---|---|---|---|
| 1 | 0.478737 | 0.389469 | 0.066588 | 0.229189 |
| 2 | 0.343906 | 0.242626 | 0.076170 | 0.118271 |
| 3 | 0.226400 | 0.114507 | 0.081959 | 0.051257 |

Eigenvalues of INV(E)*H
= CanRsq/(1-CanRsq)

|   | Eigenvalue | Difference | Proportion | Cumulative |
|---|---|---|---|---|
| 1 | 0.2973 | 0.1632 | 0.6124 | 0.6124 |
| 2 | 0.1341 | 0.0801 | 0.2763 | 0.8887 |
| 3 | 0.0540 |  | 0.1113 | 1.0000 |

Test of H0: The canonical correlations in the
current row and all that follow are zero

|   | Likelihood Ratio | Approx F | Num DF | Den DF | Pr > F |
|---|---|---|---|---|---|
| 1 | 0.64480972 | 1.5801 | 36 | 355.2811 | 0.0211 |
| 2 | 0.83653444 | 1.0268 | 22 | 242 | 0.4318 |
| 3 | 0.94874326 | 0.6591 | 10 | 122 | 0.7601 |

Total Canonical Structure

|  | CAN1 | CAN2 | CAN3 |
|---|---|---|---|
| VOCABLRY | 0.593274 | 0.311258 | -0.054219 |
| VFLUENCY | 0.521887 | 0.238670 | 0.352998 |
| READCOMP | 0.306846 | 0.386048 | 0.158447 |
| ARTHMTIC | -0.366718 | -0.093762 | -0.084633 |
| NFLUENCY | -0.175489 | 0.010702 | 0.350915 |
| WORDPROB | -0.446918 | 0.242132 | 0.545117 |
| FFLUENCY | 0.573573 | -0.023286 | 0.374173 |
| PERSPEED | 0.539279 | -0.444014 | 0.435625 |
| SPATIAL | 0.418702 | 0.120643 | 0.364582 |
| VMEMORY | -0.358273 | -0.250032 | 0.321674 |
| NUMSPANF | -0.527471 | -0.071390 | 0.388592 |
| NUMSPANB | -0.356067 | -0.182415 | 0.376879 |

correlation analysis between group membership and a set of variables. For now, the reader should interpret canonical to mean discriminant. With four groups, three discriminant functions are possible of which only the first discriminant function is significant (according to the $F$ transformation of Wilks's $\Lambda$, $p < .021$). This one discriminant function can account for 61% of the group separation.

Table 5.4 includes the within-canonical-structure matrix that can be used for interpreting the single significant function. As can be seen from Table 5.4, a majority of the cognitive tests have respectable ($>\pm.3$) coefficients on the first function and provide probably the most reasonable interpretation that this function represents general intelligence. Table 5.5 provides the raw coefficients and the group means on the first discriminant function. Examining the group means on the first function provided in Table 5.5, we see that the groups are ordered as expected on an intelligence function—the high IQ group has the highest mean (1.016) and the low IQ group the lowest (–0.449).

## FACTORIAL DESCRIPTIVE DISCRIMINANT
## ANALYSIS DESIGNS

Until this point, our discussion has assumed that the discriminant analysis was based on a single-factor (independent) variable design. However, this is an unnecessary restriction because discriminant analyses can also be performed on multifactor variable designs. In fact, just as the one-factor MANOVA model (chapter 3) was extended to the multifactor MANOVA (chapter 4), a single-factor descriminant analysis can be extended to a multifactorial analysis. In the unifactor case, the purpose is to identify linear combinations of variables that best separate the groups. In the multifactorial case, the purpose is to identify linear combinations of variables that provide the best separation within each of the MANOVA effects. For example, assume that a MANOVA was conducted with two independent variables, $A$ and $B$. Three significant effects are possible from the design: a main effect for $A$, a main effect for $B$, and the $A \times B$ interaction. Assume further that each of the three effects is significant. For each effect, it is possible to identify a separate set of linear discriminant functions—with the number of functions for each effect based on either $p$ (the number of variables) or $df_{Between} - 1$ (the degrees of freedom minus one specific to the effect), whichever is smaller. It is entirely possible (if not probable) that the discriminant functions that provide the best separation for one effect will be different from the discriminant functions that provide the best separation for another effect. Nevertheless, interpreting the functions and determining their significance proceeds exactly as described earlier in this chapter. The only complexity added by the factorial design is that now the process is conducted separately for each effect (in this case, three times).

TABLE 5.4
SAS System Canonical Discriminant Analysis, Structure

Between Canonical Structure

|  | CAN1 | CAN2 | CAN3 |
|---|---|---|---|
| VOCABLRY | 0.934984 | 0.352380 | -0.040409 |
| VFLUENCY | 0.909000 | 0.298625 | 0.290762 |
| READCOMP | 0.730014 | 0.659773 | 0.178268 |
| ARTHMTIC | -0.977930 | -0.179616 | -0.106732 |
| NFLUENCY | -0.726208 | 0.031813 | 0.686739 |
| WORDPROB | -0.820833 | 0.319463 | 0.473473 |
| FFLUENCY | 0.955190 | -0.027857 | 0.294681 |
| PERSPEED | 0.817654 | -0.483609 | 0.312354 |
| SPATIAL | 0.908188 | 0.187981 | 0.373976 |
| VMEMORY | -0.835770 | -0.418995 | 0.354869 |
| NUMSPANF | -0.940374 | -0.091428 | 0.327623 |
| NUMSPANB | -0.849415 | -0.312600 | 0.425176 |

Pooled Within Canonical Structure

|  | CAN1 | CAN2 | CAN3 |
|---|---|---|---|
| VOCABLRY | 0.546705 | 0.306769 | -0.055431 |
| VFLUENCY | 0.476550 | 0.233090 | 0.357605 |
| READCOMP | 0.275023 | 0.370071 | 0.157556 |
| ARTHMTIC | -0.327281 | -0.089497 | -0.083797 |
| NFLUENCY | -0.155113 | 0.010117 | 0.344114 |
| WORDPROB | -0.406425 | 0.235504 | 0.549975 |
| FFLUENCY | 0.525766 | -0.022829 | 0.380519 |
| PERSPEED | 0.498992 | -0.439410 | 0.447191 |
| SPATIAL | 0.376897 | 0.116148 | 0.364094 |
| VMEMORY | -0.321389 | -0.239887 | 0.320136 |
| NUMSPANF | -0.480755 | -0.069591 | 0.392934 |
| NUMSPANB | -0.319104 | -0.174845 | 0.374716 |

Total-Sample Standardized Canonical Coefficients

|  | CAN1 | CAN2 | CAN3 |
|---|---|---|---|
| VOCABLRY | 0.330199759 | 0.252423720 | -0.351535013 |
| VFLUENCY | 0.201072578 | -0.046090043 | 0.779310556 |
| READCOMP | -0.050268014 | 0.371395937 | -0.106217108 |
| ARTHMTIC | -0.362103695 | -0.219649635 | -0.393613025 |
| NFLUENCY | 0.387857400 | 0.109684514 | 0.406775068 |
| WORDPROB | -0.515928643 | 0.437712789 | 0.407136838 |
| FFLUENCY | 0.245950487 | 0.376047485 | 0.103606665 |
| PERSPEED | 0.210528125 | -1.280758104 | 0.246313114 |
| SPATIAL | 0.174516548 | 0.589878626 | -0.034049159 |
| VMEMORY | 0.122551448 | -0.216938029 | -0.084074156 |
| NUMSPANF | -0.351142518 | 0.029780372 | 0.276873012 |
| NUMSPANB | -0.214802822 | -0.170345561 | 0.333215705 |

TABLE 5.5

SAS System Canonical Discriminant Analysis, Raw Coefficients
and Group Means on First Discriminant Function

Pooled Within-Class Standardized Canonical Coefficients

|  | CAN1 | CAN2 | CAN3 |
|---|---|---|---|
| VOCABLRY | 0.318177888 | 0.243233509 | -0.338736371 |
| VFLUENCY | 0.195529325 | -0.044819413 | 0.757826194 |
| READCOMP | -0.049800387 | 0.367940962 | -0.105229005 |
| ARTHMTIC | -0.360276594 | -0.218541328 | -0.391626934 |
| NFLUENCY | 0.389639528 | 0.110188493 | 0.408644118 |
| WORDPROB | -0.503764755 | 0.427392972 | 0.397537900 |
| FFLUENCY | 0.238250748 | 0.364274921 | 0.100363148 |
| PERSPEED | 0.202032470 | -1.229074377 | 0.236373392 |
| SPATIAL | 0.172150760 | 0.581882089 | -0.033587580 |
| VMEMORY | 0.121308607 | -0.214737978 | -0.083221528 |
| NUMSPANF | -0.342096515 | 0.029013181 | 0.269740313 |
| NUMSPANB | -0.212828851 | -0.168780139 | 0.330153558 |

Raw Canonical Coefficients

|  | CAN1 | CAN2 | CAN3 |
|---|---|---|---|
| VOCABLRY | 0.281074568 | 0.214869594 | -0.299235688 |
| VFLUENCY | 0.166268523 | -0.038112225 | 0.644418132 |
| READCOMP | -0.050000622 | 0.369420358 | -0.105652104 |
| ARTHMTIC | -0.093459905 | -0.056692142 | -0.101592545 |
| NFLUENCY | 0.133086365 | 0.037636289 | 0.139577626 |
| WORDPROB | -0.253918964 | 0.215424322 | 0.200376090 |
| FFLUENCY | 0.334596798 | 0.511583797 | 0.140948931 |
| PERSPEED | 0.220300706 | -1.340210080 | 0.257746812 |
| SPATIAL | 0.150674592 | 0.509291078 | -0.029397459 |
| VMEMORY | 0.113243490 | -0.200461276 | -0.077688604 |
| NUMSPANF | -0.291089649 | 0.024687292 | 0.229521814 |
| NUMSPANB | -0.174141985 | -0.138100207 | 0.270140047 |

Class Means on Canonical Variables

| GRP | CAN1 | CAN2 | CAN3 |
|---|---|---|---|
| 1 | -0.449135346 | -0.058582215 | 0.062488493 |
| 2 | 0.226312828 | 0.602420859 | -0.315929157 |
| 3 | 0.690536721 | -0.494228010 | -0.123754881 |
| 4 | 1.016103083 | 0.486263403 | 0.671652253 |

## PREDICTIVE DISCRIMINANT ANALYSIS

As indicated in the introduction to this chapter, a second goal of discriminant analysis is classification. In this situation, there are $k$ preexisting groups and we wish to classify a new person into one of them. For example, for purposes of hiring a job applicant, the applicant can be placed into one of two groups: hired or not hired. Because predictive and descriptive discriminant analysis have different goals, the techniques may be conducted independent of each other.

## CLASSIFICATION BASED ON GENERALIZED DISTANCE

A reasonable criterion for placing an individual into a group is to consider how far that individual's observation vector is from the mean score vector within each of the groups. The individual is placed within the group that has the vector of mean scores "closest" to the individual's observation vector. In multivariate analysis, closeness is typically measured by the generalized distance formula ($D^2$) introduced in chapter 3. This generalized distance $D^2$ is also known as the Mahalanobis distance:

$$D_{ij}^2 = (x_i - \overline{x}_j)' S^{-1}(x_i - \overline{x}_j)$$

where $x_i$ is the observation vector of the variables for the $i$th person, $\overline{x}_j$ is the mean vector of observations for the $j$th group, and $S$ is the pooled within-groups covariance matrix.

As an example, assume the values of $D^2$ for an individual were computed for three groups and found to be 1.5, 2.5, and 3.0. Based on the results of the Mahalanobis distance the individual would be placed into the first group because the $D^2$ value for that group (1.5) is the lowest. Although many other classification procedures have been proposed in the literature, to date most depend on or are a function of $D^2$. We next offer an overview of the most popular classification procedures.

## POSTERIOR PROBABILITY OF GROUP MEMBERSHIP

In chapter 3 it was indicated that $D^2$ was distributed as a chi-square variate with $p$ degrees of freedom. Assuming a multivariate normal distribution within each of the groups, the estimated probability that an individual belongs to the $j$th group, $P(X|G_j)$, can be determined:

$$P(X|G_j) \equiv f(X|G_j) = \frac{1}{\sqrt{(2\pi)^p}\sqrt{|S_g|}}\, e^{[(-1/2)(X-\overline{X}_g)'S_g^{-1}(X-\overline{X}_g)]} = (2\pi)^{-p/2}\, |S|^{-1/2} e^{(-1/2)D_{ij}^2}$$

where $f(X|G_j)$ is the multivariate normal probability density function. Thus, $P(X|G_j)$ is a function of $D_{ij}^2$. The probability $P(X|G_j)$ should be read as "the probability that an individual has an observation vector $X$ given membership in group $j$." Huberty (1994) referred to $P(X | G_j)$ as a *typicality probability*, and that is the term we adopt here.

If the typicality probability for a person is computed for each of the groups, then the probability that individual $I$ is a member of group $j$ is:

$$P(G_j | X) = \frac{P(X | G_j)}{\sum\limits_{j=1}^{k} P(X | G_j)}$$

$P(G_j | X)$ is called a *posterior probability*; an individual's posterior probabilities should sum to one across the groups. $P(G_j | X)$ is read as "the probability that an individual belongs to group $j$ given observation vector $X$." It is very important to distinguish the meaning of $P(X | G_j)$ from $P(G_j | X)$. The posterior probability gives the probability that a person with a specific observation vector $X$ belongs to group $j$; the typicality probability gives the probability that a person with membership in group $j$ will have observation vector $X$. If posterior probabilities are used to assign a person group membership, the person's largest posterior probability is used. For instance, if the posterior probabilities of membership to each of four groups were .4, .3, .2, and .1, the group for which the probability equaled .4 would be selected.

Therefore, group membership can be assigned based on either the minimum chi-square or the maximum typicality or posterior probability. There are two qualifications, however, that should be made to these two assignment schemes. First, note it is the pooled within-groups covariance matrix $S$ that is used to calculate $D_{ij}^2$. This assumes that no significant differences exist among the groups on $S$.[2] When this assumption can be made, and the matrix $S$ common to the groups is used, we are employing a *linear classification rule* to the assignment of a person to one of the groups. On the other hand, when evidence exists that $S$ is not the same in each group, we must use each group's specific $S$ for calculating the probability of membership in a specific group. When group-specific covariance matrices are used, we are employing a *quadratic classification rule*. The second qualification concerns the fact that we have been assuming that the prior probability of a person's membership in each of the groups is the same. For example, if there are two groups, we assume there is a 50% prior probability of membership in either group; if there are four groups, we assume a 25% prior probability. In the sample, if not the population, this is unlikely to be strictly true. Particularly if there is a large difference in the prior likelihood of membership among the groups in the population, it is not only beneficial but necessary to take these differences into account when assigning group membership.

---

[2]This issue can be thought of in the same way as the homogeneity of variance assumption is made in ANOVA.

## AN OVERVIEW OF THE DIFFERENT TYPES
## OF CLASSIFICATION RULES

There are four possible types of classification rules: (a) linear/equal prior probability rules; (b) linear/unequal prior probability rules; (c) quadratic/equal prior probability rules; or (d) quadratic/unequal prior probability rules. Each of these rules is discussed further. In addition, a number of nonparametric classification rules have been developed that do not assume multivariate normality. Thus, the classification rules could be further subdivided into parametric and nonparametric types. All of the classification rules discussed in this book are parametric; the nonparametric rules are beyond the scope of this book, but the interested reader may refer to Hand (1982). As a final point, the "cost" of misclassification could be factored into the classification rule. Cooley and Lohnes (1971) and Tatsuoka (1971) provide further details.

### LINEAR/EQUAL PRIOR PROBABILITY
### CLASSIFICATION RULES

When assuming equality of group covariance matrices and equal prior probability of group membership, the classification rule for $D_{ij}^2$ is:

$$(x_i - \overline{x}_j)' S^{-1} (x_i - \overline{x}_j)$$

and the classification rule based on $P(G_j | X)$ is

$$\frac{P(X | G_j)}{\sum\limits_{j=1}^{k} P(X | G_j)}$$

These expressions for $D_{ij}^2$ and $P(G_j | X)$ are identical to those given earlier: An individual is placed within a group for which $D_{ij}^2$ is a minimum or, equivalently, for which $P(G_j | X)$ is a maximum.

Although the linear/equal prior probability rules are the simplest classification rules, they do require homogeneity of the covariance matrices $S$ in order to be appropriate. Furthermore, one must also have good reason to believe that the groups are equally numerous in the population. Unfortunately, the latter assumption of equal numerosity is a difficult assumption to verify in practice, except in special situations. For example, it is impossible for an employer to know what proportion of the population would be qualified or what proportion would not be for a clerical job. On the other hand, if the position were one for a teacher of gifted children, it can safely be assumed that the major part of the population would not fall within the qualified category.

Testing the homogeneity of covariance matrices assumption is much more straightforward (relatively speaking), and may be accomplished by using Bartlett's test of homogeneity of within-covariance matrices (Morrison, 1976). Bartlett's test of homogeneity is provided as standard output by most computer programs along with its computational definition (e.g., the information provided in Table 5.12 is output generated by PROC DISCRIM in SAS). Although Bartlett's test is generally quite useful, a serious problem exists with the test. In particular, the null hypothesis of the homogeneity of covariance is often rejected when the data are not multivariate normal (Olson, 1974). As such, if ever the multivariate normality status of the data are in doubt, we recommend that Bartlett's test should be evaluated with a conservation $\alpha$—perhaps $\alpha < .001$.

## LINEAR/UNEQUAL PRIOR PROBABILITY RULES

If $q_j$ is the prior probability of membership in group $j$, then the classification rule for $D_{ij}^2$ becomes:

$$D_{ij}^2 - 2 \ln q_j$$

and the classification rule based on $P(G_j | X)$ is:

$$\frac{q_j P(X | G_j)}{\sum_{j=1}^{k} q_j P(X | G_j)}$$

where ln represents the natural logarithm, and $q_j$ serves as a corrective factor for basing group membership solely on group differences on the variables.

Even if the set of variables provides an appreciable degree of discrimination among the groups, it must be remembered that some of the group separation is sample specific. In particular, if the variability of $q_j$ among the groups is large, some of the chance group separation can be tempered by the application of prior probabilities. Thus, if one has strong reason to believe that $q_j$ in the population is .90 for one group and .10 for a second group, classification without using prior probabilities would be foolhardy. Because $q_j$ must always be positive, it serves to increase the probability of membership in a group. The primary question to address here is how, if one believes the prior probabilities differ among the groups, is $q_j$ determined. In the absence of information concerning the representation of the groups in the population, there are primarily two ways one can estimate $q_j$. The first approach is much the simpler: One assumes that the population proportion ($\pi_j$, which $q_j$ represents) can be estimated by the sample proportion, $n_j/n$, which asymptotically converges to (i.e., is the maximum likelihood estimator of) $\pi_j$ under large $n$. Nonetheless, even when sample size is "large," if the sampling plan of the study was not intentionally designed to represent the relative sizes of the groups in the population, it would be difficult to place much confidence in $n_j/n$, as an estimator of $\pi_j$. The second approach is more complicated, and is based on analyzing a "mixture" distribution, in which one estimates the proportions $\pi_1, \ldots, \pi_k$ from a sample mixture of $k$ groups (for a complete discussion on how this is accomplished see Macdonald, 1975).

## QUADRATIC/EQUAL PRIOR PROBABILITY RULES

Assuming one has rejected the null hypothesis of equality of covariance matrices across groups (by using Bartlett's test), classification may proceed by using each of the group covariance matrices $S_j$ instead of $S$ pooled across groups. The classification rule for $D^2_{ij}$ is then:

$$\ln \ |S_j| + D^2_{ij}$$

and the classification rule for $P(G_j | X)$ is:

$$\frac{|S_j|^{(-1/2)} \ e^{(-1/2)D^2_{ij}}}{\sum_{j=1}^{k} |S_j|^{(-1/2)} \ e^{(-1/2)D^2_{ij}}}$$

An examination of the two preceding equations reveals that the probability of group membership is conditioned on the specific group's covariance matrix: For each person to be classified, $D^2_{ij}$ and $P(G_j | X)$ are computed for each of the groups, and then the person is assigned to the group for which $D^2_{ij}$ is a minimum or $P(G_j | X)$ is a maximum. Less apparent is the origin of the term *quadratic*. Although not obvious, the two classification rules contain a quadratic term denoting the difference between the group covariance matrices (Lachenbruch, 1975).

Given the added analytic complexity introduced by using a quadratic rule, one may ask how important it is to use quadratic instead of linear classification when the homogeneity of covariance assumption is violated. McLachlan (1992) concluded that if the ratio of sample size to the number of variables is small then a linear rule should be used, even under significant covariance heterogeneity. Further, quadratic rules appear less robust against departures from normality than linear rules (Johnson & Wichern, 1982). Moreover, especially with small $n$, the results of linear classification are more stable across samples than the results of quadratic classification (Huberty & Curry, 1978). When these reservations concerning quadratic rules are joined with the observation that in many circumstances the results from linear and quadratic rules are very similar (Tatsuoka, 1971), caution should be exercised in applying a quadratic rule unless there is extreme covariance heterogeneity. With extreme covariance heterogeneity, one would find the largest differences in classification between linear and quadratic rules.

## QUADRATIC/UNEQUAL PRIOR PROBABILITY RULES

Quadratic rules employing unequal prior probabilities are a direct extension of the rules with equal prior probabilities. The classification rule for $D^2_{ij}$ is:

$$\ln |S_j| + D^2_{ij} - 2 \ln q_j$$

and for $P(G_j | X)$ is:

$$\frac{q_j |S_j|^{(-1/2)} e^{(-1/2)D^2_{ij}}}{\sum_{j=1}^{k} q_j |S_j|^{(-1/2)} e^{(-1/2)D^2_{ij}}}$$

The limitations and difficulties that quadratic rules and nonequal prior probability rules bring with them are, of course, even more magnified when the two are combined together. The general message is the same: *If* multivariate normality is present, and *if* covariance heterogeneity is extreme, and *if* sample size is large relative to the number of variables, and *if* there is good basis for estimating $q_j$, then quadratic/nonequal probability rules can (should) be used. Otherwise, it is much better to stay within the linear classification, equal prior probability framework.

## TYPE OF DATA USED FOR CLASSIFICATION

To this point, we have consistently referred to the data used for classification as the observation vector $X$, consisting of the *original* observed variables. As it turns out, it makes no difference whether classification is done using $X$ or the *complete* set of linear discriminant function scores—the classification results will be the same. Nevertheless, considering the general interest in parsimony, preference should be given to classification on the fewer linear discriminant scores as opposed to the larger number of $X$. However, it is important to note that if fewer than the maximum number of linear discriminant functions are retained, classification results can differ between the two. Of course, if the absent functions have been removed for good reason (e.g., lack of significance, negligible variance explained, lack of interpretability), classification based on the reduced set of linear discriminant functions is more desirable.

## THE FISHER TWO-GROUP CLASSIFICATION FUNCTION

R. A. Fisher (1936) was the first to develop the linear discriminant function for the classification of two groups. It is of some interest to examine the two-group case, for as becomes apparent, the calculations for the single linear discriminant function used for classification are much simplified and provide insight into the discriminant analysis procedures (recall that with $k - 1$ groups, there is only one discriminant function).

Consider the formula for Hotelling's $T^2$ first presented in chapter 3:

$$\frac{n_1 n_2}{n_1 + n_2}(\bar{x}_1 - \bar{x}_2)'S^{-1}(\bar{x}_1 - \bar{x}_2) = \frac{(\bar{x}_1 - \bar{x}_2)^2 n_1 n_2/(n_1 + n_2)}{S}$$

Recall from the discussion earlier in this chapter that a solution for the discriminant coefficients $a$ is one that maximizes the ratio of between to within sums of squares. This ratio can also be written as:

$$\frac{(a'\bar{x}_1 - \bar{x}_2)^2 n_1 n_2/(n_1 + n_2)}{a'Sa}$$

The vector of coefficients $a$ is obtained by finding the eigenvalues and eigenvectors of the following equation:

$$\left[\frac{n_1 n_2}{n_1 + n_2}(\bar{x}_1 - \bar{x}_2)(\bar{x}_1 - \bar{x}_2)' - \lambda S\right]a = 0$$

By definition, there is only one solution for $a$:

$$CS^{-1}(\bar{x}_1 - \bar{x}_2)$$

where $C$ is a scalar.

Because $C$ is a constant, we may remove $C$ and write:

$$a \propto S^{-1}(\bar{x}_1 - \bar{x}_2)$$

This implies that the linear discriminant function score ($Y$) for each person is then proportional to:

$$(\bar{x}_1 - \bar{x}_2)' S^{-1} x$$

The mean discriminant score $(\bar{Y}_1)$ for Group 1 is proportional to $(\bar{x}_1 - \bar{x}_2)'$ $S^{-1}\bar{x}_1$ and the mean for Group 2 $(\bar{Y}_2)$ is proportional to $(\bar{x}_1 - \bar{x}_2)' S^{-1}\bar{x}_2$. The center of these two means (the midpoint between Groups 1 and 2) is $\frac{1}{2}(\bar{x}_1 - \bar{x}_2)' S^{-1}(\bar{x}_1 + \bar{x}_2)$. This implies that if a person's discriminant score is greater than $\frac{1}{2}(\bar{x}_1 - \bar{x}_2)' S^{-1}(\bar{x}_1 + \bar{x}_2)$, the person's score is closer to the mean of Group 1, and therefore is assigned to Group 1. On the other hand, if a person's discriminant score is less than $\frac{1}{2}(\bar{x}_1 - \bar{x}_2)' S^{-1}(\bar{x}_1 + \bar{x}_2)$, the person's score is closer to the mean of Group 2, and therefore is assigned to Group 2.

Referring back to Fig. 5.1a, recall that the discriminant function means for Groups 1 and 2 are connected by a discriminant function axis. The point on the discriminant axis, $\frac{1}{2}(\bar{x}_1 - \bar{x}_2)' S^{-1}(\bar{x}_1 + \bar{x}_2)$, would be located at the origin, the point (0,0) precisely in between the means for Group 1 and for Group 2. If the discriminant score for a person is greater than 0, that person is closer to Group 1; if the discriminant score is less than 0, that person is closer to Group 2.

## EVALUATING CLASSIFICATION ACCURACY

The prototypic method for assessing the accuracy of classification is first to construct the discriminant functions, then use these functions to assign group membership to the individuals in the data—the same people who contributed to constructing the functions in the first place! A good measure of the accuracy of the classification scheme is called the *hit rate*, or the proportion of the sample correctly classified. From earlier discussions of multiple regression analysis (and see chapter 6), the sample squared multiple correlation coefficient $R^2$ is a positively biased estimator of its population counterpart. In other words, the value of the variables for predicting the criterion (as expressed by $R^2$) is probably overestimated in the sample from which $R^2$ was determined. And understandably so, because the idiosyncracies of the sample that contributed to the maximization of the correlation between the predictors and the criterion will never be repeated in another sample. Therefore, the preference is for the shrunken $\check{R}^2$, the $R^2$ corrected for positive bias. As it turns out, an identical problem plagues the hit-rate estimate in discriminant analysis. It too is positively biased, and in most cases gives a too optimistic impression of how accurately the discriminant function will classify in future samples. Unfortunately, unlike $R^2$ in multiple regression, no direct correction for positive bias in the hit rate for discriminant analysis to date has been proposed.

In general, two strategies are used to obtain a more realistic estimate of the hit rate. The first strategy is very similar to the cross-validation procedure used in regression. In summary, before conducting the discriminant analysis, the sample is divided into two random halves, a *training sample* and a *test sample.* The training sample is used to develop the classification rule, and then this rule is applied to the test sample. The hit-rate estimate is the proportion of subjects classified correctly in the test sample. Schaafsma and van Vark (1979) recommended that the test sample be one-third of the total sample. The second strategy is referred to as the "leave-one-out" method (the leave-one-out method is also used in regression analysis under the guise of the PRESS statistic to estimate $R^2$; see Allen, 1971). With the leave-one-out method, the discriminant functions are calculated $N$ times, each time leaving out one subject. The discriminant functions calculated without using this subject are used to classify the subject. The hit rate is the proportion of times deleted cases are correctly assigned.

Assuming one has obtained a good estimate of the hit rate, how can the magnitude of the hit rate be evaluated? For example, an analyst may be very pleased with a hit rate of 60% for two groups—until the analyst realizes that by chance alone (i.e., without the assistance of a classification rule) 50% of the sample could have been correctly assigned. In this situation, the analyst is confronted with the problem of determining at what point the hit rate observed significantly surpasses the hit rate expected by chance alone. One alternative is to use the proportional chance criterion.

To use the proportional chance criterion, one must first obtain the hit-rate frequency expected by chance ($H_e$). If the sample size in each group is the same, then $H_e$ is simply $n_j$. If sample size differs, then $H_e = \sum_{j=1}^{k} \frac{1}{k} n_j$. The hit-rate frequency observed ($H_o$) is the sum of the main diagonal elements of the classification matrix. Then the difference between $H_o$ and $H_e$ may be tested by the following statistic:

$$z = \frac{H_o - H_e}{\sqrt{H_e(N - H_e)/N}}$$

where the significance of $z$ is found by comparison with a critical value from a standard normal distribution.

Instead of using the proportional chance criterion, the maximum chance criterion can also be invoked. This criterion is particularly helpful when the prior probability of group membership differs. For example, let us assume that there are three groups, with prior probabilities of .60, .30, and .10. According to the maximum chance criterion, the maximum proportion of correctly assigned subjects will be the highest prior probability value, or in this case, .60. For example, if there were 120 subjects to classify, by the maximum chance criterion we would predict .60 × 120 = 72 of them to be assigned correctly without using a classification rule. One can use the frequency predicted by the maximum chance criterion as an estimate of $H_e$ and employ the $z$ test given earlier.

The improvement over chance criterion (IOCC) describes how much better we did by using a classification rule than by relying on chance assignment. This criterion is defined as:

$$IOCC = \frac{H_o/N - H_e/N}{1 - H_e/N}$$

Based on the above example (using $H_e$ = 40, $H_o$ = 60, and $N$ =120), the IOCC would be

$$IOCC = \frac{60/120 - 40/120}{1 - 40/120} = .24$$

which indicates that our improvement over chance is not that great—actually, it is terrible! However, assuming that we wanted to reach an IOCC = .90 (90% fewer classification errors by using a classification rule), the improvement over chance criterion can be used as an effect size estimate of the hit rate, in conjunction with a $z$ test of the significance between $H_o$ and $H_e$.

### NUMERICAL CLASSIFICATION EXAMPLE

In this example, an admissions officer of a business school is trying to determine what criteria to use to admit students to the school's MBA program. To assist with this study, data were collected on 150 recent applicants who were categorized into three student groups (each $n$ = 50): Group A, students who have been admitted into college; Group B, students who are borderline admittances; and Group C, students who have been rejected. The students were measured on five variables: high school GPA, GMAT scores, an Index score (comprised of letters of recommendation ratings), socioeconomic status (SES), and age. As indicated earlier, the goal of this analysis is to assess how well, using these five variables, individuals can be correctly classified.

The SAS program instructions for PROC DISCRIM (the appropriate procedure for predictive discriminant analysis) are shown in Table 5.6. Prior to conducting a predictive discriminant analysis, it can be very informative to examine a histogram of the groups' standing on each of the variables to assess how much each variable will likely contribute to classification. A PROC CHART command, which will draw such a histogram, is invoked for each of the variables. Following the PROC CHART commands are four PROC DISCRIM commands, which will generate classifications according to four different rules: (a) a linear classification with equal prior probabilities; (b) a linear classification with unequal prior probabilities; (c) a quadratic classification with equal prior probabilities; and (d) a quadratic classification with unequal prior probabilities. The POOL=TEST option invokes Bartlett's test for homogeneity of group covariance matrices; if this test is rejected, SAS automatically

TABLE 5.6

```
data one;
infile dat1;
input student gpa gmat index ses age;
run;
proc chart;
vbar gpa / subgroup = student midpoints=1 to 4 by .5;
format student student.;
run;
proc chart;
vbar gmat / subgroup = student midpoints=200 to 800 by 50;
format student student.;
run;
proc chart;
vbar index / subgroup = student midpoints=5 to 25 by 2;
format student student.;
run;
proc chart;
vbar ses / subgroup = student midpoints=1 to 5 by 1;
format student student.;
run;
proc chart;
vbar age / subgroup = student midpoints=16 to 20 by .5;
format student student.;
run;
title 'linear classification with equal prior probabilities';
proc discrim all;
class student;
var gpa gmat index ses age;
run;
title 'linear classification with unequal prior probabilities';
proc discrim all;
class student;
var gpa gmat index ses age;
priors '1'=.1 '2'=.1 '3'=.8;
run;
title 'quadratic classification with equal prior probabilities';
proc discrim all pool=test;
class student;
var gpa gmat index ses age;
run;
title 'quadratic classification with unequal prior probabilities';
proc discrim all pool=test;
class student;
var gpa gmat index ses age;
priors '1'=.1 '2'=.1 '3'=.8; run;
```

uses quadratic classification. Because there are three groups, SAS will by default assign a prior probability of .333 to each unless the user specifies other prior probabilities. In order to assign one's own prior probability values, one can use the PRIORS command. For two of the analyses presented, we have assigned prior probabilities of .1 for the accepted group, .1 for the borderline group, and .8 for the rejected group (this is a very selective school!). The variable histograms are shown in Figs. 5.3–5.7 and the results in Tables 5.7–5.16.

Examining the histograms in Figs. 5.3 through 5.7, it is clear that the groups are highly separated on GPA and GMAT, somewhat less so on Index,

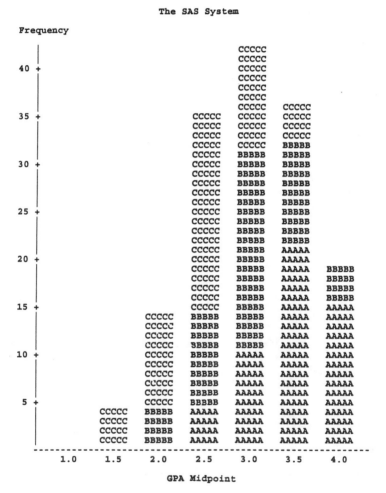

FIG. 5.3. High school gradepoint average (GPA) for students in Groups A, B, and C.

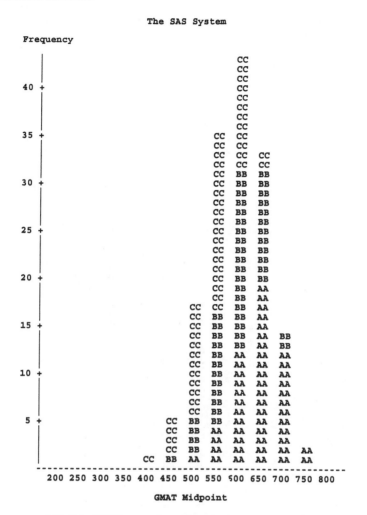

FIG. 5.4. GMAT scores for students in Groups A, B, and C.

and not separated at all on SES or age. Table 5.7 provides the estimated distances ($D^2$) between the groups. Assuming multivariate normality, according to the results of the significance tests for these distances, all the group are highly separated from one another. Table 5.8 provides the classification results using the linear, equal prior probability rule. The hit rate is equal to the sum of the diagonal elements divided by the total number of elements, $(46 + 36 + 38)/150 = 80\%$, or $1 - .80 = .20$, the total error rate reported below the classification table. Note that the group that has the highest error rate is the borderline students (Group 2)—the group we would expect to be the most difficult to classify. Table 5.9 provides the results from

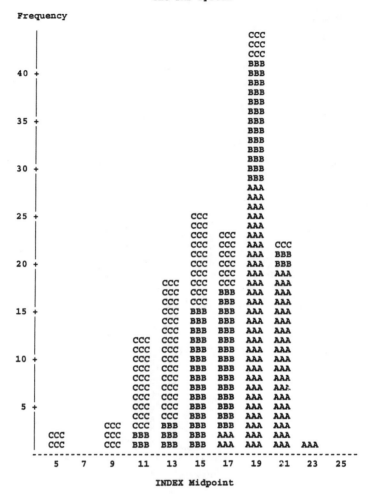

FIG. 5.5.  Index scores for students in Groups A, B, and C.

the "leave-one-out method" for estimating the hit rate; the error rate reported here is .226, implying a hit rate of .774, only slightly lower than the initial rate of .80. Given that the hit rate expected by chance is .33 × 150 = 49.5 students, and the observed hit rate is 120 students, the observed hit rate significantly exceeds the chance hit rate:

$$z = \frac{120 - 49.5}{\sqrt{\frac{49.5(150 - 49.5)}{150}}} = 12.24 \quad p < .00001$$

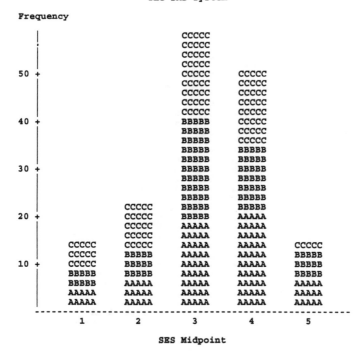

FIG. 5.6. Socioeconomic status for students in Groups A, B, and C.

As a measure of effect size, the improvement over chance criterion is: (.80 − .33)/(1 − .33), or .70, meaning that we made 70% fewer errors by using the linear classification rule.

Table 5.10 presents the results of using a linear, unequal prior probability classification rule, and Table 5.11 presents the leave-one-out validation. According to Table 5.10, we can correctly classify 90% (i.e., 1 − .104 = .90) of the students. Even using the leave-one-out method of estimating the hit rate results in a hit rate of 88% (1 − .123 = .88). Using the maximum by chance rate as our expected hit rate (.80, for the group with the highest prior probability), we find that using a classification rule still produces a significant increment in the number of people correctly classified:

$$z = \frac{135 - 120}{\sqrt{\dfrac{120(150 - 120)}{150}}} = -3.01 \quad p < .01$$

The improvement over chance criterion (using the maximum by chance rate as our expected rate) is (.90 − .80)/(1 − .80) or 50%.

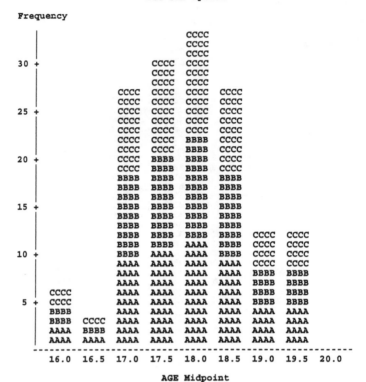

FIG. 5.7. Age (years) for students in Groups A, B, and C.

TABLE 5.7
SAS System Discriminant Analysis for Distances Between Groups

Pairwise Squared Distances Between Groups
$$D^2 (i|j) = (\overline{X}_i - \overline{X}_j)' \, COV^{-1} \, (\overline{X}_i - \overline{X}_j)$$

Squared Distance to STUDENT

| From STUDENT | 1 | 2 | 3 |
|---|---|---|---|
| 1 | 0 | 3.41032 | 13.64128 |
| 2 | 3.41032 | 0 | 3.41032 |
| 3 | 13.64128 | 3.41032 | 0 |

F Statistics, NDF=5, DDF=143 for Squared Distance to STUDENT

| From STUDENT | 1 | 2 | 3 |
|---|---|---|---|
| 1 | 0 | 16.58761 | 66.35045 |
| 2 | 16.58761 | 0 | 16.58761 |
| 3 | 66.35045 | 16.58761 | 0 |

Prob > Mahalanobis Distance for Squared Distance to STUDENT

| From STUDENT | 1 | 2 | 3 |
|---|---|---|---|
| 1 | 1.0000 | 0.0001 | 0.0001 |
| 2 | 0.0001 | 1.0000 | 0.0001 |
| 3 | 0.0001 | 0.0001 | 1.0000 |

TABLE 5.8
SAS System Discriminant Analysis, Linear Equal Prior Probability Rule

Classification Summary for Calibration Data: WORK.ALL
Resubstitution Summary using Linear Discriminant Function
Generalized Squared Distance Function:

$$D_j^2 (X) = (X-\bar{X}_j)' \text{ COV}^{-1}(X-\bar{X}_j)$$

Posterior Probability of Membership in each STUDENT:

$$\Pr(j|X) = \exp(-.5 \ D_j^2(X)) / \underset{k}{\text{SUM}} \ \exp(-.5 \ D_k^2(X))$$

Number of Observations and Percent Classified into STUDENT:

| From STUDENT | 1 | 2 | 3 | Total |
|---|---|---|---|---|
| 1 | 46 | 4 | 0 | 50 |
|  | 92.00 | 8.00 | 0.00 | 100.00 |
| 2 | 10 | 36 | 4 | 50 |
|  | 20.00 | 72.00 | 8.00 | 100.00 |
| 3 | 0 | 12 | 38 | 50 |
|  | 0.00 | 24.00 | 76.00 | 100.00 |
| Total | 56 | 52 | 42 | 150 |
| Percent | 37.33 | 34.67 | 28.00 | 100.00 |
| Priors | 0.3333 | 0.3333 | 0.3333 | |

Error Count Estimates for STUDENT:

|  | 1 | 2 | 3 | Total |
|---|---|---|---|---|
| Rate | 0.0800 | 0.2800 | 0.2400 | 0.2000 |
| Priors | 0.3333 | 0.3333 | 0.3333 | |

Table 5.12 shows the results of Bartlett's test for group homogeneity of covariance matrices. As can be seen in Table 5.12, the assumption of homogeneity is rejected; $\chi^2 = 52.051$, df = 30, $p < .01$. Because we have determined that the group covariance matrices are unequal, we employ a quadratic classification rule—first with equal prior probabilities, and then with unequal prior probabilities. Table 5.13 shows the hit rate under the quadratic rule to be $1 - .173$ or 83%; Table 5.14 shows the comparable leave-one-out hit rate to be $1 - .190 = 81\%$. If we apply unequal prior probabilities, then as Table 5.15 shows, the hit rate is $1 - .082 = 92\%$; the comparable leave-one-out hit rate from Table 5.16 is $1 - .106 = 89\%$. We may conclude from this example two important classification results. First, there was little bias in the estimation of the hit rates (the leave-one-out method hit rate did not differ much from the initial hit rate), and second, the best hit rate was obtained using a quadratic, nonequal prior probability rule.

TABLE 5.9
SAS System Discriminant Analysis, Leave-One-Out Method

Classification Results for Calibration Data: WORK.ALL
Resubstitution Results using Linear Discriminant Function
Generalized Squared Distance Function:

$$D_j^2(X) = (X-\bar{X}_j)'\ COV^{-1}(X-\bar{X}_j)$$

Posterior Probability of Membership in each STUDENT:

$$Pr(j|X) = \exp(-.5\ D_j^2(X))/\underset{k}{SUM}\ \exp(-.5\ D_k^2(X))$$

Number of Observations and Average Posterior Probabilities
Classified into STUDENT:

| From STUDENT | 1 | 2 | 3 |
|---|---|---|---|
| 1 | 46 | 4 | 0 |
|   | 0.8408 | 0.6303 | . |
| 2 | 10 | 36 | 4 |
|   | 0.6506 | 0.6673 | 0.8314 |
| 3 | 0 | 12 | 38 |
|   | . | 0.6661 | 0.8703 |
| Total | 56 | 52 | 42 |
|   | 0.8069 | 0.6641 | 0.8666 |
| Priors | 0.3333 | 0.3333 | 0.3333 |

Posterior Probability Error Rate Estimates for STUDENT:

| Estimate | 1 | 2 | 3 | Total |
|---|---|---|---|---|
| Stratified | 0.0963 | 0.3093 | 0.2721 | 0.2259 |
| Unstratified | 0.0963 | 0.3093 | 0.2721 | 0.2259 |
| Priors | 0.3333 | 0.3333 | 0.3333 | |

**122**

TABLE 5.10
SAS System Discriminant Analysis Using
Linear Unequal Prior Probability Rule

Classification Summary for Calibration Data: WORK.ALL
Resubstitution Summary using Linear Discriminant Function
Generalized Squared Distance Function:

$$D_j^2(X) = (X-\overline{X}_j)'\ COV^{-1}(X-\overline{X}_j)$$

Posterior Probability of Membership in each STUDENT:

$$Pr(j\,|\,X) = \exp(-.5\ D_j^2(X))\,/\,\underset{k}{SUM}\ \exp(-.5\ D_k^2(X))$$

Number of Observations and Percent Classified into STUDENT:

| From STUDENT | 1 | 2 | 3 | Total |
|---|---|---|---|---|
| 1 | 46 | 3 | 1 | 50 |
|  | 92.00 | 6.00 | 2.00 | 100.00 |
| 2 | 10 | 10 | 30 | 50 |
|  | 20.00 | 20.00 | 60.00 | 100.00 |
| 3 | 0 | 1 | 49 | 50 |
|  | 0.00 | 2.00 | 98.00 | 100.00 |
| Total | 56 | 14 | 80 | 150 |
| Percent | 37.33 | 9.33 | 53.33 | 100.00 |
| Priors | 0.1000 | 0.1000 | 0.8000 | |

Error Count Estimates for STUDENT:

| | 1 | 2 | 3 | Total |
|---|---|---|---|---|
| Rate | 0.0800 | 0.8000 | 0.0200 | 0.1040 |
| Priors | 0.1000 | 0.1000 | 0.8000 | |

TABLE 5.11
SAS System Discriminant Analysis with Leave-One-Out Validation

Classification Results for Calibration Data: WORK.ALL
Resubstitution Results using Linear Discriminant Function
Generalized Squared Distance Function:

$$D_j^2(X) = (X-\overline{X}_j)'\ COV^{-1}(X-\overline{X}_j)$$

Posterior Probability of Membership in each STUDENT:

$$Pr(j\,|\,X) = \exp(-.5\ D_j^2(X))\,/\,\underset{k}{SUM}\ \exp(-.5\ D_k^2(X))$$

Number of Observations and Average Posterior Probabilities

*(Continued)*

TABLE 5.11
*(Continued)*

Classified into STUDENT:

| From STUDENT | 1 | 2 | 3 |
|---|---|---|---|
| 1 | 46 | 3 | 1 |
| | 0.8327 | 0.4511 | 0.8515 |
| 2 | 10 | 10 | 30 |
| | 0.6233 | 0.4730 | 0.7303 |
| 3 | 0 | 1 | 49 |
| | . | 0.4895 | 0.9215 |
| Total | 56 | 14 | 80 |
| | 0.7953 | 0.4695 | 0.8490 |
| Priors | 0.1000 | 0.1000 | 0.8000 |

Posterior Probability Error Rate Estimates for STUDENT:

| Estimate | 1 | 2 | 3 | Total |
|---|---|---|---|---|
| Stratified | 0.1093 | 0.8000 | 0.0400 | 0.1229 |
| Unstratified | -1.9691 | 0.5618 | 0.4340 | 0.2065 |
| Priors | 0.1000 | 0.1000 | 0.8000 | |

TABLE 5.12
SAS System Discriminant Analysis, Bartlett's Test
for Homogeneity of Within-Covariance Matrices

Notation:  K   = Number of Groups
           P   = Number of Variables
           N   = Total Number of Observations - Number of Groups
           N(i) = Number of Observations in the i'th Group - 1

$$V = \Pi \frac{|\text{Within SS Matrix}(i)|^{N(i)/2}}{|\text{Pooled SS Matrix}|^{N/2}}$$

$$RHO = 1.0 - \left[ SUM \frac{1}{N(i)} - \frac{1}{N} \right] \frac{2P^2 + 3P - 1}{6(P+1)(K-1)}$$

$$DF = .5(K-1)P(P+1)$$

Under null hypothesis: $-2$ RHO ln $\left[ \dfrac{N^{PN/2} V}{\Pi\ N(i)^{PN(i)/2}} \right]$

is distributed approximately as chi-square(DF)

Test Chi-Square Value = 52.050872
with 30 DF Prob > Chi-Sq = 0.0075

Since the chi-square value is significant at the 0.1 level,
the within covariance matrices will be used in the discriminant function.

*Note.* Reference, Morrison (1976), p. 252.

TABLE 5.13
SAS System Discriminant Analysis, Hit Rate Under the Quadratic Rule

Classification Summary for Calibration Data: WORK.ALL
Resubstitution Summary using Quadratic Discriminant Function
Generalized Squared Distance Function:

$$D_j^2(X) = (X-\bar{X}_j)' \, COV_j^{-1}(X-\bar{X}_j) + \ln \, |COV_j|$$

Posterior Probability of Membership in each STUDENT:

$$Pr(j\,|\,X) = \exp(-.5 \, D_j^2(X)) / \underset{k}{SUM} \, \exp(-.5 \, D_k^2(X))$$

Number of Observations and Percent Classified into STUDENT:

| From STUDENT | 1 | 2 | 3 | Total |
|---|---|---|---|---|
| 1 | 47 | 2 | 1 | 50 |
|  | 94.00 | 4.00 | 2.00 | 100.00 |
| 2 | 7 | 39 | 4 | 50 |
|  | 14.00 | 78.00 | 8.00 | 100.00 |
| 3 | 1 | 11 | 38 | 50 |
|  | 2.00 | 22.00 | 76.00 | 100.00 |
| Total | 55 | 52 | 43 | 150 |
| Percent | 36.67 | 34.67 | 28.67 | 100.00 |
| Priors | 0.3333 | 0.3333 | 0.3333 | |

Error Count Estimates for STUDENT:

| Estimate | 1 | 2 | 3 | Total |
|---|---|---|---|---|
| Rate | 0.0600 | 0.2200 | 0.2400 | 0.1733 |
| Priors | 0.3333 | 0.3333 | 0.3333 | |

TABLE 5.14
SAS System Discriminant Analysis, Leave-One-Out Hit Rate

Classification Results for Calibration Data: WORK.ALL
Resubstitution Results using Quadratic Discriminant Function
Generalized Squared Distance Function:

$$D_j^2(X) = (X-\bar{X}_j)' \ COV_j^{-1}(X-\bar{X}_j) + \ln \ |COV_j|$$

Posterior Probability of Membership in each STUDENT:

$$Pr(j|X) = \exp(-.5 \ D_j^2(X)) / \underset{k}{SUM} \ \exp(-.5 \ D_k^2(X))$$

Number of Observations and Average Posterior Probabilities
Classified into STUDENT:

| From STUDENT | 1 | 2 | 3 |
|---|---|---|---|
| 1 | 47<br>0.8851 | 2<br>0.6032 | 1<br>0.5365 |
| 2 | 7<br>0.7804 | 39<br>0.7309 | 4<br>0.8303 |
| 3 | 1<br>0.5415 | 11<br>0.6513 | 38<br>0.8713 |
| Total | 55<br>0.8656 | 52<br>0.7091 | 43<br>0.8597 |
| Priors | 0.3333 | 0.3333 | 0.3333 |

Posterior Probability Error Rate Estimates for STUDENT:

| Estimate | 1 | 2 | 3 | Total |
|---|---|---|---|---|
| Stratified | 0.0479 | 0.2625 | 0.2607 | 0.1904 |
| Unstratified | 0.0479 | 0.2625 | 0.2607 | 0.1904 |
| Priors | 0.3333 | 0.3333 | 0.3333 | |

## TABLE 5.15
SAS System Discriminant Analysis, Hit Rate with Unequal Prior Probabilities

Classification Summary for Calibration Data: WORK.ALL
Resubstitution Summary using Quadratic Discriminant Function
Generalized Squared Distance Function:

$$D_j^2(X) = (X-\bar{X}_j)'\ COV_j^{-1}(X-\bar{X}_j) + \ln\ |COV_j|$$

Posterior Probability of Membership in each STUDENT:

$$Pr(j\,|X) = \exp(-.5\ D_j^2(X))/\underset{k}{SUM}\ \exp(-.5\ D_k^2(X))$$

Number of Observations and Percent Classified into STUDENT:

| From STUDENT | 1 | 2 | 3 | Total |
|---|---|---|---|---|
| 1 | 47 | 0 | 3 | 50 |
| | 94.00 | 0.00 | 6.00 | 100.00 |
| 2 | 7 | 12 | 31 | 50 |
| | 14.00 | 24.00 | 62.00 | 100.00 |
| 3 | 0 | 0 | 50 | 50 |
| | 0.00 | 0.00 | 100.00 | 100.00 |
| Total | 54 | 12 | 84 | 150 |
| Percent | 36.00 | 8.00 | 56.00 | 100.00 |
| Priors | 0.1000 | 0.1000 | 0.8000 | |

Error Count Estimates for STUDENT:

| | 1 | 2 | 3 | Total |
|---|---|---|---|---|
| Rate | 0.0600 | 0.7600 | 0.0000 | 0.0820 |
| Priors | 0.1000 | 0.1000 | 0.8000 | |

Classification Results for Calibration Data: WORK.ALL
Resubstitution Results using Quadratic Discriminant Function
Generalized Squared Distance Function:

$$D_j^2(X) = (X-\bar{X}_j)'\ COV_j^{-1}(X-\bar{X}_j) + \ln\ |COV_j|$$

Posterior Probability of Membership in each STUDENT:

$$Pr(j|X) = \exp(-.5\ D_j^2(X))/\underset{k}{SUM}\ \exp(-.5\ D_k^2(X))$$

Number of Observations and Average Posterior Probabilities
Classified into STUDENT:

| From STUDENT | 1 | 2 | 3 |
|---|---|---|---|
| 1 | 47 | 0 | 3 |
|   | 0.8553 | . | 0.7295 |
| 2 | 7 | 12 | 31 |
|   | 0.7039 | 0.5794 | 0.7560 |
| 3 | 0 | 0 | 50 |
|   | . | . | 0.9236 |
| Total | 54 | 12 | 84 |
|   | 0.8357 | 0.5794 | 0.8549 |
| Priors | 0.1000 | 0.1000 | 0.8000 |

Posterior Probability Error Rate Estimates for STUDENT:

| Estimate | 1 | 2 | 3 | Total |
|---|---|---|---|---|
| Stratified | 0.0974 | 0.8609 | 0.0123 | 0.1057 |
| Unstratified | -2.0086 | 0.5365 | 0.4016 | 0.1741 |
| Priors | 0.1000 | 0.1000 | 0.8000 | |

## EXERCISES

1. A country is interested in resuming its selective service program for individuals under age 30 but only wants to select those individuals who would make good inductees. In order to determine what dimensions separate good from bad inductees, the military examines the profiles of three groups of soldiers: a group of soldiers who are still in the service and have been promoted (Group 1); a group of soldiers who are still in the service but have not been promoted (Group 2); and a group of soldiers who have been dishonorably discharged (Group 3). Six variables were selected: X1, the average number of traffic tickets in a given year; X2, highest level of high school education completed; X3, number of years been out of school; X4, daily frequency of alcohol intake; X5, score from a compliance to authority scale; X6, score from an openness to experience scale. Ten soldiers were randomly selected from each group; their data are presented here:

| Group | X1 | X2 | X3 | X4 | X5 | X6 |
|-------|----|----|----|----|----|----|
| 1 | 4 | 4 | 4 | 0 | 0 | 0 |
| 1 | 4 | 4 | 4 | 0 | 0 | 0 |
| 1 | 4 | 3 | 4 | 0 | 0 | 1 |
| 1 | 3 | 4 | 4 | 1 | 0 | 1 |
| 1 | 4 | 2 | 3 | 0 | 0 | 0 |
| 1 | 4 | 4 | 4 | 0 | 0 | 0 |
| 1 | 4 | 4 | 4 | 1 | 0 | 0 |
| 1 | 3 | 3 | 3 | 0 | 0 | 0 |
| 1 | 3 | 3 | 4 | 0 | 0 | 2 |
| 1 | 3 | 4 | 3 | 0 | 0 | 1 |
| 2 | 4 | 4 | 4 | 0 | 0 | 0 |
| 2 | 4 | 1 | 4 | 0 | 0 | 0 |
| 2 | 4 | 2 | 2 | 2 | 3 | 2 |
| 2 | 4 | 3 | 4 | 0 | 0 | 5 |
| 2 | 3 | 3 | 3 | 0 | 2 | 5 |
| 2 | 4 | 3 | 4 | 0 | 2 | 2 |
| 2 | 2 | 2 | 2 | 0 | 1 | 5 |
| 2 | 2 | 3 | 3 | 0 | 2 | 1 |
| 2 | 4 | 4 | 1 | 0 | 2 | 3 |
| 2 | 4 | 4 | 4 | 0 | 0 | 1 |
| 3 | 4 | 3 | 4 | 1 | 3 | 3 |
| 3 | 2 | 2 | 2 | 3 | 3 | 3 |
| 3 | 4 | 4 | 4 | 0 | 2 | 1 |
| 3 | 4 | 4 | 3 | 0 | 1 | 6 |
| 3 | 3 | 3 | 4 | 0 | 0 | 6 |
| 3 | 3 | 3 | 3 | 0 | 0 | 6 |
| 3 | 4 | 3 | 4 | 0 | 0 | 0 |
| 3 | 3 | 3 | 3 | 3 | 3 | 3 |
| 3 | 4 | 3 | 4 | 1 | 1 | 2 |
| 3 | 2 | 2 | 2 | 0 | 2 | 1 |

Perform a descriptive discriminant analysis to assess on what dimensions the groups differ.

2. Use discriminant analysis to construct a linear classification rule for assigning people to one of three groups in question 1, assuming equal prior probabilities. What is the hit rate? Does the hit rate differ significantly than what would have been obtained from classification based on chance? What is the effect size of this hit rate? Does the hit rate appear to be too optimistic? Is a quadratic classification rule necessary? Redo the linear classification analysis, assuming unequal prior probabilites: .25 for Group 1; .50 for Group 2; .25 for Group 3. Have the results changed in any way?

3. The dean of a large university would like to be able to screen applicants based on their potential for completing a BA or BS degree. The following data are available for 20 students receiving their bachelor's degree ("Successful") and 10 dropout students ("Failing"):

| "Successful" Students | | "Failing" Students | |
|---|---|---|---|
| GPA | SAT score | GPA | SAT score |
| 3.90 | 1200 | 2.40 | 900 |
| 3.75 | 1050 | 1.90 | 800 |
| 3.31 | 1050 | 2.70 | 950 |
| 2.40 | 1000 | 3.00 | 1000 |
| 2.75 | 1050 | 3.10 | 1000 |
| 3.40 | 1100 | 2.60 | 950 |
| 3.45 | 950 | 2.20 | 900 |
| 3.33 | 1000 | 3.30 | 1100 |
| 2.80 | 900 | 2.10 | 900 |
| 3.40 | 950 | 2.20 | 1000 |
| 3.10 | 1000 | | |
| 3.65 | 1050 | | |
| 3.70 | 1250 | | |
| 3.25 | 1100 | | |
| 2.60 | 900 | | |
| 3.90 | 1250 | | |
| 4.00 | 1300 | | |
| 3.65 | 1200 | | |
| 3.70 | 1200 | | |
| 2.50 | 1000 | | |

(a) Perform a discriminant analysis of the two groups. Which variable appears to be the best predictor for completing the degree requirements? (b) An applicant has a GPA of 3.30 and an SAT score of 1050. Should this student be admitted to the university? Why?

4. For a recent study by Marcoulides, Mills, and Unterbrink (1993) a discriminant analysis was used to improve preemployment drug screening. The following information on five variables is available for a sample of 20

applicants: age, sex, Drug Avoidance Scale (DAS) score, years of education, and drug use (Yes or No).

| Age | Sex | DAS | Years of Education | Use Drugs |
|-----|-----|-----|--------------------|-----------|
| 22 | M | 50 | 13 | Y |
| 41 | M | 98 | 16 | N |
| 36 | M | 90 | 14 | N |
| 32 | F | 95 | 16 | N |
| 29 | M | 60 | 12 | Y |
| 28 | M | 55 | 12 | Y |
| 27 | F | 92 | 14 | N |
| 33 | F | 78 | 12 | N |
| 25 | F | 79 | 15 | N |
| 34 | M | 60 | 13 | Y |
| 25 | M | 70 | 11 | Y |
| 44 | M | 80 | 12 | N |
| 37 | F | 94 | 14 | N |
| 38 | M | 92 | 13 | N |
| 50 | F | 90 | 16 | N |
| 43 | F | 60 | 10 | Y |
| 48 | M | 87 | 12 | N |
| 42 | M | 85 | 14 | N |
| 29 | F | 50 | 12 | Y |
| 31 | M | 75 | 12 | Y |

(a) Perform discriminant analysis for two groups, based on whether or not the applicant is a drug user. (b) Would you hire a 27-year-old male applicant, with a DAS score of 80 and 13 years of education? Why?

5. The California Basic Educational Skills Test (CBEST) is a standardized teacher proficiency exam used to establish a minimum competency requirement for hiring teachers in California. A study by Bruno and Marcoulides (1985) examined the type of knowledge that best classifies applicants who pass the exam from applicants who fail. The exam consists of three parts—math, writing, and reading—and the minimum passing score is 123. Scores for a sample of 15 teachers are presented:

| Math | Writing | Reading | Race | Employed (Yes or No) |
|------|---------|---------|------|----------------------|
| 65 | 50 | 53 | White | Yes |
| 45 | 53 | 47 | White | Yes |
| 30 | 37 | 35 | Black | Yes |
| 50 | 50 | 45 | Asian | No |
| 27 | 33 | 35 | Black | Yes |
| 43 | 32 | 40 | Black | No |
| 25 | 33 | 20 | Black | No |
| 36 | 50 | 43 | Asian | Yes |
| 70 | 65 | 63 | White | Yes |
| 40 | 45 | 53 | White | No |
| 27 | 33 | 40 | Black | No |
| 60 | 70 | 59 | White | No |
| 48 | 65 | 70 | Black | Yes |
| 29 | 40 | 46 | Asian | No |
| 50 | 63 | 71 | White | Yes |

(a) Perform a discriminant analysis for two groups (employed vs. not employed). (b) Which variable is the best predictor of whether or not a teacher will find employment?

# Canonical Correlation

One of the primary goals of multivariate statistical analysis is to describe the relationships among a set of variables. Recall that the Pearson product–moment correlation coefficient is a simple measure of relationship. The Pearson correlation coefficient represents the magnitude of a relation between two variables and is commonly referred to as a bivariate correlation analysis. This chapter is concerned with presenting the multivariate generalization of bivariate correlation. The multivariate generalization is called the canonical correlation. In canonical correlation analysis, the relations between two sets of variables are examined. Although this definition of canonical correlation may sound very similar to multiple regression, there are some important differences between the two procedures. In particular, in multiple regression analysis the concern is with examining the relationship between a single criterion variable and a set of predictor variables. For this reason, multiple regression analysis is not a true multivariate technique as we have defined multivariate in chapter 1. Nonetheless, many of the procedures and much of the rationale of canonical correlation analysis are direct extensions of multiple regression analysis. Therefore, the first part of this chapter briefly reviews multiple regression; for much greater details the reader should consult Cohen and Cohen (1983), Dielman (1996), and Pedhazur and Schmelkin (1991). To some extent, even discriminant analysis may be thought of as a special case of regression analysis in which the discriminant functions describe the relation between a single criterion variable (the categorical grouping variable) and a set of predictors. In fact, recall from the previous chapter that discriminant functions are frequently referred to as "canonical discriminant functions." Thus, in a rather reversed way, the reader has already been exposed to a restricted form of canonical correlation in the guise of a discriminant analysis.

## MULTIPLE CORRELATION/REGRESSION

Multiple regression analysis is concerned with the prediction of a dependent (criterion) variable $Y$ by a set of independent (predictor) variables $X$. The multiple regression model is generally defined as:

$$Y_i = \beta_0 + \beta_1 X_i + \beta_2 X_i + \cdots + \beta_k X_i + e_i$$

where $\beta_0$ is the regression intercept, $\beta_1, \ldots, \beta_k$ are the partial regression coefficients (i.e., the effect of each predictor holding constant the effects of the other predictors), and $e_i$ are the errors of prediction or residuals.

In what is commonly referred to as the *normal regression model*, where there is interest in both estimating and testing the parameters of the model, the $e_i$ are assumed normal, homoskedastic, and independent. The parameters $\beta$ are estimated so as to minimize the sum of the squared errors of prediction:

$$\min \sum_i [Y_i - (\beta_0 + \beta_1 X_i + \beta_2 X_i + \cdots + \beta_k X_i)]^2 = \min \sum_i (Y_i - \hat{Y})^2 = \sum_i e_i^2 = S$$

where $\hat{Y}$ is the predicted value of $Y$. The estimates of the parameters $\beta$, when substituted into this expression, produce the least possible value for $S$; this is a least-squares criterion. In the single predictor case, the least-squares criterion constructs a line that best fits the points $(X, Y)$ for each subject, as illustrated in Fig. 6.1.

The parameters $\beta$ of the regression model are found by differentiating $S$ with respect to each of the parameters and setting the result equal to zero. For the single predictor case, the solutions to $\beta_0$ and $\beta_1$ are:

$$\sum_i (Y_i - \beta_0 - \beta_1 X_i) = 0$$

$$\sum_i X_i (Y_i - \beta_0 - \beta_1 X_i) = 0$$

Through additional algebraic manipulation, it can be shown that the estimates of $\beta_0$ and $\beta_1$ are:

$$b_0 = \overline{Y} - \beta_1 \overline{X}$$

$$b_1 = r_{XY} \frac{S_Y}{S_X}$$

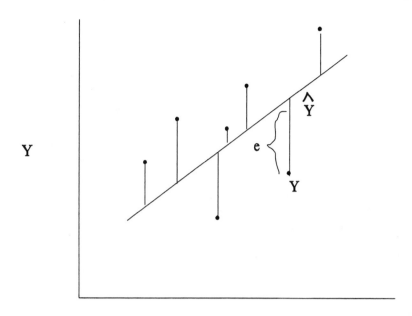

**X**

FIG. 6.1.

The estimate for $b_1$ also may be expressed as:

$$b_1 = \frac{\text{Cov}_{XY}}{S_X^2}$$

Substituting the values of $b_0$ and $b_1$ and into the multiple regression model leads to the estimation equation:

$$\hat{Y} = \overline{Y} + b_1(X - \overline{X})$$

The least-squares solution for the parameter estimates is a mathematical maximization procedure; no other method will produce values of the parameters that will predict $Y$ more closely. Unfortunately, the algebraic expressions for the parameter estimates become increasingly complex as the number of predictors rises, and their solution only really becomes feasible with the assistance of a computer.

Similar to the analysis of variance presented in chapter 3, the multiple regression model is centered on the partitioning of the $Y$ variance. To gain some insight into this partitioning, first consider the following identity:

$$Y_i - \hat{Y} = (Y_i - \overline{Y}) - (\hat{Y} - \overline{Y})$$

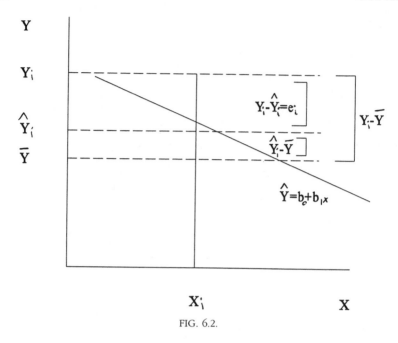

FIG. 6.2.

This identity is illustrated geometrically in Fig. 6.2. As shown in Fig. 6.2, the residual $e_i = Y_i - \hat{Y}_i$ is the difference between two quantities: (a) the deviation of $Y_i$ from the mean $\overline{Y}$, and (b) the deviation of $\hat{Y}$ from the mean $\overline{Y}$. The identity can also be written as:

$$(Y_i - \overline{Y}) = (\hat{Y} - \overline{Y}) + (Y_i - \hat{Y})$$

Squaring both sides of this identity and summing, we obtain:

$$\sum(Y_i - \overline{Y})^2 = \sum(\hat{Y} - \overline{Y})^2 + \sum(Y_i - \hat{Y})^2$$

The term $\sum(Y_i - \overline{Y})^2$ is the sum of squares about the mean ($SS_{tot}$); $\sum(Y_i - \hat{Y}_i)^2$ is the sum of squares about the regression line ($SS_{res}$); and $\sum(\hat{Y}_i - \overline{Y})^2$ is the sum of squares due to the regression line ($SS_{reg}$). These results may be organized in an analysis of variance source table as shown in Table 6.1.

From this partitioning of the $Y$ variance it should be clear to the reader that some of the variation can be attributed to the regression line (that which is predictable from $X$) and some of the variation cannot be attributed to the regression line (that which is not predictable from $X$). This latter source of variation is, of course, the residual or error variation. Therefore, the goal of regression analysis is to maximize $SS_{reg}$ relative to $SS_{res}$—in other words, to ensure that as much as possible of the total variance ($SS_{tot}$) is due to regression

TABLE 6.1
Analysis of Variance Source Table

| Source | SS | df | MS | F |
|--------|-----|-----|-----|-----|
| Regression | $SS_{reg}$ | $k$ | $SS_{reg}/k$ | |
| Residual (error) | $SS_{res}$ | $N - k - 1$ | $SS_{res}/(N - k - 1)$ | $MS_{reg}/MS_{res}$ |

($SS_{reg}$). As it turns out, the ratio of these two sources of variation is the squared multiple correlation ($R^2$) between $Y$ and the set of $X$ predictors:

$$R^2 = \frac{\Sigma(\hat{Y} - \overline{Y})^2}{\Sigma(Y_i - \overline{Y})^2} = \frac{SS_{reg}}{SS_{tot}}$$

Therefore $R^2$ is the proportion of the $Y$ variation due to the set of predictors $X$. Because the regression parameters have been constructed so as to produce values of $\hat{Y}_i$ as close as possible to $Y_i$, $R^2$ may also be interpreted as the square of the correlation between $Y_i$ and an optimally weighted set of predictors $X$. From Table 6.1, because $F = \frac{MS_{reg}}{MS_{res}}$, $F$ may also be written in terms of $R^2$:

$$F = \frac{R^2/k}{(1 - R^2)/(N - k - 1)}$$

which is evaluated with $df = k$, $N - k - 1$. It should be emphasized that for a given $Y$ and a set of predictors $X$, the variance partitioning of $Y$ will be the same whether it is conducted through the analysis of variance or multiple regression analysis. Historically, multiple regression analysis came first but posed an extreme computational burden when the number of predictors was greater than three. The complexity of the regression calculations arise primarily from the presence of correlations among the predictors. To remove the computational burden associated with correlated predictors, Fisher (1936) developed the analysis of variance model for equal $n$ at each level of the predictor variables with uncorrelated predictors. If the predictors are not correlated (i.e., $r_{ij}^2 = 0$, for any $i$ and $j$), then it can be shown that for $k$ predictors the $R^2$ is equivalent to the sum of the individual squared correlations:

$$R^2 = r_{y1}^2 + r_{y2}^2 + \cdots + r_{yk}^2$$

With the present accessibility of computers for conducting multiple regression analyses, there is actually little justification for the continued use of the analysis of variance, especially when the analytic disadvantages associated with it (e.g., loss of continuity in the predictors) is considered.

The goal of multiple regression analysis is to maximize the correlation between the criterion and set of predictors. Unfortunately, this maximization leads to bias in the sample $R^2$ as an estimate of the population $R^2$. This is because every aspect of the sample data is used to construct estimates of the regression parameters, even aspects that are idiosyncratic to a specific sample. Thus, even if $R^2$ were zero in the population, the expected value of $R^2$ in the sample is actually equal to:

$$\frac{k}{N-1}$$

where $k$ is the number of predictors. As the number of predictors increases relative to sample size, the positive bias in $R^2$ increases. In order to adjust for this positive bias, a correction for the upward bias in $R^2$ is obtained by using the equation:

$$\overset{\lor}{R}{}^2 = 1 - (1 - R^2)\frac{N-1}{N-k-1}$$

where $\overset{\lor}{R}{}^2$ is referred to as the *adjusted* or *shrunken* $R^2$ and is an estimate of the population squared multiple correlation coefficient. However, because of capitalization on chance characteristics of samples, this estimate will tend not to generalize very well to other samples. Instead, Herzberg (1969) proposed $R_c^2$ as a correction to $R^2$ that would reflect how well it would generalize to other samples:

$$R_c^2 = 1 - \left(\frac{N-1}{N-k-1}\right)\left(\frac{N-2}{N-k-2}\right)\left(\frac{N+1}{N}\right)(1 - R^2)$$

and, by definition, $R_c^2 < \overset{\lor}{R}{}^2 < R^2$. Therefore, $R_c^2$ can be used as a measure of how well one would expect an obtained $R^2$ to cross-validate or generalize across other samples. Despite the availability of the $R_c^2$ as an estimate, we highly recommend that the process of cross-validation should still be done— either with randomly splitting one sample, and using one half to develop the model and the other half to test it, or by obtaining a completely new sample.

## AN EXAMPLE OF REGRESSION ANALYSIS USING SAS

In order to illustrate a multiple regression analysis with SAS, a simple example is presented with two predictors: the number of drivers and the maximum temperature in January, used to predict the number of auto fatalities in

TABLE 6.2
SAS Regression Model Setup

```
DATA ONE;
INFILE DAT1;
INPUT DEATHS DRIVERS TEMP;
LABEL DEATHS = 'NUMBER OF AUTOMOBILE FATALITIES IN JANUARY'
      DRIVERS = 'NUMBER OF DRIVERS'
      TEMP='MAXIMUM NORMAL TEMPERATURE FOR JANUARY';
RUN;
PROC REG;
MODEL DEATHS = DRIVERS TEMP;
RUN;
```

January across the 50 states (Andrews, 1973). In order to run a regression analysis in SAS, the PROC REG procedure must be used and the regression model must be defined as shown in Table 6.2. The output from this SAS program are presented in Table 6.3.

As can be seen by examining the results in Table 6.3, the proportion of variance in auto deaths that is attributable to the number of drivers and temperature is approximately 94% ($R^2 = .934$). In addition, the adjusted $R^2$ (i.e., $\tilde{R}^2 = .931$) is only slightly less than $R^2$ (due to the large subject to variable ratio of 25:1). Although not given, the $R_c^2$ can easily be computed as .923. Based on these results, we can expect the $R^2$ to generalize to other months of the year. Individually, both predictors have a significant relation to auto deaths. An examination of the parameter estimates in Table 6.3 reveals that because the variable drivers is measured in units of $1/1000$, the regression coefficient implies that for every $1/1000$ increase in the number of drivers, there is an increase of 4.178 deaths. In addition, for every degree increase in temperature, there is an increase of 11.097 deaths.

As mentioned previously, because $R^2$ tends to be inflated as the number of predictors relative to sample size increases, care should be taken to limit the number of predictors in the model. Unfortunately, many analysts find this very difficult to do, often incorporating many more variables in the regression analysis than the sample size can reasonably support. When a large number of predictors is available, there are three general strategies that can be used to select some subset of the predictors for regression analysis. The three strategies are (a) stepwise selection, (b) hierarchical selection, and (c) stepwise selection within hierarchical selection.

The least desirable selection strategy is stepwise selection. Stepwise selection refers to a family of selection strategies, whose common characteristic is to select variables for inclusion in the regression equation based solely on statistical criteria. One variant of stepwise regression is referred to as *forward selection*. In this method, one begins with a set of $k$ predictors, and

TABLE 6.3
SAS System Output, MODEL1, Dependent Variable: DEATHS

Analysis of Variance

| Source | DF | Sum of Squares | Mean Square | F Value | Prob>F |
|--------|-----|----------------|-------------|---------|--------|
| Model | 2 | 36213259.888 | 18106629.944 | 335.015 | 0.0001 |
| Error | 47 | 2540221.2318 | 54047.260251 | | |
| C Total | 49 | 38753481.12 | | | |

| | | | | |
|--------|----------|-----------|----------|--------|
| Root MSE | 232.48067 | R-square | 0.9345 | |
| Dep Mean | 926.76000 | Adj R-sq | 0.9317 | |
| C.V. | 25.08532 | | | |

Parameter Estimates

| Variable | DF | Parameter Estimate | Standard Error | T for H0: Parameter=0 |
|----------|-----|--------------------|----------------|-----------------------|
| INTERCEP | 1 | -331.685550 | 122.75320283 | -2.702 |
| DRIVERS | 1 | 4.177700 | 0.17150840 | 24.359 |
| TEMP | 1 | 11.096755 | 2.87969568 | 3.853 |

| Variable | DF | Prob > |T| |
|----------|-----|-----------|
| INTERCEP | 1 | 0.0096 |
| DRIVERS | 1 | 0.0001 |
| TEMP | 1 | 0.0004 |

examines each of the $k$ bivariate correlations with the dependent variable ($Y$) to determine which predictor is most highly related to $Y$. The predictor with the highest correlation is allowed to enter the equation first. Subsequently, from the remaining $k - 1$ predictors, the predictor that when added to the equation increases $R^2$ the most with the first variable already in the equation is added. This process of comparing each predictor's contribution to $R^2$ with previously selected predictors in the equation continues until all the predictors have been selected, or until the addition of none of the remaining predictors results in a significant increase in $R^2$. It is important to note that if, during the process of adding predictors, a previously entered predictor becomes nonsignificant, it is still allowed to remain in the equation. *Backward selection* proceeds in a fashion analogous to forward selection but in the opposite direction. One begins the model with all $k$ predictors already in the regression equation. Then predictors are successively removed from the equation until the decrease in $R^2$ is considered statistically significant. *Stepwise selection* itself incorporates both a forward and backward selection strategy. Stepwise selection begins as in forward selection by incrementally building up the regression equation, but now predictors are removed (backward selection) if their significance falls below a certain level.

The PROC REG procedure in SAS incorporates eight stepwise methods (e.g., forward selection, backward elimination, stepwise, maximum $R^2$ improvement, minimum $R^2$ improvement, $R^2$ selection, adjusted $R^2$ selection, Mallows $C_p$ selection). Although the details of these stepwise strategies differ, they all have in common the selection of variables based on statistical criteria alone. Therefore, all of the preceding strategies are subject to the same three flaws, namely, (a) capitalization on chance sample characteristics, (b) an enormously inflated Type I error rate, and (c) the frequent missing out of important variables for trivial reasons. This last flaw is particularly bothersome. For example, consider a model confronted with an increase in $R^2$ of .110 for one variable and an increase in $R^2$ of .111 for a second variable. Unfortunately, it is the latter variable that is selected for inclusion in the equation even though its contribution to the $R^2$ is trivially greater than the other variable's contribution. The nonentered variable may never be included in the equation, especially if its correlation with the other predictor is significant.

Stepwise strategies are nonetheless very popular, and one sees them applied in any number of multivariate contexts. For example, the selection of variables in descriptive discriminant is often done through a stepwise strategy, wherein the criterion to be minimized is Wilks's $\Lambda$. Note that as in most contexts in which stepwise regression is conducted, adding additional variables can never decrease our prediction of the criterion. We say most, however, because one of the rare exceptions is predictive discriminant analysis, where the criterion to be maximized is the hit rate: The hit rate can actually decrease with the addition of variables.

A second, more desirable strategy is to select and order variables in terms of their theoretical importance. Hierarchical regression, as this is called, builds up the regression equation by including variables that are viewed as having theoretically important relations with the criterion. The ordering is logical. For example, one does not include an effect before its probable cause in an equation and one also does not ordinarily select variables that could never possibly be effects (e.g., sex) to enter in after variables that could never possibly be causes of these effects (e.g., occupation). The result of this substantively guided selection of variables is typically the estimation of many, many fewer regression equations than stepwise selection, with a concomitant smaller increase in the Type I error rate and capitalization on chance sample characteristics. A regression model constructed through a hierarchical strategy is more likely to cross-validate well to other samples than one constructed through a stepwise strategy.

Nonetheless, if confronted by a multitude of predictors, many of them tapping into the same domain, a combination hierarchical–stepwise selection strategy may be feasible. The researcher should at least know the relative theoretical importance of different types of predictors, even if each type is represented by more than one variable in the data. For example, let us assume there are three different variable domains in the data: cognitive, personality, and attitudinal. The researcher should enter the set of variables represented by a particular domain first if it is believed that domain most contributes to the criterion. This is a hierarchical strategy. But within this domain or set of variables, those variables that contribute significantly to $R^2$ can be retained, and those that do not are deleted. This is a stepwise strategy. Thus, the domain ordering is theoretically determined but the variable ordering within domains is statistically determined.

## DISCRIMINANT ANALYSIS AND REGRESSION ANALYSIS

Informally, one may think of discriminant analysis as an exercise in regression analysis. In other words, a set of variables is used to predict group membership. We can go even further though and state that the relation between the two is mathematically identical when the number of groups is equal to two. In fact, when the number of groups is two, the regression coefficients are identical within a constant of proportionality to the raw discriminant function coefficients. And when both the predictors and the criterion (consisting of the two groups) have been standardized to unit variance, the regression and discriminant function coefficients are identical.

Consider again the solution for the raw discriminant function coefficients $a$ given in chapter 5 when there are two groups:

$$CS^{-1}(\bar{x}_1 - \bar{x}_2)$$

Assuming that all variables have been standardized to unit variance, then the solution for $a$ becomes:

$$\frac{n_1 n_2}{n_1 + n_2}(\bar{x} - \bar{x}_2)$$

It can be shown that $C$ is equal to 2 when there are two groups and all variables have been standardized to unit variance. Now consider the formula for the correlation ($r_{xy}$) between a continuous predictor and a dichotomous criterion when the variables have been standardized to unit variance as given by:

$$r_{xy} = (\bar{x}_1 - \bar{x}_2) \frac{\sqrt{n_1 n_2}}{n_1 + n_2}$$

Because the variables have been standardized to unit variance, the regression coefficient $(B_{yx})$ for the regression of the criterion onto the predictor is:

$$B_{yx} = r_{xy} \frac{s_y}{s_x} = r_{xy} \frac{1}{1} = r_{xy}$$

Therefore,

$$B_{yx} = r_{yx} = (\bar{x}_1 - \bar{x}_2) \frac{\sqrt{n_1 n_2}}{n_1 + n_2}$$

The regression coefficient is a function of the sample size weighted mean difference between the two groups. Thus, when comparing the discriminant coefficient $a$ with $B_{yx}$ we have:

$$a = \frac{\sqrt{n_1 n_2}}{n_1 + n_2} (\bar{x}_1 - \bar{x}_2) = B_{yx} = (\bar{x}_1 - \bar{x}_2) \frac{\sqrt{n_1 n_2}}{n_1 + n_2}$$

which demonstrates that the regression coefficient and discriminant function coefficient are identical.

This identity is also illustrated in the following consumer behavior example. In this example a marketing researcher is attempting to predict whether a person will purchase a new product or not based on level of income. For purposes of analysis, purchasers are scored as 1 ($n_1 = 77$; $\bar{x}_1 = .32$), and nonpurchasers as 0 ($n_2 = 114$; $\bar{x}_2 = -.24$); all variables are standardized to unit variance. There is a total of $n_1 + n_2 = 191$ persons. Regressing the purchase status groups onto income, we find that the regression coefficient is:

$$B_{yx} = .25 = (.32 - -.24) \times \frac{\sqrt{114 \times 77}}{191}$$

Next a discriminant analysis is performed, again with the two purchase status groups. The solution for the single eigenvalue is .06, the eigenvector of which is equal to $\sqrt{.06} = .25$. Now the raw discriminant coefficient is obtained by multiplying the eigenvector by the square root of the variable's variance, but because the variance is equal to unity, the raw discriminant coefficient $a = .25$. Further, the mean discriminant function scores for the groups are .32 and $-.24$, as they should be, given the identity of $a$ and $B_{yx}$. Thus, it is clear that the two procedures provide identical results when the number of groups is two.

## CANONICAL CORRELATION ANALYSIS

Canonical correlation analysis is concerned with examining the linear relationship between two sets of variables. Traditionally, one set of variables is denoted by $X$ and the second set of variables is denoted by $Y$. In canonical correlation, the relationship is examined by finding a linear composite of the $Y$ variables and a linear composite of the $X$ variables such that the scores derived from the $Y$ composite are maximally correlated with the scores derived from the $X$ composite. Each linear composite created is referred to as a canonical variate. As such, canonical correlation is the correlation between the $Y$ and $X$ canonical variate scores. For ease of presentation, the canonical variate scores for the $Y$ composite are usually denoted by $V$ and the canonical variate scores for the $X$ composite are denoted by $U$. It may be easy to consider the predicted value of $y$ ($\hat{y}$) obtained from the linear composite of $X$ predictors in multiple regression analysis as analogous (and under restricted circumstances, identical) conceptually to a canonical variate score. Similarly, one may think of linear discriminant function scores as identical to canonical variate scores when the variables in one set represent membership in one or another group. However, unlike multiple regression or discriminant analysis, in canonical correlation no distinction is made with respect to either $Y$ or $X$ as criterion or predictor variables. Canonical correlations are symmetric measures of relationship and it is arbitrary which set is labeled $Y$ and which set is labeled $X$.

Typically, more than one pair of canonical variates can be extracted from the data. For example, if there are $q$ $Y$ variables and $p$ $X$ variables then there can be a minimum of $q$ or $p$ pairs of variables. The canonical variates are formed as follows:

$$U_1 = a_{11}X_1 + a_{12}X_2 + \cdots + a_{1p}X_p$$
$$U_2 = a_{21}X_1 + a_{22}X_2 + \cdots + a_{2p}X_p$$
$$\cdot$$
$$\cdot$$
$$\cdot$$
$$U_r = a_{r1}X_1 + a_{r2}X_2 + \cdots + a_{rp}X_p$$

$$V_1 = b_{11}Y_1 + b_{12}Y_2 + \cdots + b_{1q}Y_q$$
$$V_2 = b_{21}Y_1 + b_{22}Y_2 + \cdots + b_{2q}Y_q$$
$$\cdot$$
$$\cdot$$
$$\cdot$$
$$V_r = b_{r1}Y_1 + b_{r2}Y_2 + \ldots + b_{rq}Y_q$$

where the $U$ are the canonical variate scores for the $X$ composites, the $V$ are the canonical variate scores for the $Y$ composites, and the $r$ is the minimum of $q$ or $p$.

The linear composites are constructed so that the correlation between $U_1$ and $V_1$ is a maximum, the correlation between $U_2$ and $V_2$ is a maximum (subject to the condition that $U_1$ and $V_1$ are uncorrelated with neither $U_2$ nor $V_2$), and the correlation between $U_r$ and $V_r$ is a maximum (subject to the condition that $U_r$ and $V_r$ are uncorrelated with neither $U_1$, $U_2$, . . . , $U_{r-1}$ nor $V_1$, $V_2$, . . . , $V_{r-1}$. Each canonical variate can be thought of as a latent variable, with each of the (canonical) correlations $R(U_1, V_1)$ through $R(U_r, V_r)$ representing the relationship between latent variables, $R(U_1, V_1) > R(U_2, V_2) > \cdots > R(U_r, V_r)$. The canonical correlations themselves are subject to interpretation as different dimensions of the $X$, $Y$ relationship.

The example presented in Fig. 6.3 illustrates the meaning of canonical correlation. In this example, there are two $X$ variables ($X_1$ and $X_2$), and there are two $Y$ variables ($Y_1$ and $Y_2$). Both variables have been standardized to unit variance and zero mean (i.e., transformed to $z$ scores). Figure 6.3a displays the scatterplot between $X_1$ and $X_2$, and Fig. 6.3b the scatterplot between $Y_1$ and $Y_2$. The regression line predicting one variable from the

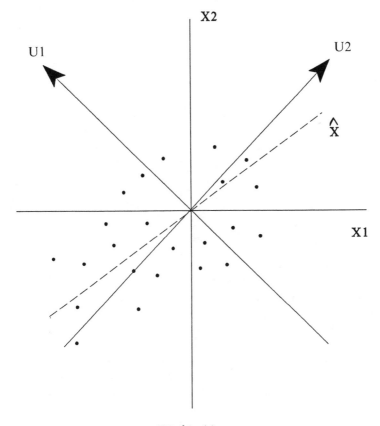

FIG. 6.3. (a).

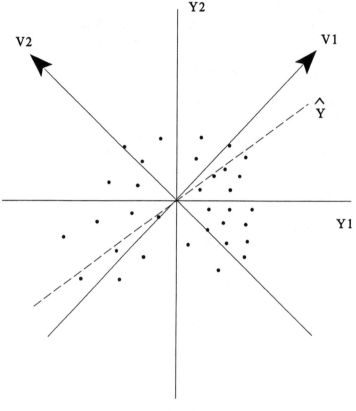

FIG. 6.3. (b).

other has been drawn in each scatterplot (recall from an earlier discussion that the regression slope is also equal to the correlation between the variables when the variables are $z$ transformed). Also drawn in the scatterplot are the vectors $U_1$ and $U_2$ (see Fig. 6.3a) and the vectors $V_1$ and $V_2$ (see Fig. 6.3b). These vectors are the canonical variates, and the projection of the scatterplot points onto these canonical variates will generate canonical variate scores. Figure 6.3c presents the newly created canonical variate scores $U_1$ and $V_1$ plotted against each other, and Fig. 6.3d presents the canonical variate scores $U_2$ and $V_2$ plotted against each other. For both Fig. 6.3c and Fig. 6.3d, the regression line relating the scores is also drawn. These regression lines (slopes) are the two possible canonical correlations relating the $X$ and $Y$ sets of variables. As can be seen by examining the regression lines, the canonical correlation between the $X$ and $Y$ variables depicted in Fig. 6.3c is the maximum obtainable value of the canonical correlation. The canonical correlation depicted in Fig. 6.3d is somewhat smaller than (and uncorrelated with) the canonical correlation depicted in Fig. 6.3c. As it turns out, the

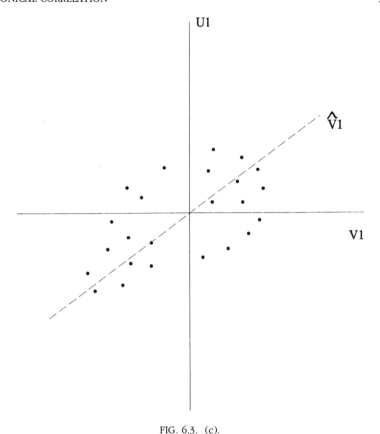

FIG. 6.3. (c).

correlations displayed in Fig. 6.3c and Fig. 6.3d can be considered latent variable correlations—correlations between two unobserved variables that have been constructed from two linear combinations of observed variables such that the latent variable correlation is at a maximum.

## OBTAINING THE CANONICAL CORRELATIONS

As mentioned previously, canonical correlations are the correlations between canonical variates that have been constructed from weighted combinations of observed variables. The central problem is to find the values of the canonical coefficients (i.e., the values of $a$ and $b$). Not surprisingly, once again the eigenvalue method is used for calculating the coefficients. In fact, the eigenvalue method used is almost identical to the method used in discriminant analysis. In order to obtain the canonical coefficients we first define a "super" matrix that represents the correlations among the set of $X$

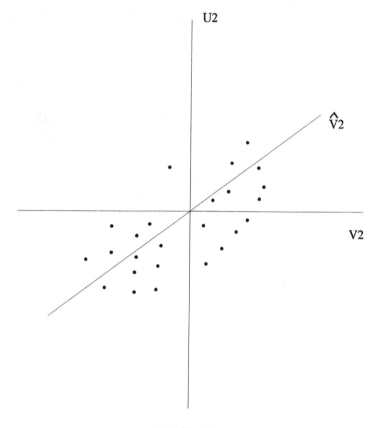

FIG. 6.3. (d).

variables ($R_{xx}$), the correlations among the set of $Y$ variables ($R_{yy}$), and the correlations between the set of $X$ and set of $Y$ variables ($R_{xy} = R_{yx}$). This "super" matrix looks like

$$\begin{bmatrix} R_{YY} | R_{YX} \\ \hline R_{XY} | R_{XX} \end{bmatrix}$$

Using this matrix, the eigenvalue decomposition of the matrix product

$$R_{XX}^{-1} R_{XY} R_{YY}^{-1} R_{YX}$$

is found. The eigenvalue decomposition that provides a solution to the coefficients $a$ and $b$ is given by

$$(R_{xx}^{-1} R_{xy} R_{yy}^{-1} R_{yx} - \lambda_j I_j) b_j = 0$$

and

$$a_j = \frac{R_{YY}^{-1}R_{YX}b_j}{\sqrt{\lambda_j}}$$

Fortunately, once the reader understands what the solutions involve, the actual computations are best left to a computer procedure.

Although we mentioned earlier that the eigenvalue method used in canonical correlation is almost identical to that used in discriminant analysis, it turns out that there is one important difference between the solutions for the eigenvectors of discriminant and canonical correlation analyses. Recall from the discussion in chapter 2 that eigenvectors are usually derived under the unit norm condition (i.e., $x_1'x_2 = 0$). This is the scaling condition that is imposed in discriminant analysis. However, the scaling condition imposed instead in canonical correlation analysis is

$$b_j' R_{XX} b_j = 1$$

This restriction ensures that the canonical variate scores will have unit standard deviations. The canonical variate scores are equal to

$$u_{i1} = x_i' a_1$$

$$v_{i1} = y_i' b_1$$

Therefore, the $j$th canonical correlation may be defined as:

$$R_j = \frac{1}{n-1} \sum_{i=1}^{n} u_{ij}v_{ij}$$

Thus, it can be seen that the canonical correlation is the average cross-product of standardized scores. It is also the case that the canonical correlation ($R_j$) is equal to the square root of the $j$th eigenvalue. This is the reason the canonical coefficients are easily obtained via the eigenvalue decomposition method.

## INTERPRETATION OF CANONICAL VARIATES

To a great extent, the interpretation of canonical variates proceeds in a similar manner to how discriminant functions were interpreted in discriminant analysis. First, one examines the degree to which each variable is associated with a particular variate. Those variables most closely related to the variate are used as a guide to identify and label the latent variable the variate may represent. Finally, once the nature of the two canonical variates of a pair have been identified, the analyst must determine the dimension that links them. The following simple example illustrates the interpretation of canonical variates. Assume we have two sets of observed variables: One set consists of four cognitive variables and the other consists of four socio-economic variables. We extract four canonical correlations (four pairs of canonical variates) with the "defining" variables (defined by large, positive coefficient values) in bold type:

### First Canonical Correlation

| Socioeconomic Variable Variate: | Cognitive Variable Variate: |
|---|---|
| **Income** | **Verbal fluidity** |
| **Occupational status** | **Spatial fluidity** |
| **Education level** | **Delayed memory** |
| **Number of children** | **Mechanical drawings** |

### Second Canonical Correlation

| Socioeconomic Variable Variate: | Cognitive Variable Variate: |
|---|---|
| **Income** | **Verbal fluidity** |
| Occupational status | Spatial fluidity |
| Education level | Delayed memory |
| Number of children | Mechanical drawings |

### Third Canonical Correlation

| Socioeconomic Variable Variate: | Cognitive Variable Variate: |
|---|---|
| Income | Verbal fluidity |
| **Occupational status** | **Spatial fluidity** |
| Education level | Delayed memory |
| Number of children | **Mechanical drawings** |

### Fourth Canonical Correlation

| Socioeconomic Variable Variate: | Cognitive Variable Variate: |
|---|---|
| Income | **Verbal fluidity** |
| Occupational status | Spatial fluidity |
| **Education level** | **Delayed memory** |
| Number of children | Mechanical drawings |

For the first canonical correlation, each of the observed variables is related highly to its respective variate. Our interpretation here may be that higher socioeconomic status (as determined by higher income, higher occupational status, higher education level, and higher number of children) is related to higher overall cognitive ability (obtaining higher scores on all four tests). The second canonical correlation defines a relation between high income and high verbal fluidity (i.e., more verbally proficient people have higher income generating jobs). The third canonical correlation shows a relation between higher occupational status and higher spatial ability (mechanical drawings is largely a measure of spatial ability). As for the fourth canonical correlation, there is a relation between higher education level and higher verbal fluidity/delayed memory. Perhaps the latter two variables are the most important for educational advancement.

Two types of canonical variate coefficients are generally reported: standardized coefficients and structure coefficients, both comparable in meaning to their discriminant analysis counterparts. The standardized coefficients may be interpreted as standardized partial regression coefficients, showing the unique relation between the variable and the canonical variate (partialing out the effects of the other variables). They are obtained by multiplying the raw canonical coefficients by the appropriate variable's standard deviation. The structure coefficients are simple bivariate correlations between variables and their variates, and are given as

$$s_X = R_{XX}A$$
$$s_Y = R_{YY}B$$

where $A$ and $B$ refer to matrices of raw canonical variate coefficients for $X$ and $Y$, respectively. The advantages and disadvantages of standardized versus structure coefficients were discussed extensively in chapter 5, along with our reasons for preferring to base our canonical variate interpretations on the structure coefficients.

## TESTS OF SIGNIFICANCE
## OF CANONICAL CORRELATIONS

Testing the significance of canonical correlations follows a residualization procedure identical to that of discriminant analysis. One first tests whether there is any association between the $r$ pairs of canonical variates using Rao's $F$ approximation to Wilks's $\Lambda$. If the first $F$ is significant, it is clear that there is at least one significant canonical correlation. After removing the first canonical correlation, the first residual (consisting of the second through $r$th canonical correlations) is tested, and one continues to test successive residuals until the residuals are no longer significant. Although the minimum of $p,q$ canonical correlations are possible, substantially fewer residuals may be significant. Note that the significance of residuals is being tested—not individual canonical correlations. As argued earlier, testing the significance of individual canonical correlations is an uncertain affair at best because the size ordering of the canonical correlations in the sample may not match their ordering in the population.

## THE PRACTICAL IMPORTANCE
## OF CANONICAL CORRELATIONS

How may the practical, as opposed to the statistical, significance of the canonical correlations be evaluated? The canonical correlation squared ($R^2$) is directly interpretable as a measure of effect size. Nonetheless, $R^2$ in canonical correlation has a very different interpretation than $R^2$ in multiple regression. The multiple regression $R^2$ represents the proportion of criterion variance *accounted* for by the set of predictors. In contrast, canonical correlation $R^2$ represents the amount of variance *shared* between the two sets of variables. What this implies is that the canonical correlation could be very large yet be based on a relatively small proportion of each set's variance. Remember: Canonical correlation maximizes the correlation between two sets of variables. This is a much different goal than maximizing the amount of variance accounted for within each set. Although a large $R^2$ may be found between two sets of variables, if either one or both sets are contributing little variance to this correlation, one has to wonder about how well the constituent canonical variates represent their respective latent variables.

Stewart and Love (1968) proposed a "redundancy index" to assess how much of the variance in one variable set is accounted for by the other. The Stewart–Love index ($R_{j_{n}x}$) for the *j*th canonical correlation, if one were interested in how much variance in set *Y* is accounted for by set *X*, is given by:

$$R_{j_{n}x} = R_j^2 \frac{\sum\limits_i s_i^2}{q}$$

where the $s_i$ are structure coefficients. Adding $R_{j_{n}x}$ across all the canonical correlations also provides overall how much of the *Y* set variance is accounted for by the *X* set. Of course, the Stewart–Love index is not without its critics. In fact, some authors (e.g., Cramer & Nicewander, 1979, who also report some alternatives) have criticized the index for not being truly multivariate because it fails to take into account the intercorrelations among the *Y* variables. This objection is somewhat warranted because the redundancy coefficient is not symmetric; rarely will $R_{j_{n}x} = R_{j_{x} y}$.

## NUMERICAL EXAMPLE USING SAS

Sadri and Marcoulides (1994) recently conducted a study to examine the relationship between a set of organizational characteristic variables and a set of worker psychological characteristic variables. In order to conduct a canonical correlation, the PROC CANCORR procedure must be used. Table 6.4 presents the example SAS setup used for this analysis. The five organizational variables included in the study are quality of communication, job

TABLE 6.4
Example SAS Setup

```
data one;
infile dat1;
input qualcom jobresp orgeff manchar manrel
      prsgrwth relatshp sysmaint lifesat jobsat;
label qualcom='quality of communication'
      jobresp='job responsibility'
      orgeff='organizational effectiveness'
      manchar='manager characteristics'
      manrel='manager relations'
      prsgrwth='personal growth'
      relatshp='relationship'
      sysmaint='systems maintenance'
      lifesat='life satisfaction'
      jobsat='job satisfaction';
run;
proc cancorr all;
var qualcom jobresp orgeff manchar manrel;
with prsgrwth relatshp sysmaint lifesat jobsat;
run;
```

responsibility, organizational effectiveness, manager characteristics, and manager relations. The five psychological variables used in this study are personal growth, relationship, systems maintenance (three scales from an organizational climate scale), life satisfaction, and job satisfaction. In order to specify in SAS the differences between the two sets of variables, one set is arbitrarily designated as the VAR set (the organizational variables) and the other set as the WITH set (the psychological variables).

The results of the analysis are presented in Tables 6.5 through 6.7. As can be seen by examining Table 6.5, there are maximally five possible canonical correlations. According to the $F$ test presented in Table 6.5, only the first two should be retained (i.e., $F = .0001$ and $F = .0402$). The magnitude of the first canonical correlation is equal to .654 and the second is equal to .351. Combined, the two correlations account for 55% of the covariance between the two sets of variables. Table 6.6 provides the structure coefficients between each canonical variate and its variables. Because only two canonical correlations were retained as important, interpretation is limited to the pairs V1,W1 and V2,W2. Four of the five organizational variables are associated with V1 (with the exception of job responsibility) and four of the five psychological variables are associated with W1 (with the exception of life satisfaction). One interpretation of the first canonical correlation could be that it represents the association between emotions specific to work (W1) and managerial interactions (V1). This latter interpretation on V1 seems plausible because job responsibility is the only variable in the set that is not directly concerned with

## TABLE 6.5
### SAS System Canonical Correlation Analysis

| | Canonical Correlation | Adjusted Canonical Correlation | Approx Standard Error | Squared Canonical Correlation |
|---|---|---|---|---|
| 1 | 0.653949 | 0.629505 | 0.047207 | 0.427649 |
| 2 | 0.350657 | 0.293084 | 0.072337 | 0.122961 |
| 3 | 0.224492 | 0.181658 | 0.078322 | 0.050397 |
| 4 | 0.093506 | . | 0.081757 | 0.008743 |
| 5 | 0.000923 | . | 0.082479 | 0.000001 |

Eigenvalues of INV(E)*H
= CanRsq/(1-CanRsq)

| | Eigenvalue | Difference | Proportion | Cumulative |
|---|---|---|---|---|
| 1 | 0.7472 | 0.6070 | 0.7871 | 0.7871 |
| 2 | 0.1402 | 0.0871 | 0.1477 | 0.9348 |
| 3 | 0.0531 | 0.0443 | 0.0559 | 0.9907 |
| 4 | 0.0088 | 0.0088 | 0.0093 | 1.0000 |
| 5 | 0.0000 | | 0.0000 | 1.0000 |

Test of H0: The canonical correlations in the
current row and all that follow are zero

| | Likelihood Ratio | Approx F | Num DF | Den DF | Pr > F |
|---|---|---|---|---|---|
| 1 | 0.47250842 | 4.5989 | 25 | 514.1492 | 0.0001 |
| 2 | 0.82555697 | 1.7213 | 16 | 425.2896 | 0.0402 |
| 3 | 0.94129970 | 0.9532 | 9 | 340.8738 | 0.4789 |
| 4 | 0.99125575 | 0.3103 | 4 | 282 | 0.8710 |
| 5 | 0.99999915 | 0.0001 | 1 | 142 | 0.9912 |

Multivariate Statistics and F Approximations

S=5   M=-0.5   N=68

| Statistic | Value | F | Num DF | Den DF | Pr > F |
|---|---|---|---|---|---|
| Wilks' Lambda | 0.472508 | 4.5989 | 25 | 514.15 | 0.0001 |
| Pillai's Trace | 0.60975 | 3.9444 | 25 | 710 | 0.0001 |
| Hotelling-Lawley Trace | 0.949272 | 5.1792 | 25 | 682 | 0.0001 |
| Roy's Greatest Root | 0.747179 | 21.22 | 5 | 142 | 0.0001 |

*Note.* F statistic for Roy's greatest root is an upper bound.

## TABLE 6.6
### SAS System Canonical Structure

Correlations Between the 'VAR' Variables
and Their Canonical Variables

|          | V1      | V2      |                                     |
|----------|---------|---------|-------------------------------------|
| QUALCOM2 | 0.8218  | 0.0031  | quality of communication squared    |
| JOBRESP2 | 0.1662  | 0.6285  | job responsibility squared          |
| ORGEFF   | 0.7006  | -0.0187 | organizational effectiveness        |
| MANCHAR  | 0.7471  | 0.5437  | manager characteristics             |
| MANREL2  | 0.7398  | 0.3511  | relationship with manager squared   |

|          | V3      | V4      |                                     |
|----------|---------|---------|-------------------------------------|
| QUALCOM2 | -0.5281 | 0.2131  | quality of communication squared    |
| JOBRESP2 | -0.5005 | 0.1115  | job responsibility squared          |
| ORGEFF   | 0.2426  | 0.1327  | organizational effectiveness        |
| MANCHAR  | -0.0324 | 0.3477  | manager characteristics             |
| MANREL2  | -0.1015 | -0.5557 | relationship with manager squared   |

|          | V5      |                                     |
|----------|---------|-------------------------------------|
| QUALCOM2 | 0.0199  | quality of communication squared    |
| JOBRESP2 | -0.5608 | job responsibility squared          |
| ORGEFF   | -0.6575 | organizational effectiveness        |
| MANCHAR  | -0.1558 | manager characteristics             |
| MANREL2  | 0.1015  | relationship with manager squared   |

Correlations Between the 'WITH' Variables
and Their Canonical Variables

|          | W1      | W2      |                                        |
|----------|---------|---------|----------------------------------------|
| PRSGRWTH | 0.7964  | -0.3555 | wes personal growth                    |
| RELATSHP | 0.8586  | 0.0557  | wes relationship                       |
| SYSMAINT | 0.7282  | -0.4663 | wes system maintenance and change      |
| EIGHT2   | -0.1604 | 0.0024  | faces - life satisfaction              |
| SATTOT   | 0.8963  | 0.3344  | sum of 90 satisfaction items           |

|          | W3      | W4      |                                        |
|----------|---------|---------|----------------------------------------|
| PRSGRWTH | -0.4591 | -0.1388 | wes personal growth                    |
| RELATSHP | 0.1157  | -0.4894 | wes relationship                       |
| SYSMAINT | 0.4679  | -0.1179 | wes system maintenance and change      |
| EIGHT2   | 0.2689  | 0.1892  | faces - life satisfaction              |
| SATTOT   | 0.1955  | 0.2146  | sum of 90 satisfaction items           |

|          | W5      |                                        |
|----------|---------|----------------------------------------|
| PRSGRWTH | 0.0968  | wes personal growth                    |
| RELATSHP | 0.0821  | wes relationship                       |
| SYSMAINT | -0.1393 | wes system maintenance and change      |
| EIGHT2   | 0.9307  | faces - life satisfaction              |
| SATTOT   | -0.0215 | sum of 90 satisfaction items           |

TABLE 6.7

SAS System Canonical Redundancy Analysis

### Raw Variance of the 'VAR' Variables Explained by

| | Their Own Canonical Variables | | | The Opposite Canonical Variables | |
|---|---|---|---|---|---|
| | Proportion | Cumulative Proportion | Canonical R-Squared | Proportion | Cumulative Proportion |
| 1 | 0.3963 | 0.3963 | 0.4276 | 0.1695 | 0.1695 |
| 2 | 0.1867 | 0.5829 | 0.1230 | 0.0230 | 0.1924 |
| 3 | 0.1746 | 0.7575 | 0.0504 | 0.0088 | 0.2012 |
| 4 | 0.1239 | 0.8815 | 0.0087 | 0.0011 | 0.2023 |
| 5 | 0.1185 | 1.0000 | 0.0000 | 0.0000 | 0.2023 |

### Raw Variance of the 'WITH' Variables Explained by

| | Their Own Canonical Variables | | | The Opposite Canonical Variables | |
|---|---|---|---|---|---|
| | Proportion | Cumulative Proportion | Canonical R-Squared | Proportion | Cumulative Proportion |
| 1 | 0.3677 | 0.3677 | 0.4276 | 0.1572 | 0.1572 |
| 2 | 0.0609 | 0.4286 | 0.1230 | 0.0075 | 0.1647 |
| 3 | 0.0893 | 0.5179 | 0.0504 | 0.0045 | 0.1692 |
| 4 | 0.0527 | 0.5705 | 0.0087 | 0.0005 | 0.1697 |
| 5 | 0.4295 | 1.0000 | 0.0000 | 0.0000 | 0.1697 |

### Standardized Variance of the 'VAR' Variables Explained by

| | Their Own Canonical Variables | | | The Opposite Canonical Variables | |
|---|---|---|---|---|---|
| | Proportion | Cumulative Proportion | Canonical R-Squared | Proportion | Cumulative Proportion |
| 1 | 0.4599 | 0.4599 | 0.4276 | 0.1967 | 0.1967 |
| 2 | 0.1628 | 0.6227 | 0.1230 | 0.0200 | 0.2167 |
| 3 | 0.1199 | 0.7426 | 0.0504 | 0.0060 | 0.2227 |
| 4 | 0.1010 | 0.8437 | 0.0087 | 0.0009 | 0.2236 |
| 5 | 0.1563 | 1.0000 | 0.0000 | 0.0000 | 0.2236 |

### Standardized Variance of the 'WITH' Variables Explained by

| | Their Own Canonical Variables | | | The Opposite Canonical Variables | |
|---|---|---|---|---|---|
| | Proportion | Cumulative Proportion | Canonical R-Squared | Proportion | Cumulative Proportion |
| 1 | 0.5462 | 0.5462 | 0.4276 | 0.2336 | 0.2336 |
| 2 | 0.0918 | 0.6380 | 0.1230 | 0.0113 | 0.2449 |
| 3 | 0.1107 | 0.7487 | 0.0504 | 0.0056 | 0.2504 |
| 4 | 0.0709 | 0.8196 | 0.0087 | 0.0006 | 0.2511 |
| 5 | 0.1804 | 1.0000 | 0.0000 | 0.0000 | 0.2511 |

*(Continued)*

TABLE 6.7
*(Continued)*

Squared Multiple Correlations Between the 'VAR' Variables and
the First 'M' Canonical Variables of the 'WITH' Variables

| M | 1 | 2 | |
|---|---|---|---|
| QUALCOM2 | 0.2888 | 0.2888 | quality of communication squared |
| JOBRESP2 | 0.0118 | 0.0604 | job responsibility squared |
| ORGEFF | 0.2099 | 0.2100 | organizational effectiveness |
| MANCHAR | 0.2387 | 0.2751 | manager characteristics |
| MANREL2 | 0.2341 | 0.2492 | relationship with manager squared |

| M | 3 | 4 | |
|---|---|---|---|
| QUALCOM2 | 0.3028 | 0.3032 | quality of communication squared |
| JOBRESP2 | 0.0730 | 0.0731 | job responsibility squared |
| ORGEFF | 0.2129 | 0.2131 | organizational effectiveness |
| MANCHAR | 0.2751 | 0.2762 | manager characteristics |
| MANREL2 | 0.2498 | 0.2525 | relationship with manager squared |

| M | 5 | |
|---|---|---|
| QUALCOM2 | 0.3032 | quality of communication squared |
| JOBRESP2 | 0.0731 | job responsibility squared |
| ORGEFF | 0.2131 | organizational effectiveness |
| MANCHAR | 0.2762 | manager characteristics |
| MANREL2 | 0.2525 | relationship with manager squared |

Squared Multiple Correlations Between the 'WITH' Variables and
the First 'M' Canonical Variables of the 'VAR' Variables

| M | 1 | 2 | |
|---|---|---|---|
| PRSGRWTH | 0.2712 | 0.2868 | wes personal growth |
| RELATSHP | 0.3153 | 0.3157 | wes relationship |
| SYSMAINT | 0.2268 | 0.2535 | wes system maintenance and change |
| EIGHT2 | 0.0110 | 0.0110 | faces - life satisfaction |
| SATTOT | 0.3436 | 0.3573 | sum of 90 satisfaction items |

the worker's manager. The second canonical correlation (W2) may be considered a psychological stasis variate. High scorers on this variate have low personal growth, low change, yet adequate satisfaction. It seems that V2 is most defined by job responsibility and manager characteristics, leading to the speculation that the variate may represent to what degree the manager controls the worker's responsibilities. Thus, the second canonical variate may represent the relation between managerial demands and psychological equilibrium. A Stewart–Love redundancy analysis is presented in Table 6.7 (simply labeled Canonical Redundancy Analysis). By examining the standardized results for the VAR set (organizational variables), it can be seen that the amount of variance explained in the organizational set by the psychological set for the first two canonical correlations is .217 or 22% (i.e., the sum of .1967 and .02). Similarly, by examining the standardized results for the WITH set (psychological variables), it can be seen that the amount of variance explained in the psychological set by the organizational set for the first two canonical correlations is .245 or 25%. Thus, although the combination of the first two squared canonical correlations explains 55% of the covariance between the two sets, less than a third of each set's variance is included in these correlations. Clearly, great care should be taken when making interpretations of canonical correlations, and looking at the redundancy index at least provides insight into this interpretation.

## EXERCISES

1. Ganster, Fusilier, and Mayes (1986) performed a canonical correlation analysis between a set of six job stressors and a set of four job outcomes. The correlation matrix that follows was used for the analysis ($N = 326$):

```
STD            1.18 0.94 0.62 0.84 1.13 1.11 0.89 1.36 0.45 0.47
CORR    V1     1.00 ........
CORR    V2     0.33 1.00 ........
CORR    V3     -.20 0.00 1.00 .......
CORR    V4     -.20 0.01 0.56 1.00 ......
CORR    V5     0.18 0.26 0.24 0.35 1.00 .....
CORR    V6     0.20 -.21 -.24 -.35 0.20 1.00 ....
CORR    V7     0.09 0.16 0.09 0.07 0.14 -.09 1.00 ...
CORR    V8     0.25 0.26 0.26 0.23 0.45 -.09 0.30 1.00 ..
CORR    V9     0.19 0.15 -.08 -.07 0.14 0.00 0.14 0.13 1.00 .
CORR    V10    0.21 0.26 0.21 0.09 0.24 -.12 0.51 0.25 0.43 1.00
V1=ROLE CONFLICT
V2=ROLE AMBIGUITY
V3=WORK UNDERLOAD
V4=LACK OF VARIABILITY
V5=UNDERUTILIZATION
V6=RESPONSIBILITY
V7=LIFE DISSATISFACTION
V8=JOB DISSATISFACTION
V9=SOMATIC COMPLAINTS
V10=DEPRESSION
```

V1–V6 are the job stressors; V7–V10 are the job outcomes. Perform a canonical correlation analysis of these data. How many significant canonical correlations are there? Can a reasonable interpretation be applied to them? How much of the variance from one set is *predictable* from the other?

2. The weekly rates of return for five stocks were recorded for the period January 1995 through March 1995. The data provided here are for 12 weekly rates of returns. The stocks observed are X1, Allied Chemical; X2, Dow Chemical; X3, Union Chemical; X4, Standard Oil Company; and X5, Texas Oil.

| X1 | X2 | X3 | X4 | X5 |
|------|------|------|------|------|
| 2.12 | 1.92 | 2.25 | 5.15 | 6.42 |
| 3.98 | 3.45 | 2.78 | 4.20 | 5.30 |
| 1.03 | 1.86 | 2.81 | 5.90 | 4.80 |
| 2.56 | 2.90 | 2.33 | 4.76 | 4.23 |
| 2.37 | 2.98 | 2.41 | 4.82 | 5.88 |
| 1.75 | 1.22 | 2.79 | 4.25 | 5.88 |
| 2.82 | 4.09 | 2.98 | 4.71 | 6.21 |
| 4.01 | 3.69 | 3.04 | 4.76 | 5.10 |
| 2.25 | 2.82 | 2.21 | 5.04 | 6.16 |
| 2.44 | 1.73 | 2.18 | 5.12 | 4.81 |
| 2.81 | 2.16 | 2.74 | 4.98 | 4.72 |
| 1.32 | 1.99 | 2.36 | 4.24 | 4.96 |
| 4.12 | 3.91 | 2.31 | 4.26 | 5.62 |
| 2.40 | 2.56 | 2.29 | 4.31 | 5.75 |
| 3.63 | 3.00 | 2.18 | 4.18 | 5.01 |

Perform a canonical correlation analysis using the chemical companies as one set and the oil companies as a second set.

3. Height and weight data are presented here for a sample of 20 father and adult son dyads (height is given in inches and weight in pounds).

| Father | | Son | |
|--------|--------|--------|--------|
| Height | Weight | Height | Weight |
| 74 | 205 | 73 | 195 |
| 72 | 195 | 73 | 190 |
| 69 | 185 | 70 | 193 |
| 70 | 197 | 71 | 210 |
| 71 | 176 | 69 | 180 |
| 76 | 245 | 72 | 250 |
| 70 | 231 | 69 | 225 |
| 72 | 216 | 71 | 197 |
| 73 | 194 | 73 | 199 |
| 69 | 250 | 71 | 215 |
| 75 | 263 | 71 | 247 |
| 74 | 201 | 70 | 211 |
| 73 | 189 | 72 | 175 |
| 74 | 190 | 69 | 179 |
| 73 | 217 | 71 | 199 |
| 71 | 200 | 70 | 187 |
| 72 | 233 | 75 | 230 |
| 75 | 249 | 75 | 234 |
| 69 | 187 | 72 | 193 |
| 73 | 204 | 75 | 186 |

Perform a canonical correlation analysis on the father and son dyads and interpret your results.

4. This data set includes six predictor scores ($x$) for student success on three measures ($y$) in a statistics class, for a sample of 25 students:

| $x_1$ | $x_2$ | $x_3$ | $x_4$ | $x_5$ | $x_6$ | $y_1$ | $y_2$ | $y_3$ |
|---|---|---|---|---|---|---|---|---|
| 1 | 0 | 3.45 | 470 | 0 | 27 | 48 | 61 | 119 |
| 2 | 1 | 2.71 | 490 | 1 | 10 | 60 | 42 | 68 |
| 1 | 0 | 3.50 | 510 | 0 | 24 | 47 | 78 | 119 |
| 1 | 1 | 2.81 | 480 | 0 | 13 | 24 | 44 | 100 |
| 3 | 1 | 3.10 | 600 | 0 | 26 | 48 | 60 | 79 |
| 2 | 0 | 3.69 | 610 | 1 | 28 | 57 | 49 | 89 |
| 2 | 1 | 3.17 | 620 | 0 | 14 | 52 | 61 | 98 |
| 3 | 1 | 3.57 | 560 | 0 | 20 | 42 | 77 | 107 |
| 3 | 1 | 2.76 | 710 | 1 | 28 | 67 | 83 | 158 |
| 2 | 0 | 3.81 | 460 | 0 | 10 | 48 | 67 | 110 |
| 2 | 1 | 3.60 | 580 | 0 | 28 | 69 | 64 | 116 |
| 3 | 0 | 3.10 | 500 | 1 | 13 | 21 | 40 | 69 |
| 2 | 1 | 3.18 | 410 | 0 | 24 | 52 | 73 | 107 |
| 2 | 1 | 3.50 | 440 | 1 | 25 | 45 | 47 | 125 |
| 1 | 1 | 3.43 | 210 | 1 | 26 | 35 | 57 | 54 |
| 2 | 0 | 3.49 | 610 | 0 | 16 | 59 | 58 | 100 |
| 2 | 0 | 3.76 | 510 | 0 | 25 | 68 | 56 | 138 |
| 3 | 1 | 3.71 | 500 | 0 | 33 | 36 | 58 | 68 |
| 1 | 1 | 4.00 | 470 | 1 | 5 | 45 | 24 | 82 |
| 3 | 1 | 3.69 | 700 | 0 | 21 | 54 | 100 | 132 |
| 1 | 0 | 3.14 | 610 | 0 | 13 | 55 | 83 | 87 |
| 2 | 0 | 3.46 | 480 | 0 | 9 | 31 | 70 | 79 |
| 1 | 0 | 3.29 | 470 | 1 | 13 | 39 | 48 | 89 |
| 2 | 1 | 3.80 | 670 | 1 | 32 | 67 | 35 | 119 |
| 2 | 0 | 2.76 | 580 | 0 | 10 | 45 | 14 | 100 |

$x_1$ = socioeconomic status (SES) (1 = high, 2 = middle, 3 = low)
$x_2$ = sex (1 = male, 2 = female)
$x_3$ = gradepoint average (GPA)
$x_4$ = Scholastic Aptitude Test (SAT)
$x_5$ = previous statistics class (0 = no, 1 = yes)
$x_6$ = pretest score
$y_1$ = Exam 1 score
$y_2$ = Exam 2 score
$y_3$ = Exam 3 score

(a) Perform a canonical correlation analysis on two sets of variables, defined as $x$ = predictors and $y$ = statistics course measures. (b) Test the canonical variates for statistical significance.

5. You have just collected data on a class of 25 students who took a class that required extensive computer work. Before the class, each student completed a questionnaire that included: (1) major (collapsed into four categories—science [S], fine arts [F], behavioral science [B], and manage-

ment/business [M]), (2) gender, (3) computer experience (number of computer courses taken and whether or not they have a home computer), (4) a computer anxiety score (53–265 with higher scores indicating more anxiety), (5) a computer attitudes score (26–130 with higher scores indicating more positive attitudes), and (6) a math anxiety score (90–450 with higher scores indicating more math anxiety). You also have collected some information concerning their course performance in this class (A, B, C, D, F).

| Student | Major | Gender | Number of Courses | Home | Anxiety | Attitude | Math | Grade |
|---------|-------|--------|-------------------|------|---------|----------|------|-------|
| 1 | S | M | 5 | Y | 75 | 120 | 101 | A |
| 2 | S | M | 4 | Y | 65 | 109 | 99 | A |
| 3 | S | M | 3 | Y | 72 | 99 | 110 | A |
| 4 | S | M | 7 | Y | 89 | 97 | 109 | A |
| 5 | F | F | 0 | N | 109 | 52 | 207 | C |
| 6 | F | F | 1 | N | 125 | 68 | 189 | C |
| 7 | F | M | 0 | N | 155 | 56 | 222 | B |
| 8 | F | F | 1 | Y | 111 | 88 | 117 | B |
| 9 | F | F | 0 | N | 122 | 70 | 127 | C |
| 10 | F | F | 0 | N | 133 | 71 | 133 | B |
| 11 | B | F | 1 | N | 129 | 101 | 111 | B |
| 12 | B | F | 2 | N | 88 | 98 | 99 | A |
| 13 | B | M | 3 | Y | 75 | 100 | 93 | A |
| 14 | B | F | 0 | N | 135 | 74 | 189 | C |
| 15 | B | F | 0 | N | 146 | 62 | 222 | D |
| 16 | B | F | 0 | N | 135 | 80 | 200 | D |
| 17 | B | M | 1 | N | 176 | 82 | 188 | C |
| 18 | B | M | 0 | N | 142 | 69 | 176 | D |
| 19 | M | M | 5 | Y | 53 | 125 | 101 | B |
| 20 | M | M | 7 | Y | 55 | 120 | 100 | A |
| 21 | M | F | 8 | Y | 60 | 119 | 99 | A |
| 22 | M | M | 4 | Y | 72 | 121 | 109 | A |
| 23 | M | F | 5 | Y | 75 | 108 | 121 | B |
| 24 | M | M | 3 | N | 101 | 89 | 135 | B |
| 25 | M | M | 6 | Y | 80 | 109 | 98 | A |

Conduct a canonical correlation using demographics as one set of variables and test results as the other.

# Principal Components
# and Factor Analysis

Principal component analysis (PCA) and factor analysis (FA) are multivariate statistical techniques that can be used to reduce a large number of interrelated variables to a smaller set of variables. For example, a survey questionnaire with 100 items may be completed by several thousand respondents. Such an extensive survey questionnaire will usually contain a lot of information that cannot be easily understood without some sort of summarization. Although simple descriptive statistics can help one to examine the frequency distribution of responses or correlations on particular items or pairs of items, visualizing relationships among items is generally limited to two or three dimensions. As such, it seems almost natural to try to reduce the dimensionality of the data for descriptive purposes but still preserve as much of the underlying structure of the relationships as possible.

There is no doubt that some items on survey questionnaires often ask questions about the same issues. For example, one's responses to questions concerning the revamping of the welfare system in the United States and the change in the taxation laws for high-income earners can both be considered indicators of a particular political party or position. The concern, therefore, is how to reduce this number of interrelated variables to a much smaller (and more manageable) number of variables without losing the information in the original responses. PCA and FA are two related multivariate techniques that can be used to accomplish this reduction.

The basic approach to this reduction first involves gathering observations from a sample of $n$ people on some set of $p$ variables of interest. This provides a simple data matrix containing observations on $p$ variables from the $n$ sample. In matrix form, this collection of information can be represented by a matrix $X$ as follows:

$$
X = \text{Persons} \quad
\begin{array}{c}
 \\
1 \\
2 \\
\cdot \\
\cdot \\
\cdot \\
n
\end{array}
\begin{array}{c}
\overset{\displaystyle \text{Variables}}{\overset{\displaystyle 1 \quad 2 \ \ldots \ p}{\left[
\begin{array}{ccccc}
x_{11} & x_{12} & \cdot & \cdot & x_{1p} \\
x_{21} & \cdot & \cdot & \cdot & \cdot \\
\cdot & \cdot & \cdot & \cdot & \cdot \\
\cdot & \cdot & \cdot & \cdot & \cdot \\
\cdot & \cdot & \cdot & \cdot & \cdot \\
x_{n1} & \cdot & \cdot & \cdot & x_{np}
\end{array}
\right]}}
\end{array}
$$

The concern, therefore, is whether the matrix $X$ can be replaced by another matrix with far less columns than $p$, and yet represent the information in $X$ as closely as possible. For example, if the variables in the matrix $X$ were scores obtained on a series of different tests, then $X$ could be summarized by simply adding the rows (i.e., this would provide a total score for each person). This summed or total score can be considered a one-dimensional representation of the original collection of $p$ variables. This is the approach that PCA uses, except that in most observed data matrices the results will not be simple one-dimensional representations but multidimensional ones.

The factor analytic approach was originally developed by psychologists as a paradigm that was meant to represent hypothetically existing entities (also referred to as constructs or latent variables). Although no direct measuring instrument existed for these entities, psychologists still wanted somehow to handle the entities as if they were measurable. For example, business climate is commonly discussed as if it is a real observable variable. Nevertheless, it is a construct in organizational theory that is generally regarded as a way of describing the perceptions of workers about a variety of conditions concerning the work environment. The same is true for organizational values, socioeconomic status, quality of life, and numerous others.

The idea behind the factor analytic paradigm goes back to Karl Pearson (1901), but it was developed as a multivariate technique by Hotelling (1933). It is also generally acknowledged that Charles Spearman first applied this method to study the structure of human abilities. Spearman formulated the idea that an individual's ability scores were manifestations of some underlying general ability called general intelligence or $g$. However, scores on different items that were not perfectly correlated were explained by other specific abilities or factors (such as verbal, mechanical, or numerical). The general and specific factors combined to produce the actual performance. This was labeled the two-factor theory of human abilities. However, as more researchers got involved in this line of study, they found Spearman's theory too restrictive (e.g., Thurstone and his associates at the University of Chicago in the 1930s; Thurstone, 1935, 1947) and established what has since come to be known as the common-factor theory.

In simple terms, factor analysis (FA) is a technique for expressing in the language of mathematics hypothetical variables (factors) by using a variety of common indicators that can be measured. The analysis is considered exploratory when the concern is with determining how many factors (or latent variables) are necessary to explain the relationships among the indicators. The analysis is considered confirmatory when a preexisting theory (definition of interrelationships among variables) directs the search (the confirmatory factor analytic model is discussed in the next chapter).

## PCA VERSUS FA

Principal component analysis (PCA) and factor analysis (FA) are generally seen as competing strategies. In fact, there is a good deal of disagreement with regard to which is the appropriate method to use. Mathematical statisticians prefer to use the PCA method. Applied researchers prefer FA. We believe that PCA and FA constitute distinct models with different goals. As such, users should first examine the goals of the analysis in order to decide which method is most appropriate. PCA is a data reduction method. The goal is to arrive at a relatively small number of components that will extract most of the variance of a relatively large set of variables. FA is aimed at explaining common variance. Common variance is the variance shared by the observed variables.

The literature on PCA and FA is vast and usually mathematically complex. Some easier presentations can be found in Comrey (1973), Comrey and Lee (1992), Harman (1976), and Gorsuch (1983). The major treatment of FA within the statistical tradition is given by Lawley and Maxwell (1963, 1971) and Mulaik (1972), whose books remain unrivaled as reference sources. The information presented in this chapter is intended to provide the reader with the necessary background to understand and conduct PCA and FA analyses. However, it is by no means complete!

## PRINCIPAL COMPONENT ANALYSIS

Principal component analysis (PCA) is a method used to reduce a set of observed variables into a relatively small number of components that account for most of the observed variance. This is accomplished by mathematical linear transformations of the observed variables under two conditions. The first condition is that the first component (the principal component) accounts for the maximum amount of variance possible, the second the next, and so on and so forth. The second condition is that all components are uncorrelated with each other.

PCA is similar to other multivariate techniques such as discriminant analysis (DA; see chapter 5) and canonical correlation (CANCORR; see chapter

6). The techniques all involve linear combinations of variables on the basis of maximizing some statistical property. DA focuses on differences between groups and determines the weights for the linear combination based on maximizing group differences. CANCORR focuses on relationships between sets of variables and derives linear combinations of sets that have maximum correlation. PCA maximizes the variance in the observed variables.

## AN EXAMPLE OF PCA

There is no doubt that the easiest way to illustrate PCA is by way of a simple bivariate case. Assume that two variables $X_1$ and $X_2$ are observed from a large sample, and that, for ease of presentation, their distribution is bivariate normal. Figure 7.1 presents a bivariate normal relationship with a positive correlation. Recall from chapter 2 that it is customary to represent a bivariate normal distribution by drawing cross sections of the surface of a bell-shaped mound, called isodensity contours (i.e., taking cross sections of the density surface at various elevations generates a family of concentric ellipses).

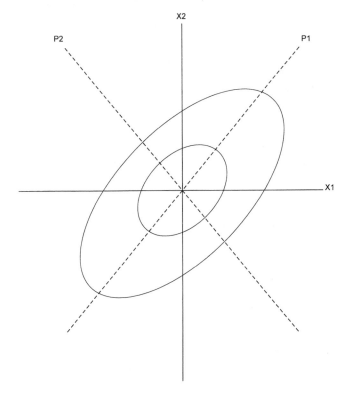

FIG. 7.1. Bivariate normal distribution with a positive correlation.

The contours presented in Fig. 7.1 show that higher values of $X_1$ tend to be associated with higher values of $X_2$ (and vice versa). This is the reason most observed cases are piled up along the first and third quadrants. As it turns out, the axes of the observed contours can be graphed as a principal axis (P1) representing the line on which most of the data points are located. Of course, there is a second axis (P2), which can be used to represent the data with the fewest observed cases.

In general terms, PCA can be conceptualized as a technique that attempts to represent the relative position of each observed case on the best dimension or axis. Obviously, for this example case, the best choice is the first axis (P1). In the case where the two variables $X_1$ and $X_2$ are perfectly correlated (see Fig. 7.2), the first principal axis contains all the information necessary to describe each case. Similarly, if the two variables are uncorrelated (see Fig. 7.3), there is no principal axis that can be used to describe the observed cases. Thus, the principal component analysis is no more than a representation of cases along some principal axes.

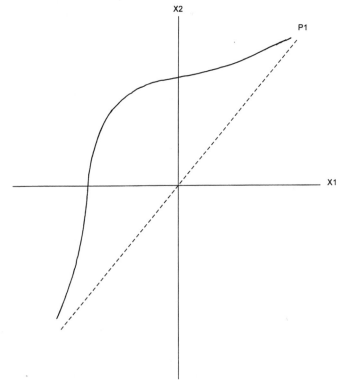

FIG. 7.2. Bivariate normal distribution with a perfect correlation between $X_1$ and $X_2$.

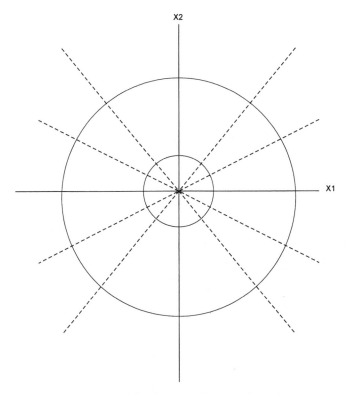

FIG. 7.3. Bivariate normal distribution with no correlation between $X_1$ and $X_2$.

It is important to note that the concept of PCA is not confined to relationships that are normal. The example case was used for ease of presentation. In addition, when there are more than two variables, although the basis for defining the principal components is similar, the mathematics are relatively cumbersome.

## THE PRINCIPAL COMPONENTS

Principal components are mathematical functions of observed variables. The PCA functions do not attempt to explain the intercorrelations among the variables, but to account for the variability observed in the data. Consider an example where eight variables of interest are observed from a sample (i.e., $x_1, x_2, \ldots, x_8$), and found to have a correlation matrix $R$. Using these eight observed variables, a PCA will attempt to construct linear combinations as follows:

$$Y_1 = \Gamma_1'X = \gamma_{11}x_1 + \gamma_{21}x_2 + \cdots + \gamma_{81}x_8$$
$$Y_2 = \Gamma_2'X = \gamma_{12}x_1 + \gamma_{22}x_2 + \cdots + \gamma_{82}x_8$$
$$Y_3 = \Gamma_3'X = \gamma_{13}x_1 + \gamma_{23}x_2 + \cdots + \gamma_{83}x_8$$
$$\cdot$$
$$\cdot$$
$$\cdot$$
$$Y_8 = \Gamma_8'X = \gamma_{18}x_1 + \gamma_{28}x_2 + \cdots + \gamma_{88}x_8$$

or in matrix notation

$$Y = \Gamma'X$$

The $Y$s in these equations are the computed principal components, and the $\gamma$s are the coefficients of the observed variables. Thus, the principal components are the uncorrelated linear combinations of $Y$s whose variances are as large as possible. The first principal component is the linear combination with maximum variance. This basically is the component that maximizes the variance of $Y_1$, written as VAR($Y_1$). However, because it is clear that VAR($Y_1$) can be increased by choosing any $\Gamma_1$ (i.e., any set of values for $\gamma_{11}$, $\gamma_{21}$, $\gamma_{31}$, $\gamma_{41}$, $\gamma_{51}$, $\gamma_{61}$, $\gamma_{71}$, and $\gamma_{81}$), a restriction that the product be equal to unity (i.e., $\Gamma'\Gamma = I$) is also imposed. Thus, the first principal component is that selection of $\gamma$s (i.e., $\Gamma_1'X$) that maximizes VAR($Y_1$) subject to $\Gamma_1'\Gamma_1 = 1$. The second principal component is the linear combination $\Gamma_2'X$ that maximizes VAR($Y_2$) subject to $\Gamma_2'\Gamma_2 = 1$, and assures that the two combinations are uncorrelated [i.e., COV($\Gamma_1'X,\Gamma_2'X$) = 0]. The analysis continues until all principal components are generated (e.g., in this case there will be eight components).

## PRINCIPAL COMPONENT LOADINGS

The actual coefficients of the observed variables (i.e., the $\gamma$s), commonly referred to as principal component loadings, are determined by multiplying the eigenvectors of the observed correlation matrix by the square roots of the respective eigenvalues. The results for the component loadings are easily determined by expanding on the following relationship between any correlation matrix and its associated eigenvalues and eigenvectors:

$$R = VLV'$$

where $R$ is the correlation matrix, $V$ the matrix of eigenvectors, and $L$ a diagonal matrix with eigenvalues.

The correlation matrix can also be represented as:

$$R = V(\sqrt{L})(\sqrt{L})V'$$

where $\sqrt{L}$ represents the square root of the matrix with eigenvalues (note that $\sqrt{L}\sqrt{L} = L$). This matrix can then be written as the product of two matrices, each a combination of eigenvectors and square roots of eigenvalues. Thus, the correlation matrix can be written as:

$$R = (V\sqrt{L})(\sqrt{L}V') = PP'$$

where $P = V\sqrt{L}$ is the product of eigenvectors and square roots of the eigenvalues.

The mathematical technique for finding the matrices $V$ and $L$ is the eigenvalue equation. The eigenvalue equation was discussed extensively in chapter 2. Nevertheless, because the eigenvalues of a matrix play such an important role in PCA and FA, it is essential to review some of the material first presented in chapter 2.

Consider an example with two variables of interest ($X_1$ and $X_2$) observed from a large sample and found to have the following correlation matrix $A$:

$$A = \begin{bmatrix} 1 & .6 \\ .6 & 1 \end{bmatrix}$$

Recall from chapter 2 that by definition the eigenvalue of any matrix (e.g., $A$) is a number (e.g., $\lambda$) such that

$$Ax = \lambda x$$

and the value of $x$ satisfying this equation is the corresponding eigenvector. Rewriting this equation one can determine that

$$Ax - \lambda x = 0$$

and from this (by collecting similar terms) that

$$(A - \lambda I)x = 0$$

which is a system of linear equations each having a right-hand side equal to zero.

Expanding this system of linear equations using the correlation matrix $A$ yields:

$$\left( \begin{bmatrix} 1 & 0.6 \\ 0.6 & 1 \end{bmatrix} - \lambda \begin{bmatrix} 1 & 0 \\ 0 & 1 \end{bmatrix} \right) \begin{bmatrix} x_1 \\ x_2 \end{bmatrix} = 0$$

or

$$\left( \begin{bmatrix} 1 & 0.6 \\ 0.6 & 1 \end{bmatrix} - \begin{bmatrix} \lambda & 0 \\ 0 & \lambda \end{bmatrix} \right) \begin{bmatrix} x_1 \\ x_2 \end{bmatrix} = 0$$

This yields

$$\begin{bmatrix} 1 - \lambda & 0.6 \\ 0.6 & 1 - \lambda \end{bmatrix} \begin{bmatrix} x_1 \\ x_2 \end{bmatrix} = 0$$

It is obvious that a trivial solution exists when $\begin{bmatrix} x_1 \\ x_2 \end{bmatrix}$ is zero. A nontrivial solution, however, exists if and only if the determinant of the $\begin{bmatrix} 1 - \lambda & 0.6 \\ 0.6 & 1 - \lambda \end{bmatrix}$ matrix is zero. Thus, to obtain a nontrivial solution one must find the values of $\lambda$ for which the determinant of $\begin{bmatrix} 1 - \lambda & 0.6 \\ 0.6 & 1 - \lambda \end{bmatrix}$ is zero. Thus, the eigenvalues are found by solving the equation

$$\left| \begin{bmatrix} 1 - \lambda & 0.6 \\ 0.6 & 1 - \lambda \end{bmatrix} \right| = 0$$

where | | denotes the determinant of the matrix.

The determinant of the matrix can be written out as:

$$(1 - \lambda)(1 - \lambda) - (0.6)(0.6) = 0$$

or simply

$$(1 - \lambda)(1 - \lambda) - r^2 = 0$$

(because $r = .6$).

This equation can also be written as:

$$\lambda^2 - 2\lambda + (1 - r^2) = 0$$

and solving for $\lambda$, the eigenvalues are $\lambda_1 = 1 + r = 1.6$ and $\lambda_2 = 1 - r = 0.4$. It is important to note that if the correlation between the variables is perfect, one of the eigenvalues is 2, the other is 0. This implies that the first principal component accounts for all the variance (as shown in Fig. 7.2). If the correlation is zero, both eigenvalues are 1. This implies that there is no principal component (see Fig. 7.3).

The associated eigenvectors for the matrix $A$ are

$$x_1 = \begin{bmatrix} 0.707 \\ 0.707 \end{bmatrix}$$

which is the eigenvector associated with $\lambda_1 = 1.6$, and

$$x_2 = \begin{bmatrix} 0.707 \\ -0.707 \end{bmatrix}$$

which is the eigenvector associated with $\lambda_2 = 0.4$.

An examination of the first eigenvector reveals that it fulfills the unit-norm condition (i.e., $x_1'x_1 = 1$). Similarly, the two eigenvectors are orthogonal (i.e., $x_1'x_2 = 0$).

There is no doubt that the computation of eigenvalues and eigenvectors is best left to computers. Nevertheless, it is important to examine how these are used in a PCA analysis. In the case of PCA, it was shown that an observed correlation matrix can be written as the product of eigenvalues and eigenvectors. For example, the correlation matrix $A$ can be written as

$$A = (V\sqrt{L})(\sqrt{L}V') = PP'$$

or simply as

$$\begin{bmatrix} 1 & 0.6 \\ 0.6 & 1 \end{bmatrix} = \begin{bmatrix} 0.89 & 0.45 \\ 0.89 & -0.45 \end{bmatrix} \begin{bmatrix} 0.89 & 0.89 \\ 0.45 & -0.45 \end{bmatrix}$$

where

$$V = \begin{bmatrix} 0.707 & 0.707 \\ 0.707 & -0.707 \end{bmatrix} \quad \text{and} \quad L = \begin{bmatrix} 1.6 & 0 \\ 0 & 0.4 \end{bmatrix}$$

Thus, the principal component loadings determined by the product of the eigenvalues and eigenvectors are given by the matrix

$$P = \begin{bmatrix} 0.89 & 0.45 \\ 0.89 & -0.45 \end{bmatrix}$$

It is important to note that the sum of the eigenvalues of a correlation matrix is always equal to the number of variables (in this case, 2). In fact, this property holds for correlation matrices of any size. As such, because the first eigenvalue represents the amount of variance explained by the first principal axis, dividing the first eigenvalue by the number of variables ($p$) provides the proportion of variance explained by the first principal component or axis. For example, the proportion of variance explained by the first component is:

$$\frac{\lambda_1}{p} = \frac{1.6}{2} = 0.8 \times 100\% = 80\%$$

The proportion of variance explained by the second principal component is obtained by dividing the second eigenvalue by the number of variables (i.e., $0.4/2 = 0.2 \times 100\% = 20\%$). This method is used in a similar manner when more components are determined. It is important to note that if a covariance matrix is used, the proportion of variance explained by each component is found by dividing the eigenvalue by the sum of the eigenvalues [i.e., $\lambda_1/(\lambda_1 + \lambda_2)$].

## HOW MANY COMPONENTS?

Unfortunately, the question of how many axes or principal components are important or needed to account for the variance in the data is not straightforward. In other words, given the eigenvalues of a correlation or covariance matrix, how does one decide how many to use in order to account for the observed variance? This is a controversial question and there is really no "correct" answer. Several different methods have been proposed in the literature, but the following three methods appear to be the most popular.

The first method involves examining the proportion of variance accounted for by the principal components. In general, if the ratio is close to or exceeds 90%, the remaining components are considered of no practical value. The second method, proposed by Kaiser (1960), and perhaps the most widely used method, is known as the Kaiser eigenvalue criterion. This criterion involves retaining only those components whose eigenvalues of the correlation matrix are greater than one. A modification to this criterion was proposed by Jolliffe (1972). Jolliffe (1972) suggested that Kaiser's rule tends to throw away too much information, and on the basis of simulation studies suggested a cutoff point of 0.7 be used for determining the number of important components obtained from correlation matrices (for covariance matrices his rule is $0.7\bar{\lambda}$, where $\bar{\lambda}$ = average eigenvalue of the covariance matrix). Finally, another popular method is Cattell's scree test (1966). This method consists of plotting the eigenvalues of the observed correlation matrix against their ordinal number and noting the point at which the plot becomes nearly horizontal. This is considered the point from which all else is mere "scree" (see Fig. 7.3 for an example of a scree plot). The term *scree* is used in the geographical sense to refer to the rubble at the foot of a steep slope. Unfortunately, although Cattell's scree test is simple to obtain, it is rather subjective because it involves a decision of just how level is level.

It should be clear that the criteria concerning how many principal components are important should be applied with caution. In general, look for consensus! For example, when the three criteria are applied to the simple correlation matrix *A* all three indicate that one principal component is sufficient. In this case, there is no question how many components should be retained. However, this will not always be the case. The best strategy is to examine all three criteria. If the results are similar, choosing the criterion to use is just a matter of personal taste. If the results are different, look for some consensus among criteria. However, always keep in mind that the more principal components that are retained relative to the number of variables, the less parsimonious one's description of the data will be.

## LOOSE ENDS

There are also two other important issues that have caused a lot of confusion in the literature. The first issue relates to the rotation of principal components. In general, those that propose the use of rotation indicate that it should be used to find simpler and more easily interpretable principal components. The basic approach proposed is to examine the patterns of variables and then rotate the axis or define new axes in such a way that the components are more interpretable. For example, in the case of the eight variables observed, proponents of this method might use rotation (if there was a way) to make the first component strictly a function of the first four variables and the second component a function of the other variables. Nevertheless, the majority of mathematical statisticians indicate that there is really no such a thing as rotation of principal components. The principal component decomposition achieved by solving the eigenvalue equation is unique, in that each succeeding component has maximum variance. This desirable mathematical property precludes the need for rotation of principal components. However, as becomes clearer in the next section, the confusion concerning rotation exists because of the similarities between principal component analysis and factor analysis.

The second issue that has caused some confusion in the literature concerns the use of covariance or correlation matrices when conducting a PCA. In general, the solution of a PCA is not invariant with respect to changes in the scales of observed variables. "Not invariant" implies that the principal components of a covariance matrix have no direct relationship with those of a corresponding correlation matrix. In other words, the eigenvalues and eigenvectors of a covariance matrix are not the same as those of a correlation matrix. If the variables are all measured on the same metric and the variances do not differ widely from one to another, then it might be reasonable to conduct a PCA on the covariance matrix rather than the correlation matrix. However, in general, variables should be standardized if they are measured on scales that have widely differing ranges or if the measurement units are not similar. Therefore, the use of a correlation matrix is recommended when conducting a principal component analysis.

## AN EXAMPLE OF PCA USING SAS

Marcoulides and Drezner (1993) presented a reanalysis of Olympic Decathlon results (the study was originally conducted by Linden, 1977). Altogether the results from 139 different athletes were collected. An important concern of this study was to determine what combination of variables can account for the observed differences among athletes. A PCA was conducted because it can be used to construct a single variable that summarizes the athletes' performances.

The observed correlation matrix representing the interrelationships among the 10 Olympic events is presented in Table 7.1. This correlation matrix can be used as input to a SAS program. The SAS program can read either a correlation matrix or raw data. The SAS deck setup to perform a PCA on the observed correlation matrix is presented in Table 7.2.

The DATA and INPUT statements indicate that a correlation matrix is used as input. It is important to note that one does not have to provide the entire correlation matrix as input. Only the diagonal and lower or upper off-diagonal elements of the matrix must be provided. The other elements can be represented using a period.

To perform a PCA on the correlation matrix, two SAS procedures can be used. The PROC PRINCOMP or the PROC FACTOR procedures can be used. The main difference between the two procedures is that PROC FACTOR can be used to conduct either a principal component analysis or a factor analysis (see next section). PROC PRINCOMP, however, was specifically designed for conducting only component analysis. Nevertheless, both procedures produce identical results when used to conduct a principal component analysis. The usefulness of the PROC FACTOR is that it can be used to provide a scree plot. The scree plot is provided in the PROC FACTOR because factor analysis uses the same criterion for determining how many factors to retain (see next section).

The results of the analysis are presented in Table 7.3. As can be seen, the choice concerning how many components are important is not straightforward. An examination of the scree plot suggests that either three or four components should be used. Using the Kaiser eigenvalue criterion it appears that three components are important. However, when Jolliffe's rule is used four components are deemed important. Finally, when examining the proportion of variance accounted for by the principal components it would appear that at least four components should be used (accounting for 73% of the variance). It is important to note that PROC FACTOR uses Kaiser's rule as the default value for determining how many components are important (see Table 7.3). In this example, PROC FACTOR determined the component loadings (called factor pattern in the output) using three components (factors). To change the default value one can use the command NFACTOR

TABLE 7.1

Correlation Matrix for Decathlon Events

| | 100-m Run | Long jump | Shot-put | High jump | 400-m Run | 110-m Hurdle | Discus | Pole vault | Javelin | 1500-m Run |
|---|---|---|---|---|---|---|---|---|---|---|
| 100-m Run | 1.00 | | | | | | | | | |
| Long jump | 0.59 | 1.00 | | | | | | | | |
| Shot-put | 0.35 | 0.42 | 1.00 | | | | | | | |
| High jump | 0.34 | 0.51 | 0.38 | 1.00 | | | | | | |
| 400-m Run | 0.63 | 0.49 | 0.19 | 0.29 | 1.00 | | | | | |
| 110-m Hurdle | 0.40 | 0.52 | 0.36 | 0.46 | 0.34 | 1.00 | | | | |
| Discus | 0.28 | 0.31 | 0.73 | 0.27 | 0.17 | 0.32 | 1.00 | | | |
| Pole vault | 0.20 | 0.36 | 0.24 | 0.39 | 0.23 | 0.33 | 0.24 | 1.00 | | |
| Javelin | 0.11 | 0.21 | 0.44 | 0.17 | 0.13 | 0.18 | 0.34 | 0.24 | 1.00 | |
| 1500-m Run | -.07 | 0.09 | -0.08 | 0.18 | 0.39 | 0.00 | -0.02 | 0.17 | 0.00 | 1.00 |

TABLE 7.2
SAS Desk Setup for PCA

```
DATA CORREL(TYPE=CORR);
_TYPE_='CORR';
INPUT_NAME_ $ a b c d e f g h i j;

CARDS;
a    1.0 .........
b    0.59 1.0 ........
c    0.35 0.42 1.0 .......
d    0.34 0.51 0.38 1.0 ......
e    0.63 0.49 0.19 0.29 1.0 .....
f    0.40 0.52 0.36 0.46 0.34 1.0 ....
g    0.28 0.31 0.73 0.27 0.17 0.32 1.0 ...
h    0.20 0.36 0.24 0.39 0.23 0.33 0.24 1.0 ..
i    0.11 0.21 0.44 0.17 0.13 0.18 0.34 0.24 1.0 .
j    -.07 0.09 -.08 0.18 0.39 0.00 -.02 0.17 -.00 1.0
;
PROC FACTOR SCREE;
```

= N in the PROC FACTOR statement. Table 7.3 presents the results of the statement "PROC FACTOR NFACTOR=4;". When the NFACTOR command is used, PROC FACTOR determines the component loadings using the specified number of factors. As can be seen by examining the loadings in Table 7.3, the values for the first three components do not change when more components are used. To verify that the results are correct, one can always use the eigenvector values that are printed as standard output for PROC PRINCOMP.

The results of this study indicate that a component analysis can be used to summarize the differences between the performances of Decathlon athletes. In fact, the results can even be used to rank order the performances of the observed athletes. Unfortunately, there is an enormous amount of output that is generated by a principal component analysis. It is essential, therefore, that at a minimum one report the eigenvalues and eigenvectors, and the component loadings for the important number of components. This information will enable others to evaluate the results and, hopefully, arrive at similar results.

## FACTOR ANALYSIS

Factor analysis (FA) is somewhat similar to principal component analysis (PCA) in that both methods can be used for data reduction. There are, however, some distinct differences between the two methods, especially with regard to the goals of analysis. The simplest way to differentiate the

## TABLE 7.3
### SAS Initial Factor Method: Principal Components

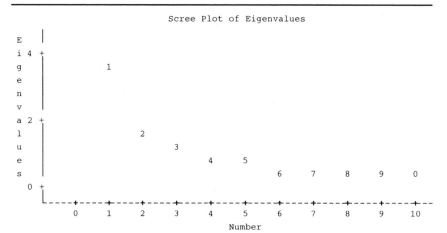

Scree Plot of Eigenvalues

```
E   |
i 4 +
g   |      1
e   |
n   |
v   |
a 2 +
l   |         2
u   |             3
e   |                4    5
s   |                          6    7    8    9    0
  0 +
    L - - -+- - - -+- - - -+- - - -+- - - -+- - - -+- - - -+- - - -+- - - -+- - - -+- -
        0    1    2    3    4    5    6    7    8    9    10
                            Number
```

Prior Communality Estimates: ONE

Eigenvalues of the Correlation Matrix:    Total = 10    Average = 1

|  | 1 | 2 | 3 | 4 | 5 |
|---|---|---|---|---|---|
| Eigenvalue | 3.786608 | 1.517302 | 1.114410 | 0.913368 | 0.720109 |
| Difference | 2.269306 | 0.402893 | 0.201041 | 0.193259 | 0.125127 |
| Proportion | 0.3787 | 0.1517 | 0.1114 | 0.0913 | 0.0720 |
| Cumulative | 0.3787 | 0.5304 | 0.6418 | 0.7332 | 0.8052 |

|  | 6 | 7 | 8 | 9 | 10 |
|---|---|---|---|---|---|
| Eigenvalue | 0.594982 | 0.526717 | 0.383705 | 0.235312 | 0.207487 |
| Difference | 0.068265 | 0.143013 | 0.148392 | 0.027825 |  |
| Proportion | 0.0595 | 0.0527 | 0.0384 | 0.0235 | 0.0207 |
| Cumulative | 0.8647 | 0.9173 | 0.9557 | 0.9793 | 1.0000 |

3 factors will be retained by the MINEIGEN criterion.

Factor Pattern

|  | FACTOR1 | FACTOR2 | FACTOR3 |
|---|---|---|---|
| A | 0.69052 | 0.21701 | -0.52025 |
| B | 0.78854 | 0.18360 | -0.19260 |
| C | 0.70187 | -0.53462 | 0.04699 |
| D | 0.67366 | 0.13401 | 0.13875 |
| E | 0.61965 | 0.55112 | -0.08376 |
| F | 0.68689 | 0.04206 | -0.16102 |
| G | 0.62121 | -0.52112 | 0.10946 |
| H | 0.53848 | 0.08698 | 0.41090 |
| I | 0.43405 | -0.43903 | 0.37191 |
| J | 0.14660 | 0.59611 | 0.65812 |

*(Continued)*

TABLE 7.3
(*Continued*)

Prior Communality Estimates: ONE

Eigenvalues of the Correlation Matrix:    Total = 10    Average = 1

|            | 1 | 2 | 3 | 4 | 5 |
|------------|----------|----------|----------|----------|----------|
| Eigenvalue | 3.786608 | 1.517302 | 1.114410 | 0.913368 | 0.720109 |
| Difference | 2.269306 | 0.402893 | 0.201041 | 0.193259 | 0.125127 |
| Proportion | 0.3787 | 0.1517 | 0.1114 | 0.0913 | 0.0720 |
| Cumulative | 0.3787 | 0.5304 | 0.6418 | 0.7332 | 0.8052 |

|            | 6 | 7 | 8 | 9 | 10 |
|------------|----------|----------|----------|----------|----------|
| Eigenvalue | 0.594982 | 0.526717 | 0.383705 | 0.235312 | 0.207487 |
| Difference | 0.068265 | 0.143013 | 0.148392 | 0.027825 | |
| Proportion | 0.0595 | 0.0527 | 0.0384 | 0.0235 | 0.0207 |
| Cumulative | 0.8647 | 0.9173 | 0.9557 | 0.9793 | 1.0000 |

4 factors will be retained by the NFACTOR criterion.

Factor Pattern

|   | FACTOR1 | FACTOR2 | FACTOR3 | FACTOR4 |
|---|---------|---------|---------|---------|
| A | 0.69052 | 0.21701 | -0.52025 | 0.20603 |
| B | 0.78854 | 0.18360 | -0.19260 | -0.09249 |
| C | 0.70187 | -0.53462 | 0.04699 | 0.17534 |
| D | 0.67366 | 0.13401 | 0.13875 | -0.39590 |
| E | 0.61965 | 0.55112 | -0.08376 | 0.41873 |
| F | 0.68689 | 0.04206 | -0.16102 | -0.34462 |
| G | 0.62121 | -0.52112 | 0.10946 | 0.23437 |
| H | 0.53848 | 0.08698 | 0.41090 | -0.43955 |
| I | 0.43405 | -0.43903 | 0.37191 | 0.23451 |
| J | 0.14660 | 0.59611 | 0.65812 | 0.27866 |

Eigenvalues of the Correlation Matrix

|        | Eigenvalue | Difference | Proportion | Cumulative |
|--------|------------|------------|------------|------------|
| PRIN1  | 3.78661 | 2.26931 | 0.378661 | 0.37866 |
| PRIN2  | 1.51730 | 0.40289 | 0.151730 | 0.53039 |
| PRIN3  | 1.11441 | 0.20104 | 0.111441 | 0.64183 |
| PRIN4  | 0.91337 | 0.19326 | 0.091337 | 0.73317 |
| PRIN5  | 0.72011 | 0.12513 | 0.072011 | 0.80518 |
| PRIN6  | 0.59498 | 0.06826 | 0.059498 | 0.86468 |
| PRIN7  | 0.52672 | 0.14301 | 0.052672 | 0.91735 |
| PRIN8  | 0.38370 | 0.14839 | 0.038370 | 0.95572 |
| PRIN9  | 0.23531 | 0.02783 | 0.023531 | 0.97925 |
| PRIN10 | 0.20749 | | 0.020749 | 1.00000 |

(*Continued*)

TABLE 7.3
*(Continued)*

| | PRIN1 | PRIN2 | PRIN3 | PRIN4 | PRIN5 |
|---|---|---|---|---|---|
| | | | Eigenvectors | | |
| A | 0.354853 | 0.176177 | -.492820 | 0.215584 | 0.158964 |
| B | 0.405226 | 0.149055 | -.182447 | -.096776 | 0.065344 |
| C | 0.360690 | -.434018 | 0.044512 | 0.183466 | -.258331 |
| D | 0.346191 | 0.108792 | 0.131436 | -.414247 | -.273814 |
| E | 0.318437 | 0.447411 | -.079347 | 0.438141 | 0.129547 |
| F | 0.352992 | 0.034147 | -.152531 | -.360595 | -.054194 |
| G | 0.319239 | -.423060 | 0.103692 | 0.245233 | -.419702 |
| H | 0.276723 | 0.070611 | 0.389234 | -.459928 | 0.322442 |
| I | 0.223059 | -.356418 | 0.352298 | 0.245380 | 0.668947 |
| J | 0.075338 | 0.483937 | 0.623421 | 0.291577 | -.285347 |

| | PRIN6 | PRIN7 | PRIN8 | PRIN9 | PRIN10 |
|---|---|---|---|---|---|
| A | -.191316 | -.183732 | 0.238553 | 0.075623 | 0.628838 |
| B | 0.095522 | -.181075 | -.832867 | 0.128975 | -.120155 |
| C | -.034246 | -.108560 | -.053435 | -.750186 | 0.023698 |
| D | 0.380931 | -.532149 | 0.381942 | 0.133159 | -.112089 |
| E | -.073981 | 0.108404 | 0.254678 | -.098130 | -.625828 |
| F | 0.321300 | 0.769343 | 0.094699 | -.058116 | 0.102623 |
| G | -.293116 | 0.153105 | 0.033253 | 0.595014 | -.086946 |
| H | -.664657 | 0.019513 | 0.047121 | -.084030 | 0.000989 |
| I | 0.400614 | -.027140 | 0.055936 | 0.147932 | 0.071795 |
| J | 0.105392 | 0.101430 | -.142850 | -.033794 | 0.402606 |

two is to remember that PCA summarizes the data by means of a linear combination of observed variables whereas FA attempts to explain common variance. The PCA method was presented in the previous section for two reasons: (a) it simplifies the presentation of factor analysis, and (b) the statistics generated by a PCA still serve as the most widely used means for solving some important issues in FA. This is why PCA is often considered a first-stage FA. Unfortunately, it is also the reason that the two techniques are so often confused as being the same. Nevertheless, the two techniques are clearly quite different.

Factor analysis is based on the assumption that some underlying constructs (factors) exist that are responsible for the interrelationships among observed variables. In general, the underlying factors are smaller in number than the number of observed variables. For example, consider a simple case where one underlying factor is responsible for the interrelationships between three observed variables. Such a simple case is depicted as a path diagram in Fig. 7.4. The path diagram in Fig. 7.4 implies that $F_1$ is common (called a common factor) to the observed variables $X_1$, $X_2$, and $X_3$. If the observed variables $X_1$, $X_2$, and $X_3$ could be observed without error, then the common factor $F_1$

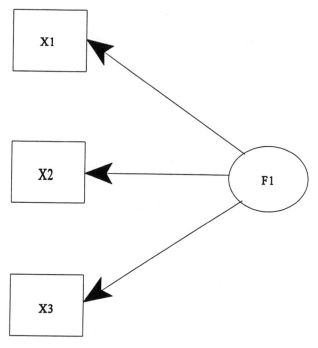

FIG. 7.4. Path model representing a one-factor model with three observed variables (without measurement error).

would perfectly account for the interrelationships among the variables. The problem is that $X_1$, $X_2$, and $X_3$ will usually not be measured without error. As such, the path diagram presented in Fig. 7.5 depicts a more realistic situation and reflects the fact that there is some error that is unique to each observed variable. For convenience, this error in the variables can also be referred to as a unique factor.

The path diagram presented in Fig. 7.5 can also be represented using the following equations:

$$X_1 = \lambda_1 F_1 + E_1$$
$$X_2 = \lambda_2 F_1 + E_2$$
$$X_3 = \lambda_3 F_1 + E_3$$

where each $\lambda$ would represent the factor loading for each observed variable. Figure 7.6 presents the same model based on the notation used in the preceding equations.

Alternatively, one may also write the above equations using matrix notation as

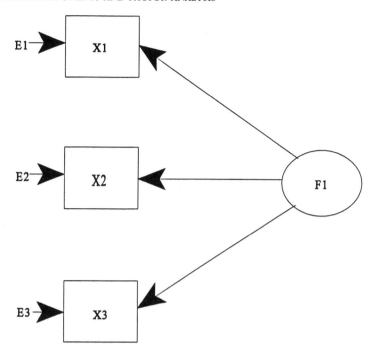

FIG. 7.5. Path model representing a one-factor model with three observed variables (with measurement error).

$$X = \lambda f + e$$

which is the same as writing

$$\begin{bmatrix} X_1 \\ X_2 \\ X_3 \end{bmatrix} = \begin{bmatrix} \lambda_1 \\ \lambda_2 \\ \lambda_3 \end{bmatrix} \begin{bmatrix} F_1 \end{bmatrix} + \begin{bmatrix} E_1 \\ E_2 \\ E_3 \end{bmatrix}$$

The assumptions of this model (based on the path diagram presented in Fig. 7.6), which are embedded in these equations, imply that there is no correlation between the factor $F$ and the errors of measurement (i.e., between the $F$ and the $E$s) or between the errors of measurement (i.e., between each of the $E$s).

With these assumptions in mind, it can be shown that the variance–covariance matrix of the observed variables (i.e., $X_1$, $X_2$, and $X_3$), represented by

$$\Sigma = XX' = \begin{bmatrix} \sigma^2_{X_1} & & \\ \sigma_{X_2 X_1} & \sigma^2_{X_2} & \\ \sigma_{X_3 X_1} & \sigma_{X_3 X_2} & \sigma^2_{X_3} \end{bmatrix}$$

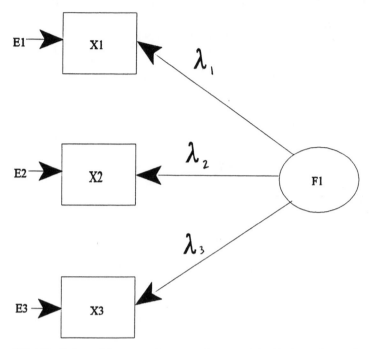

FIG. 7.6. Path model representing a one-factor model with three observed variables and factor loadings.

can also be expressed as

$$\Sigma = XX' = (\lambda f + e)(\lambda f + e)' = \Lambda \Phi \Lambda' + \Theta$$

where $\Phi$ represents the correlation among common factors and $\Theta$ refers to the errors or uniqueness.

With uncorrelated common factors (i.e., $\Phi = I$), this expression reduces to

$$\Sigma = \Lambda \Lambda' + \Theta$$

As was the case in the discussion of PCA, the observed variables are usually standardized before a factor analysis is conducted. When the observed variables are standardized, this decomposition of the variance–covariance matrix $\Sigma$ can be written as a decomposition of a correlation matrix $R$ as:

$$R = ZZ' = \Lambda \Phi \Lambda' + \Theta$$

and with uncorrelated common factors is equivalent to:

$$R = \Lambda\Lambda' + \Theta$$

with

$$R = \begin{bmatrix} 1 & & \\ r_{21} & 1 & \\ r_{31} & r_{32} & 1 \end{bmatrix}$$

Thus, the process of conducting a factor analysis begins with generating the correlation matrix between the observed variables. The correlation matrix for the observed variables is then decomposed into two other matrices: a matrix of factor loadings ($\Lambda$), and a matrix of errors ($\Theta$). Obviously, if the observed variables could be measured without error (i.e., the matrix of errors $\Theta$ were equal to zero), then the correlation matrix could be perfectly reproduced using

$$R = \Lambda\Lambda'$$

Although it is possible to carry out a factor analysis that can perfectly reproduce the observed correlation matrix $R$, such a reproduction usually requires that the $\Lambda$ matrix be the same size as $R$. This implies that in order to perfectly reproduce the correlation matrix $R$, there must be as many common factors as there are observed variables. In such a case, all the factors are common to all the variables and there is no error. Nevertheless, because we expect the observed variables to be measured with some error, in order to reproduce the correlation matrix we must take into account the error matrix $\Theta$. It is important to note that that since there is no correlation between the errors of measurement (i.e., between each of the $Es$), by definition the matrix $\Theta$ is a diagonal matrix containing values that when added to the diagonal values of the $\Lambda\Lambda'$ product matrix will always sum to one (i.e., the diagonal values of the correlation matrix $R$).

Of course, one of the major goals of factor analysis is to account for the interrelationships between the variables using fewer factors than observed variables without losing information in the observed variables. In simple terms, one may consider this goal as an attempt to reproduce the correlation matrix $R$ with a $\Lambda$ matrix that has as few factors (i.e., columns) as possible. In many cases, it would be ideal if one could reproduce the correlation matrix $R$ using only a single factor. However, this will not usually be the case. As such, the problem then becomes one of estimating the $\Lambda$ and $\Theta$ matrices that can best reproduce the observed correlation matrix. The elements of these two matrices are parameters of the factor analysis model that must be estimated from the observed data.

## METHODS OF ESTIMATION

There are many methods that have been proposed in the literature for determining the estimates of the $\Lambda$ and $\Theta$ matrices. To some extent, all of the proposed methods have the same goal in mind but are just based on different algorithms. For example, alpha factoring, least-squares factoring, principal factoring, maximum likelihood, and image analysis have all been discussed extensively in the literature as methods for extracting factor solutions (e.g., see Comrey & Lee, 1992). Obviously, it would be impossible in a one-semester multivariate text to present the mathematical computations for every factor-analytic method that has been proposed in the literature (some excellent books devoted entirely to the topic of factor analysis are available for the interested reader; see Comrey & Lee, 1992; Harman, 1976; Mulaik, 1972). Nevertheless, we believe it is essential that the mathematics of at least one of these factor extraction methods be presented in detail. Because most of the early work for obtaining factor analytic solutions is based on the principal factor method, this method is discussed extensively in this section.

The principal factor method (PFM) is very similar to the decomposition procedures of the principal component analysis (PCA) method with the exception that instead of using a correlation matrix, an "adjusted correlation matrix" whose diagonal elements have been replaced by what are called communality estimates is used. In general terms, communality is considered the variance of an observed variable that is accounted for by the common factors. In other words, communality is much like the proportion of variability in an observed variable that is accounted for by its factor. As such, when conducting a PFM analysis, the diagonal elements of the observed correlation matrix have to be replaced by communalities. Mathematically, the communality of a variable is sometimes written as

$$h^2 = 1 - \text{uniqueness}$$

where uniqueness $= \Sigma \lambda_{ij}^2$ (i.e., the sum of the squared factor loadings on each common factor). And yet, the definition of communality given earlier implies that these values can only be determined after the correlation matrix has been subjected to a factor analysis in which the number of common factors has already been decided on. As it turns out, a way around this problem is to use estimates of the communalities in the correlation matrix before subjecting it to a factor analysis. To date, the most widely used estimates of communalities are the squared multiple correlations (SMCs) of each variable with the remaining variables. Although there is still some disagreement in the literature concerning which are the best communality estimates to use, most currently available factor analytic computer programs perform iterations that improve on the initial generated estimates (whether they are SMCs or not). As such, the issue of communality estimates no longer dominates the factor-analytic literature (for a discussion on communalities and statistical packages to perform FA, see Marcoulides, 1989c). Once the communality estimates have been obtained, they are inserted in the main diagonal of the correlation matrix (thus, the name *adjusted correlation matrix*), and factors are extracted in the same way that a solution was obtained in principal components analysis.

Recall from the previous section that a principal component solution can be obtained by expanding on the following relationship between any correlation matrix and its associated eigenvalues and eigenvectors,

$$R = VLV'$$

where $R$ is the correlation matrix, $V$ the matrix of eigenvectors, and $L$ a diagonal matrix with eigenvalues. It was also shown that the correlation matrix can be represented as:

$$R = V(\sqrt{L})(\sqrt{L})V' = PP'$$

where the correlation matrix is just a product of the eigenvectors and square root of the eigenvalues. Alternatively, this correlation matrix can also be written as

$$R = \Lambda\Lambda'$$

Thus, the same relationship between the eigenvalues and eigenvectors of a correlation matrix can be used to obtain a PFM factor solution (except instead of using $PP'$, as was done in the PCA section, we use $\Lambda\Lambda'$). The only difference is that now the correlation matrix $R$ has been replaced by an "adjusted correlation matrix" and we already know that the actual number of eigenvalues in the $L$ matrix will be less than the number of observed variables (and, consequently, the number of eigenvectors will be less).

Consider an example with six observed variables of interest with the following observed correlation matrix:

$$R = \begin{bmatrix} 1 & & & & & \\ .25 & 1 & & & & \\ .25 & .25 & 1 & & & \\ 0 & 0 & 0 & 1 & & \\ 0 & 0 & 0 & .25 & 1 & \\ 0 & 0 & 0 & .25 & .25 & 1 \end{bmatrix}$$

For purposes of discussion, let us assume that the communality estimates, based on SMCs, were all determined to be equal to .25 and were inserted into the preceding correlation matrix to produce the following "adjusted correlation matrix":

$$
\text{adjusted } R =
\begin{bmatrix}
.25 & & & & & \\
.25 & .25 & & & & \\
.25 & .25 & .25 & & & \\
0 & 0 & 0 & .25 & & \\
0 & 0 & 0 & .25 & .25 & \\
0 & 0 & 0 & .25 & .25 & .25
\end{bmatrix}
$$

Solving for the eigenvalues of this "adjusted correlation matrix" $R$ provides the following eigenvalues in decreasing order: $\lambda_1 = 0.75$, $\lambda_2 = 0.75$, $\lambda_3 = 2.42 \times 10^{-18}$, $\lambda_4 = 0$, $\lambda_5 = -2.65 \times 10^{-17}$, and $\lambda_6 = -6.94 \times 10^{-17}$. Based on the values of these eigenvalues, it seems that only the first two eigenvalues appear to possess information that is useful for examining the intercorrelations in the data matrix—the other four eigenvalues are either close to or equal to zero. Thus, for the moment, let us simply assume that these first two eigenvalues are suggesting that there are really only two factors that are responsible for the interrelationships among the observed variables (the issue of how one decides on the appropriate number of factors is discussed extensively in the next section).

Using the two eigenvalues already shown, their associated eigenvectors are determined to be

$$
V =
\begin{bmatrix}
.58 & 0 \\
.58 & 0 \\
.58 & 0 \\
0 & .58 \\
0 & .58 \\
0 & .58
\end{bmatrix}
$$

and, consequently, the product of these eigenvectors with the square root of the eigenvalues is found to be

$$
\text{adjusted } R = V(\sqrt{L})(\sqrt{L})V' = \Lambda\Lambda' =
\begin{bmatrix}
.25 & & & & & \\
.25 & .25 & & & & \\
.25 & .25 & .25 & & & \\
0 & 0 & 0 & .25 & & \\
0 & 0 & 0 & .25 & .25 & \\
0 & 0 & 0 & .25 & .25 & .25
\end{bmatrix}
$$

where the eigenvalue matrix $L$ is

$$
L =
\begin{bmatrix}
.75 & 0 \\
0 & .75
\end{bmatrix}
$$

and the factor loading matrix $\Lambda$ is determined to be

$$\Lambda = V(\sqrt{L}) = \begin{bmatrix} .5 & 0 \\ .5 & 0 \\ .5 & 0 \\ 0 & .5 \\ 0 & .5 \\ 0 & .5 \end{bmatrix}$$

As can be seen, the product of the factor loading matrix (i.e., $\Lambda\Lambda'$) is able to perfectly reproduce the adjusted correlation matrix.

Finally, given that the values in the diagonal of the observed correlation matrix $R$ must be equal to 1, it should be clear that the matrix $\Theta$ is simply equal to

$$\Theta = \begin{bmatrix} .75 & & & & & \\ 0 & .75 & & & & \\ 0 & 0 & .75 & & & \\ 0 & 0 & 0 & .75 & & \\ 0 & 0 & 0 & 0 & .75 & \\ 0 & 0 & 0 & 0 & 0 & .75 \end{bmatrix}$$

Thus, the process of conducting a factor analysis based on the PFM method concludes when estimates of the $\Lambda$ and $\Theta$ matrices have been obtained via the eigenvalue decomposition. In the preceding example, the PFM method led to the extraction of two factors that are responsible for the intercorrelations among the observed variables. The first factor $(F_1)$ is common to the observed variables $X_1$, $X_2$, $X_3$, and the second factor $(F_2)$ is common to $X_4$, $X_5$, $X_6$.

## HOW MANY FACTORS SHOULD BE USED IN A FACTOR ANALYSIS?

Determining the number of factors to use in a factor analysis is somewhat similar to determining how many principal components are needed to account for the variance in an observed data matrix. As discussed previously, deciding on how many components to use in order to account for the observed variance is not straightforward and there is really no "correct" answer. Unfortunately, the same concerns apply when trying to decide on the number of factors that are responsible for the interrelationships among the observed variables (in a sense the number of factors that are compatible with the data). Several different methods have been proposed in the literature (some of which are very similar to those used in PCA, thereby adding even more confusion to the perceived overlap between the two methods), but the following four methods appear to be the most popular.

The first method involves examining the proportion of contribution by a factor. If a correlation matrix is used, the proportion of contribution can be determined by dividing each computed eigenvalue ($\lambda_i$) by the number of observed ($p$) variables in the data [i.e., ($\lambda_i/p$) × 100 = percent contribution]. If the adjusted correlation matrix is used, the proportion contribution can be determined by dividing each eigenvalue ($\lambda_i$) by the sum ($\Sigma\lambda_i$) of the eigenvalues [i.e., ($\lambda_i/\Sigma\lambda_i$) × 100 = percent contribution]. Thus, once the proportion of contribution is determined for each factor, one may set the final criterion at whatever level is considered of practical value. In general, if the ratio of the combined factors is close to or exceeds 90%, the remaining factors are considered of no practical value.

The second proposed method is the Kaiser eigenvalue criterion discussed in the PCA section. As previously mentioned, this criterion involves retaining only those factors whose eigenvalues of the correlation matrix are greater than one. If the adjusted or so-called reduced correlation matrix is used (i.e., with the estimates of communality inserted in the diagonal), then the Kaiser eigenvalue criterion involves retaining only those factors whose eigenvalues are greater than zero (e.g., as was done in the previous section that demonstrated the PFM method of extraction). It is important to note that the Kaiser eigenvalue "greater-than-one" criterion is essentially equivalent to setting the variance explained at the $1 - (100/p)\%$ level.

A third method used is Cattell's scree test (1966), discussed earlier in this chapter. As indicated, this method simply involves plotting the eigenvalues of the observed correlation matrix against their ordinal number and noting the point at which the plot becomes nearly horizontal. The number of factors corresponding to the horizontal point is considered the appropriate number to retain.

Finally, a very popular method is the large-sample $\chi^2$ test of significance that is based on the maximum likelihood factor-analytic approach. In general, the maximum likelihood approach is very similar to the PFM approach. Basically, the objective of maximum likelihood is to find the underlying factor structure, in terms of population parameters, that has the greatest likelihood of producing the observed correlation matrix. The process is accomplished iteratively until no improvement can be made to the factor structure. As such, the $\chi^2$ test of significance is a statistical test of the degree of fit of the obtained factor structure (i.e., the population values) to reproducing the observed correlation matrix.

Unfortunately, the $\chi^2$ test has been shown to be sensitive to sample size. In particular, when the sample size is very large, minor deviations may result in statistical significance. For this reason, several authors (e.g., Harman, 1976) have advised researchers not to rely too much on the $\chi^2$ test but merely to use it as a guide. Thus, it seems that although the $\chi^2$ test is a useful index to examine, it is generally accepted that it should be interpreted with caution and supplemented with other criteria. Because of these problems, we recommend that if the $\chi^2$ test is to be used, both the sample size and the number of parameters estimated should be taken into account as measures of fit. One way to accomplish this is to consider whether the ratio of the $\chi^2$ value to its degrees of freedom is less than 2. The "chi-square to the degrees of freedom ratio" ($\chi^2/df$) is a criterion that has received much attention in the confirmatory factor-analytic literature (see the discussion on fit indices in chapter 8) and continues to be popular. In order to use the ratio criterion, one simply examines whether $\chi^2/df$ is greater than 2. If so, then there exists a significant discrepancy between the observed correlation matrix and the reproduced correlation matrix (i.e., the model does not fit the data and is rejected). More factors will need to be retained in order to explain the interrelationships among the observed variables.

As with PCA, the criteria concerning how many factors to retain should be applied with caution. The best strategy is to examine as many criteria as possible. If the results are similar, determining the number of factors is easy. If the results are different, look for some consensus among the criteria. Nevertheless, one should remember that the final judgment should be based on the reasonableness of the number of factors and their potential interpretation.

## ROTATION OF FACTORS

Once the number of factors in an FA has been determined (i.e., the ability of the selected factor structure to reproduce the observed correlations has been examined), the next step is to attempt an interpretation of the retained factors. Unfortunately, a good fitting solution will not necessarily be easy to interpret. This is because of the arbitrariness or indeterminancy of the coefficients of the factor loading matrix $\Lambda$. In general, when the common factors are uncorrelated, the computed coefficients of the factor loading matrix $\Lambda$ are not unique; rather, there is an infinite number of solutions that could have easily been chosen. This is because the factor solution obtained has only determined the $m$-dimensional space containing the factors, but not the exact position of the factors (for further discussion see Harman, 1976). Thus, the factor extraction method can obtain a solution that is able to reproduce the observed correlation matrix, but the solution is not necessarily in a form that is amenable to interpretation. For this reason, most of the time, once a factor solution has been obtained and the number of factors has been determined, the factor loading matrix is rotated or transformed to a new set of coefficients that improve interpretability of the factors.

There are basically three different approaches to factor rotation that are commonly used in factor analysis: (a) graphical rotation, (b) analytic rotation, and (c) rotation to a target matrix. The three approaches are all based on Thurstone's (1947) concept of "simple structure" for improving interpretability. Although the concept of "simple structure" is not that simple to explain or achieve, it is generally characterized by cases in which observed variables included in a factor analysis have high loadings on only one factor and factors have high loadings from observed variables (for an excellent list of the criteria to follow to achieve simple structure see Mulaik, 1972). With real data, however, it is usually quite difficult to achieve a perfect simple structure. For this reason, most rotations attempt to create a solution that is as close to simple structure as possible.

Historically, the concept of simple structure was specified in terms of reference axes. For example, in Fig. 7.1 reference axes were used to represent the relative position of each observed case in a PCA. In a similar manner, let us consider the use of reference axes in a graphical representation of a two factor solution as displayed in Fig. 7.7. The graphical representation is based on the factor loading matrix

$$\Lambda = \begin{bmatrix} .7 & -.5 \\ .7 & -.5 \\ .7 & .5 \\ .7 & .5 \end{bmatrix}$$

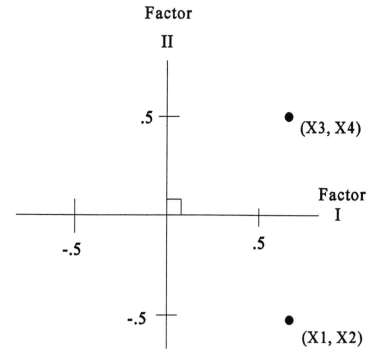

FIG. 7.7. Example graphical representation of a two-factor solution.

that was obtained by conducting a factor analysis on the following observed correlation matrix:

$$R = \begin{bmatrix} 1 & & & \\ .74 & 1 & & \\ .24 & .24 & 1 & \\ .24 & .24 & .74 & 1 \end{bmatrix}$$

As can be seen by examining the factor loading matrix $\Lambda$ and Fig. 7.7, the factors must be rotated because no clear picture of each common factor emerges.

The first approach to factor rotation commonly used is to examine the pattern of observed variables on each factor graphically and then attempt to rotate the axes or define new axes in such a way as to achieve simple structure. Figure 7.8 presents a graphical representation of new axes that

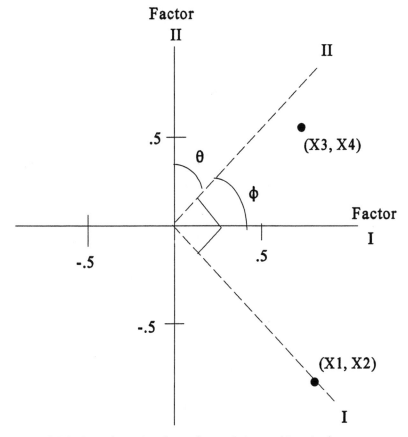

FIG. 7.8. Example rotation of a two-factor solution to achieve simple structure. (Note that $\theta + \phi = 90°$ angle.)

can be rotated to achieve simple structure. As can be seen from this figure, the rotation of the axes has led to high loadings of the observed variables $X_1$ and $X_2$ on the first factor and zero loadings on the second. This same rotation has also led to low loadings of the observed variables $X_3$ and $X_4$ on the first factor and high loadings on the second factor. In other words, as long as the two factors are uncorrelated (also referred to as orthogonal to each other—as evidenced by the 90° angle between the factors), the solution obtained appears to be as close as possible to simple structure. What is important to remember is that the basic goal of the rotation is to find a factor matrix that is closest to the simple structure as possible.

The actual values of the rotated factor matrix can be obtained mathematically by taking the factor matrix given earlier as

$$\Lambda = \begin{bmatrix} .7 & -.5 \\ .7 & -.5 \\ .7 & .5 \\ .7 & .5 \end{bmatrix}$$

and multiply it by a transformation matrix $T$ that is defined as

$$T = \begin{bmatrix} \cos\Theta & \sin\Theta \\ -\sin\Theta & \cos\Theta \end{bmatrix}$$

Recall from high-school algebra and trigonometry that the angle $\Theta$ can be determined in this example by the following steps:

1. Find the tangent of the angle $\Phi$ (given by $\tan\Phi = 0.7/0.5 = 1.4$).
2. Determine the angle $\Phi$ [given by $\Phi = \tan^{-1}(1.4) = 54.46°$].
3. Determine the angle $\Theta$ (given by $\Theta = 90° - \Phi = 35.54°$).

Using either a calculator or trigonometric tables one can quickly determine that the transformation matrix $T$ is equal to

$$T = \begin{bmatrix} .81 & .58 \\ -.58 & .81 \end{bmatrix}$$

Thus, the rotated factor matrix (symbolized by $\Lambda*$) becomes

$$\Lambda* = \Lambda T = \begin{bmatrix} .86 & 0 \\ .86 & 0 \\ .28 & .81 \\ .28 & .81 \end{bmatrix}$$

It is important to note that the transformed or rotated matrix $\Lambda^*$ will reproduce a correlation matrix $R$ that is identical to that produced using the unrotated matrix $\Lambda$. This identity can be verified using $\Lambda^*\Lambda^{*\prime} = \Lambda\Lambda'$.

The second approach to factor rotation is to use an analytic rotation method. Two analytic rotation methods have received considerable attention in the literature. These can be classified in the general categories of orthogonal techniques (for uncorrelated factors), and oblique techniques (for correlated factors). As it turns out, there are numerous (at least 15) analytic procedures that fall within these two general categories. Despite the abundance of analytic procedures, the following two methods are currently the most popular rotation methods from each category: VARIMAX (for orthogonal rotations) and PROMAX (for oblique rotations).

The VARIMAX method (Kaiser, 1958) attempts to perform a rotation by maximizing the variance of the squared loadings for each factor. This method tends to polarize the factor loadings so that they are either high or low, thereby making it easier to identify factors with specific observed variables. Mathematically, the VARIMAX rotation method seeks a transformation matrix $T$ such that when multiplied by the factor loading matrix $\Lambda$ will produce a rotated factor matrix $(\Lambda^*)$ that maximizes the variance of the squared loadings for each factor. Thus, $\Lambda^* = \Lambda T$ maximizes

$$\sum_{i=1}\sum_{j=1} = (\lambda_{ij}^2 - \overline{\lambda}_i)$$

where $\lambda_{ij}^2$ is the loading of the $i$th observed variable on the $j$th factor and $\overline{\lambda}_i$ is the mean squared loading on the $j$th factor.

The PROMAX method (Hendrickson & White, 1964) permits factors to be correlated. This method seeks an even greater polarization of the factor loadings so that the factors are more interpretable. The PROMAX method can be thought of as an approach that permits the reference axes not to be rigid, in the sense of preserving the perpendicularity of the axes. As such, the PROMAX method may be considered as an approach that includes the factor correlation matrix $(\Phi)$ in the decomposition of the observed correlation matrix (i.e., $\Phi$ is no longer an identity matrix in $R = \Lambda\Phi\Lambda' + \Theta$). The PROMAX method begins with the VARIMAX-rotated matrix $\Lambda^*$, and seeks a transformation matrix $T$ such that when multiplied by the $\Lambda^*$ matrix (i.e., the PROMAX-rotated loadings are $\Lambda^{**} = \Lambda^* T$) it minimizes

$$\text{trace } (P - \Lambda*T)'(P - \Lambda*T)$$

where *trace* denotes the sum of the diagonal elements of the product matrix and the $P$ matrix contains factor loadings that have been raised to an arbitrary power—the higher the power, the stronger the correlations between factors.

The third approach to factor rotation is called *rotation to a target matrix*. In this approach, a target matrix (in terms of factor loadings) is defined and an attempt is made to rotate the reference axes to find the factor patterns that are closest to the target matrix. Although rotation to a target matrix is used quite frequently in factor analysis, we believe that it actually fits much better into the realm of confirmatory factor-analytic techniques discussed extensively in the next chapter. This belief is based on the fact that the specification of a target matrix presumes some knowledge about the nature of the factor patterns. And yet, if the knowledge was available, why was the exploratory factor analysis undertaken? Obviously, some knowledge of the observed variables used in the study will assist one in conducting a meaningful interpretation of the factors. Even then, "when interpreting factors all factorists struggle and struggle and struggle" (McNemar, 1951, p. 357). Our humble advice is to study the observed variables and factors within the context of the theory being studied. Without doubt, the task of interpretation is made simpler by examining the magnitude of the loadings for each observed variable on each factor. Those variables that contribute most to a factor, in terms of the highest loadings, will influence the interpretation of that factor the most. To assist with interpretation, several researchers have also proposed methods for representing data in two dimensions (e.g., Andrew's plots, Anderson's stars, Chernoff's faces, and Marcoulides and Drezner's plots). Currently, the Marcoulides and Drezner (1993) approach to examining factor analytic solutions appears to be popular. In general, the approach transforms multidimensional space into a two-dimensional diagram that can be used to assist with the interpretation of the rotated factors.

As should be evidenced by the foregoing discussion on rotation methods, rotations are not easy to apply and there is really no best method for rotation. It is for this reason that we caution readers that attempts to rotate and, consequently, interpret and name factors should not be guided blindly by statistical criteria but examined within the context of the theory being studied.

## LOOSE ENDS

There are two final issues that one must be aware of when conducting a factor analysis. The first issue concerns sample size and the second concerns what are commonly called Heywood cases. Without doubt, sample size plays an important role in almost every statistical technique used. Although there is universal agreement among researchers that the larger the sample the more stable the parameter estimates, there is no agreement as to what constitutes large. As one might imagine, the factor-analytic literature is full of recommendations and rules of thumb for the determination of appropriate sample sizes. Although we believe that the sample size issue depends to a great extent on the aims of the analysis and the properties of the data, at the very least one should have a reasonable sample in order to tap in to the characteristics of the population. A rather easy to remember rule of thumb is provided by Lawley and Maxwell (1971) and suggests that the sample size ($N$) should be $N -$ (number of factors) $- 1 > 50$.

The issue of Heywood cases is also a problem that shows up frequently in factor analysis. Recall from earlier in this chapter that communality estimates (e.g., SMCs) are used in an adjusted correlation matrix in order to conduct a factor analysis. As with any correlation cofficient, one would expect the communality estimates to lie between zero and one. Unfortunately, there is a mathematical peculiarity with the common factor analysis method, and often the communality estimates may exceed one. If a communality estimate exceeds one, it is called a Heywood case and suggests that some unique factor has negative variance (i.e., some diagonal element of the matrix $\Theta$ is negative). This is a clear indication that something is wrong with the factor analysis. Because Heywood cases show up quite frequently in factor analysis, most available computer software packages have built-in algorithms for dealing the problem. For example, in the SAS package a Heywood case problem can be corrected by using the command HEYwood in the procedure statement.

Although factor analysts disagree whether or not a factor solution with a Heywood case should be completed, most feel that it is simply an indication that one of the following conditions has occurred:

1. Not enough data have been gathered from the population to provide stable estimates.
2. Too few common factors have been retained.
3. Too many common factors have been retained.
4. The computed communality estimates are incorrect.
5. No common factor model is appropriate for the observed variables.

As long as the Heywood case has occurred because of conditions 1–4, a factor analysis can be performed once the condition has been corrected. However, if the cause of the problem is condition 5, nothing can be done to correct the Heywood case.

## AN EXAMPLE OF FACTOR ANALYSIS USING SAS

Sadri and Marcoulides (1994) recently conducted a study of the dynamics of occupational stress. In this study data were collected from 437 individuals employed in diverse companies within the southern Orange County area of

Southern California. Participants in the study responded to a questionnaire consisting of variables designed to measure the impact of personality, coping, and organizational variables on stress-related outcomes. For the purposes of this example, only 15 of the variables will be examined (which represent 6 variables related to stress, 3 variables related to locus of control, and 6 variables related to coping strategies). An important concern of this study was to determine what factors might account for the intercorrelations between the observed variables.

The SAS program setup to perform a factor analysis on the observed data is presented in Table 7.4. The DATA and INPUT statements indicate that raw data are used as input (due to space limitations, only the first 20 observations are provided). To perform a factor analysis on the observed data, the PROC FACTOR procedure is used (the same procedure was presented in the PCA section of this chapter). Although 30 variables are included in the observed data, only 15 are selected in the PROC FACTOR statement for analysis by using the VAR definition statement. For simplicity, the variables are labeled STRESS1–STRESS6, LOC1–LOC3, and COPING1–COPING6.

The results of the analysis are presented in Table 7.5. As can be seen in Table 7.5a and 7.5b, determining how many factors to retain for this solution is relatively straightforward. Both the Kaiser eigenvalue criterion and the proportion criterion indicate that three factors should be retained. In addition, the $\chi^2/df$ ratio (which is close to 2) suggests that three factors are sufficient (especially considering the relatively large sample size used in the study).

The results in Table 7.5c represent the initial (and arbitrary) three-factor solution obtained for the observed data. As can be seen in Table 7.5c, these results are not amenable to a clear interpretation. By using the statement ROTATE=VARIMAX in the FA specification, an orthogonal rotation is achieved. Although the default in SAS is to perform a VARIMAX rotation, it is important that the reader become familiar with the syntax of the rotation statement in order to take advantage of the other rotation schemes within the program. Table 7.5d presents the orthogonal tranformation matrix ($T$) that is used to achieve the transformation presented in Table 7.5e. As can be seen by examining Table 7.5e, the orthogonally rotated three-factor solution is now relatively easier to interpret. Thus, one can see that the six variables labeled STRESS all have high loadings on the second factor and low loadings on the other factors. Similarly, the LOC variables have high loadings on the third factor and low on the other factors, and the COPING variables all have high loadings on the first factor and low on the other factors. Based on these results, one might provide the following labels to the three factors: Factor1, "Coping Strategies"; Factor2, "Sources of Stress"; and Factor3, "Locus of Control" (for a complete discussion see Sadri & Marcoulides, 1994).

TABLE 7.4
Example of Factor Analysis in SAS

```
data stress;
input id country stress1 stress2 stress3 stress4 stress5 stress6
typea1 typea2 typea3 tota loc1 loc2 loc3 totloc
coping1 coping2 coping3 coping4 coping5 coping6
js1 js2 js3 js4 js5 totjs mh ph ;
CARDS;
001 1 29 33 27 27 36 30 24 24 10 58 18 14 09 41 11 26 15 13 15 27 28 18 17 20 13 096 34 27
002 1 15 21 19 20 29 28 27 18 11 56 15 14 11 40 15 26 12 09 16 28 30 23 24 23 16 116 29 23
003 1 20 33 20 28 40 23 25 22 14 61 16 12 07 35 17 29 15 13 15 29 32 20 27 22 16 117 42 27
004 1 30 30 19 27 42 25 27 23 14 64 19 13 07 39 16 30 17 12 16 31 32 22 25 21 17 117 32 29
005 1 25 25 29 16 29 22 20 19 10 49 15 10 10 35 16 25 12 09 15 23 33 20 20 20 15 108 44 21
006 1 27 36 30 26 30 34 18 14 16 48 16 10 08 34 15 26 14 17 15 25 26 19 17 14 13 089 54 26
007 1 26 33 30 27 40 30 20 19 11 50 19 13 10 42 17 30 12 16 18 26 20 16 17 17 14 084 32 26
008 1 32 39 34 43 44 56 27 23 12 62 14 13 10 37 13 24 13 13 15 25 27 19 19 18 12 095 52 25
009 1 36 44 33 26 47 36 15 17 07 39 15 14 11 40 15 28 16 15 19 30 32 19 23 19 16 109 31 25
010 1 33 28 28 27 38 34 20 16 09 45 16 11 08 35 07 25 15 07 12 29 28 17 14 16 09 084 49 22
011 1 27 36 33 40 53 50 22 17 15 54 26 17 11 54 18 29 08 17 18 28 14 17 14 08 08 061 50 16
012 1 33 42 34 44 46 57 15 14 17 46 22 16 10 48 17 29 09 08 15 17 11 16 19 19 10 075 85 52
013 1 33 40 36 40 53 42 22 18 12 52 18 11 09 38 22 33 15 11 15 31 32 21 23 20 16 112 33 20
014 1 23 32 28 21 39 30 23 18 12 53 15 12 10 37 16 24 08 19 14 20 27 20 18 18 14 097 30 19
015 1 29 41 36 32 42 36 26 16 14 56 19 17 13 49 15 23 09 12 16 23 22 15 19 17 13 086 64 36
016 1 24 24 19 10 18 20 22 09 12 43 17 17 10 44 12 22 13 10 14 25 27 21 19 18 14 099 35 18
017 1 27 30 35 29 41 30 21 20 08 49 14 15 09 38 17 27 12 15 16 23 22 20 18 17 15 092 37 18
018 1 29 33 34 38 50 47 23 25 12 60 20 14 08 42 15 26 17 14 15 29 24 18 18 17 11 088 35 17
019 1 35 43 34 36 53 45 16 14 11 41 21 12 10 43 18 29 12 18 15 25 20 15 14 12 10 071 54 29
020 1 30 30 27 21 37 33 21 16 08 45 15 14 11 40 20 30 13 15 17 30 33 22 24 23 15 117 41 29
: : : : : :
: : : : : :


(THE DATA SET CONTAINS 437 OBSERVATIONS)

: : : : : :
: : : : : :
;
Proc Factor method=ml rotate=varimax;
    var stress1 stress2 stress3 stress4 stress5 stress6
        loc1 loc2 loc3
        coping1 coping2 coping3 coping4 coping5 coping6;
run;
```

**197**

TABLE 7.5
Results of SAS Factor Analysis

---

*(a) Preliminary Choice of Factors*

Initial Factor Method: Maximum Likelihood

Prior Communality Estimates: SMC

| STRESS1 | STRESS2 | STRESS3 | STRESS4 | STRESS5 |
|---|---|---|---|---|
| 0.475534 | 0.473856 | 0.553597 | 0.505023 | 0.353556 |

| STRESS6 | LOC1 | LOC2 | LOC3 | COPING1 |
|---|---|---|---|---|
| 0.276626 | 0.573769 | 0.475546 | 0.506476 | 0.469796 |

| COPING2 | COPING3 | COPING4 | COPING5 | COPING6 |
|---|---|---|---|---|
| 0.587208 | 0.629230 | 0.576955 | 0.845560 | 0.722158 |

Initial Factor Method: Maximum Likelihood

Preliminary Eigenvalues:  Total = 21.7199123   Average = 1.44799416

|  | 1 | 2 | 3 | 4 | 5 |
|---|---|---|---|---|---|
| Eigenvalue | 14.201627 | 5.195412 | 3.569391 | 0.347890 | 0.262178 |
| Difference | 9.006215 | 1.626021 | 3.221501 | 0.085712 | 0.157229 |
| Proportion | 0.6539 | 0.2392 | 0.1643 | 0.0160 | 0.0121 |
| Cumulative | 0.6539 | 0.8931 | 1.0574 | 1.0734 | 1.0855 |

|  | 6 | 7 | 8 | 9 | 10 |
|---|---|---|---|---|---|
| Eigenvalue | 0.104949 | 0.037674 | 0.010417 | −0.075930 | −0.161242 |
| Difference | 0.067274 | 0.027258 | 0.086347 | 0.085312 | 0.071366 |
| Proportion | 0.0048 | 0.0017 | 0.0005 | −0.0035 | −0.0074 |
| Cumulative | 1.0903 | 1.0920 | 1.0925 | 1.0890 | 1.0816 |

|  | 11 | 12 | 13 | 14 | 15 |
|---|---|---|---|---|---|
| Eigenvalue | −0.232608 | −0.286692 | −0.371947 | −0.405135 | −0.476071 |
| Difference | 0.054084 | 0.085255 | 0.033188 | 0.070937 |  |
| Proportion | −0.0107 | −0.0132 | −0.0171 | −0.0187 | −0.0219 |
| Cumulative | 1.0709 | 1.0577 | 1.0406 | 1.0219 | 1.0000 |

3 factors will be retained by the PROPORTION criterion.

---

*(b) Use of Three Factors*

---

Initial Factor Method: Maximum Likelihood

Significance tests based on 432 observations:

Test of H0: No common factors.
    vs HA: At least one common factor.

Chi-square = 3412.998  df = 105  Prob>chi**2 = 0.0000

Test of H0: 3 Factors are sufficient.
    vs HA: More factors are needed.

Chi-square = 159.944  df = 63  Prob>chi**2 = 0.0000

---

*(Continued)*

TABLE 7.5
*(Continued)*

Initial Factor Method: Maximum Likelihood

Squared Canonical Correlations

| FACTOR1 | FACTOR2 | FACTOR3 |
|---------|---------|---------|
| 0.960172 | 0.857435 | 0.834270 |

Eigenvalues of the Weighted Reduced Correlation Matrix:
Total = 35.1560104   Average = 2.34373402

| | 1 | 2 | 3 | 4 | 5 |
|---|---|---|---|---|---|
| Eigenvalue | 24.107762 | 6.014322 | 5.033927 | 0.420456 | 0.360670 |
| Difference | 18.093440 | 0.980395 | 4.613470 | 0.059786 | 0.171038 |
| Proportion | 0.6857 | 0.1711 | 0.1432 | 0.0120 | 0.0103 |
| Cumulative | 0.6857 | 0.8568 | 1.0000 | 1.0120 | 1.0222 |

| | 6 | 7 | 8 | 9 | 10 |
|---|---|---|---|---|---|
| Eigenvalue | 0.189632 | 0.163315 | 0.108840 | 0.002208 | -0.059821 |
| Difference | 0.026317 | 0.054475 | 0.106632 | 0.062029 | 0.036935 |
| Proportion | 0.0054 | 0.0046 | 0.0031 | 0.0001 | -0.0017 |
| Cumulative | 1.0276 | 1.0323 | 1.0354 | 1.0354 | 1.0337 |

| | 11 | 12 | 13 | 14 | 15 |
|---|---|---|---|---|---|
| Eigenvalue | -0.096756 | -0.158509 | -0.207435 | -0.331270 | -0.391331 |
| Difference | 0.061753 | 0.048927 | 0.123834 | 0.060062 | |
| Proportion | -0.0028 | -0.0045 | -0.0059 | -0.0094 | -0.0111 |
| Cumulative | 1.0310 | 1.0265 | 1.0206 | 1.0111 | 1.0000 |

*(c) Initial Three-Factor Solution*

Factor Pattern

| | FACTOR1 | FACTOR2 | FACTOR3 |
|---|---------|---------|---------|
| STRESS1 | 0.02480 | 0.59179 | 0.41010 |
| STRESS2 | 0.19687 | 0.61898 | 0.26257 |
| STRESS3 | 0.10488 | 0.63980 | 0.46182 |
| STRESS4 | 0.12043 | 0.58777 | 0.41220 |
| STRESS5 | 0.07996 | 0.45473 | 0.35235 |
| STRESS6 | 0.03735 | 0.42608 | 0.31924 |
| LOC1 | 0.25113 | 0.50894 | -0.63113 |
| LOC2 | 0.08016 | 0.47849 | -0.55164 |
| LOC3 | 0.04275 | 0.46919 | -0.64718 |
| COPING1 | 0.68928 | -0.00605 | 0.00740 |
| COPING2 | 0.77657 | -0.05833 | 0.05846 |
| COPING3 | 0.80355 | -0.06925 | -0.00507 |
| COPING4 | 0.76528 | -0.06531 | -0.02416 |
| COPING5 | 0.96748 | -0.05296 | 0.00400 |
| COPING6 | 0.85190 | -0.01082 | 0.05963 |

## TABLE 7.5
### (Continued)

### (d) Orthogonal Transformation Matrix (T)

Final Communality Estimates and Variable Weights
Total Communality: Weighted = 35.156011  Unweighted = 8.690721

|  | STRESS1 | STRESS2 | STRESS3 | STRESS4 | STRESS5 |
|---|---|---|---|---|---|
| Communality | 0.519012 | 0.490828 | 0.633614 | 0.529889 | 0.337327 |
| Weight | 2.078678 | 1.963978 | 2.728870 | 2.127498 | 1.509246 |

|  | STRESS6 | LOC1 | LOC2 | LOC3 | COPING1 |
|---|---|---|---|---|---|
| Communality | 0.284851 | 0.720412 | 0.539680 | 0.640806 | 0.475193 |
| Weight | 1.398408 | 3.576668 | 2.172410 | 2.784026 | 1.905468 |

|  | COPING2 | COPING3 | COPING4 | COPING5 | COPING6 |
|---|---|---|---|---|---|
| Communality | 0.609873 | 0.650509 | 0.590496 | 0.938833 | 0.729398 |
| Weight | 2.563271 | 2.861290 | 2.441974 | 16.348789 | 3.695438 |

Rotation Method: Varimax

Orthogonal Transformation Matrix

|  | 1 | 2 | 3 |
|---|---|---|---|
| 1 | 0.99259 | 0.08090 | 0.09065 |
| 2 | -0.11878 | 0.80320 | 0.58375 |
| 3 | 0.02558 | 0.59019 | -0.80686 |

### (e) Orthogonally Rotated Three-Factor Solution

Rotation Method: Varimax

Rotated Factor Pattern

|  | FACTOR1 | FACTOR2 | FACTOR3 |
|---|---|---|---|
| STRESS1 | -0.03518 | 0.71937 | 0.01681 |
| STRESS2 | 0.12860 | 0.66805 | 0.16732 |
| STRESS3 | 0.03992 | 0.79493 | 0.01037 |
| STRESS4 | 0.06027 | 0.72512 | 0.02144 |
| STRESS5 | 0.03437 | 0.57966 | -0.01160 |
| STRESS6 | -0.00537 | 0.53366 | -0.00547 |
| LOC1 | 0.17268 | 0.05661 | 0.82909 |
| LOC2 | 0.00862 | 0.06523 | 0.73168 |
| LOC3 | -0.02986 | -0.00165 | 0.79994 |
| COPING1 | 0.68508 | 0.05527 | 0.05298 |
| COPING2 | 0.77924 | 0.05048 | -0.01082 |
| COPING3 | 0.80569 | 0.00640 | 0.03651 |
| COPING4 | 0.76675 | -0.00480 | 0.05074 |
| COPING5 | 0.96670 | 0.03810 | 0.05356 |
| COPING6 | 0.84839 | 0.09543 | 0.02280 |

*(Continued)*

TABLE 7.5
*(Continued)*

Rotation Method: Varimax

Final Communality Estimates and Variable Weights
Total Communality: Weighted = 35.156011  Unweighted = 8.690721

|  | STRESS1 | STRESS2 | STRESS3 | STRESS4 | STRESS5 |
|---|---|---|---|---|---|
| Communality | 0.519012 | 0.490828 | 0.633614 | 0.529889 | 0.337327 |
| Weight | 2.078678 | 1.963978 | 2.728870 | 2.127498 | 1.509246 |

|  | STRESS6 | LOC1 | LOC2 | LOC3 | COPING1 |
|---|---|---|---|---|---|
| Communality | 0.284851 | 0.720412 | 0.539680 | 0.640806 | 0.475193 |
| Weight | 1.398408 | 3.576668 | 2.172410 | 2.784026 | 1.905468 |

|  | COPING2 | COPING3 | COPING4 | COPING5 | COPING6 |
|---|---|---|---|---|---|
| Communality | 0.609873 | 0.650509 | 0.590496 | 0.938833 | 0.729398 |
| Weight | 2.563271 | 2.861290 | 2.441974 | 16.348789 | 3.695438 |

*(f) Target Matrix*

Rotation Method: Promax

Target Matrix for Procrustean Transformation

|  | FACTOR1 | FACTOR2 | FACTOR3 |
|---|---|---|---|
| STRESS1 | -0.00012 | 0.99592 | 0.00001 |
| STRESS2 | 0.00620 | 0.86730 | 0.01365 |
| STRESS3 | 0.00013 | 0.99628 | 0.00000 |
| STRESS4 | 0.00057 | 0.98874 | 0.00003 |
| STRESS5 | 0.00021 | 0.99446 | -0.00001 |
| STRESS6 | -0.00000 | 1.00000 | -0.00000 |
| LOC1 | 0.00845 | 0.00030 | 0.93399 |
| LOC2 | 0.00000 | 0.00070 | 0.99006 |
| LOC3 | -0.00005 | -0.00000 | 1.00000 |
| COPING1 | 0.98467 | 0.00052 | 0.00045 |
| COPING2 | 0.99661 | 0.00027 | -0.00000 |
| COPING3 | 1.00000 | 0.00000 | 0.00009 |
| COPING4 | 0.99656 | -0.00000 | 0.00029 |
| COPING5 | 0.99626 | 0.00006 | 0.00017 |
| COPING6 | 0.98338 | 0.00140 | 0.00002 |

Procrustean Transformation Matrix

|  | 1 | 2 | 3 |
|---|---|---|---|
| 1 | 1.21027 | -0.07090 | -0.05587 |
| 2 | -0.06255 | 1.43106 | -0.05299 |
| 3 | -0.08112 | -0.07413 | 1.22577 |

*(Continued)*

## TABLE 7.5
*(Continued)*

### (g) Oblique Rotation of Previous Orthogonal Factor Solution

Inter-factor Correlations

|  | FACTOR1 | FACTOR2 | FACTOR3 |
|---|---|---|---|
| FACTOR1 | 1.00000 | 0.10841 | 0.11981 |
| FACTOR2 | 0.10841 | 1.00000 | 0.10414 |
| FACTOR3 | 0.11981 | 0.10414 | 1.00000 |

Rotation Method: Promax

Rotated Factor Pattern (Std Reg Coefs)

|  | FACTOR1 | FACTOR2 | FACTOR3 |
|---|---|---|---|
| STRESS1 | -0.07411 | 0.72576 | -0.01281 |
| STRESS2 | 0.08356 | 0.65802 | 0.13385 |
| STRESS3 | -0.00188 | 0.79849 | -0.02607 |
| STRESS4 | 0.02154 | 0.72655 | -0.01277 |
| STRESS5 | 0.00523 | 0.58299 | -0.03859 |
| STRESS6 | -0.03286 | 0.53830 | -0.02857 |
| LOC1 | 0.11514 | 0.00515 | 0.82662 |
| LOC2 | -0.04416 | 0.02711 | 0.73545 |
| LOC3 | -0.08410 | -0.04192 | 0.80906 |
| COPING1 | 0.68440 | 0.01873 | 0.01955 |
| COPING2 | 0.78391 | 0.01253 | -0.04898 |
| COPING3 | 0.80969 | -0.03568 | -0.00049 |
| COPING4 | 0.77003 | -0.04577 | 0.01616 |
| COPING5 | 0.96925 | -0.01266 | 0.00793 |
| COPING6 | 0.84904 | 0.05261 | -0.02019 |

Reference Axis Correlations

|  | FACTOR1 | FACTOR2 | FACTOR3 |
|---|---|---|---|
| FACTOR1 | 1.00000 | -0.09716 | -0.10976 |
| FACTOR2 | -0.09716 | 1.00000 | -0.09236 |
| FACTOR3 | -0.10976 | -0.09236 | 1.00000 |

Rotation Method: Promax

Final Communality Estimates and Variable Weights
Total Communality: Weighted = 35.156011  Unweighted = 8.690721

|  | STRESS1 | STRESS2 | STRESS3 | STRESS4 | STRESS5 |
|---|---|---|---|---|---|
| Communality | 0.519012 | 0.490828 | 0.633614 | 0.529889 | 0.337327 |
| Weight | 2.078678 | 1.963978 | 2.728870 | 2.127498 | 1.509246 |

|  | STRESS6 | LOC1 | LOC2 | LOC3 | COPING1 |
|---|---|---|---|---|---|
| Communality | 0.284851 | 0.720412 | 0.539680 | 0.640806 | 0.475193 |
| Weight | 1.398408 | 3.576668 | 2.172410 | 2.784026 | 1.905468 |

|  | COPING2 | COPING3 | COPING4 | COPING5 | COPING6 |
|---|---|---|---|---|---|
| Communality | 0.609873 | 0.650509 | 0.590496 | 0.938833 | 0.729398 |
| Weight | 2.563271 | 2.861290 | 2.441974 | 16.348789 | 3.695438 |

Table 7.5f presents the target matrix for an oblique solution based on the PROMAX rotation approach. This rotation can be conducted in SAS by using the ROTATE=PROMAX statement. As can be seen in Table 7.5e, the inter-factor correlations are all relatively low. This implies that there is very little overlap between the common factors included in this study. As such, an oblique rotation of the previous orthogonal factor solution (presented in Table 7.5g) does not change in any major way the interpretation of the factors (or the factor loadings themselves).

## EXERCISES

1. Five physical fitness variables were measured on 425 third-grade children, with correlations given as follows:

|      |     | Test |     |     |
|------|-----|------|-----|-----|
| Test | 2   | 3    | 4   | 5   |
| 1    | .64 | .60  | .40 | .58 |
| 2    |     | .78  | .34 | .23 |
| 3    |     |      | .44 | .50 |
| 4    |     |      |     | .89 |

Test 1 = high jump
Test 2 = long jump
Test 3 = $\frac{1}{4}$-mile race
Test 4 = sit-ups
Test 5 = pull-ups

(a) Conduct a factor analysis and decide how many factors to retain to adequately explain the data. (b) Interpret the factors.

2. During the course of one semester, an economics instructor administers two open-book exams, two closed-book exams, one research report, and one oral report to determine a semester grade. A sample of 20 students is given:

| OB | OB | CB | CB | RR | OR |
|----|----|----|----|----|----|
| 78 | 82 | 84 | 71 | 85 | 98 |
| 85 | 89 | 86 | 87 | 82 | 90 |
| 84 | 71 | 80 | 75 | 77 | 81 |
| 65 | 63 | 71 | 74 | 80 | 69 |
| 77 | 63 | 75 | 55 | 63 | 53 |
| 31 | 49 | 62 | 63 | 62 | 59 |
| 46 | 53 | 57 | 56 | 64 | 67 |
| 76 | 79 | 71 | 75 | 71 | 74 |
| 12 | 58 | 61 | 63 | 67 | 64 |
| 77 | 82 | 67 | 68 | 81 | 89 |
| 65 | 63 | 58 | 56 | 67 | 75 |
| 74 | 79 | 76 | 87 | 93 | 93 |
| 48 | 48 | 49 | 51 | 37 | 34 |
| 56 | 40 | 56 | 54 | 63 | 69 |
| 17 | 51 | 52 | 53 | 46 | 35 |
| 89 | 89 | 85 | 88 | 89 | 85 |
| 96 | 91 | 96 | 98 | 91 | 100 |
| 40 | 43 | 48 | 21 | 61 | 50 |
| 59 | 53 | 37 | 22 | 0 | 46 |
| 62 | 64 | 76 | 72 | 62 | 84 |

(a) Conduct a factor analysis. How many factors should be retained? (b) Interpret the factors.

3. During a job interview, applicants are judged on seven areas: appearance, likability, self-confidence, salesmanship, experience, ambition, and potential. A correlation matrix is presented for a sample of 15 applicants:

|   | 2 | 3 | 4 | 5 | 6 | 7 |
|---|----|----|----|----|----|----|
| 1 | .37 | .05 | .30 | .58 | .25 | .30 |
| 2 |    | .33 | .39 | .10 | .40 | .65 |
| 3 |    |    | .85 | .01 | .85 | .63 |
| 4 |    |    |    | .25 | .87 | .70 |
| 5 |    |    |    |    | .18 | .30 |
| 6 |    |    |    |    |    | .78 |

(a) Perform a factor analysis. Decide which factors to retain. (b) Interpret the factors.

4. The following correlation matrix for the 11 subtests of the WAIS was observed. The data on which the correlations are based describe the performance of 200 police officers and firefighters. This correlation matrix represents the observed relationships:

*Variable*

| Variable | 1 | 2 | 3 | 4 | 5 | 6 | 7 | 8 | 9 | 10 | 11 |
|---|---|---|---|---|---|---|---|---|---|---|---|
| 1. Information | 1.00 | | | | | | | | | | |
| 2. Comprehension | .37 | 1.00 | | | | | | | | | |
| 3. Arithmetic | .34 | .27 | 1.00 | | | | | | | | |
| 4. Similarities | .40 | .25 | .36 | 1.00 | | | | | | | |
| 5. Digit Span | .27 | .38 | .28 | .22 | 1.00 | | | | | | |
| 6. Vocabulary | .59 | .46 | .33 | .35 | .29 | 1.00 | | | | | |
| 7. Digit Symbol | .09 | .10 | .18 | .08 | .16 | .08 | 1.00 | | | | |
| 8. Picture Completion | .25 | .26 | .32 | .31 | .14 | .27 | .19 | 1.00 | | | |
| 9. Block Design | .27 | .29 | .38 | .26 | .18 | .24 | .13 | .36 | 1.00 | | |
| 10. Picture Arrangement | .22 | .22 | .29 | .25 | .15 | .28 | .22 | .36 | .30 | 1.00 | |
| 11. Object Assembly | .26 | .24 | .30 | .20 | .22 | .26 | .17 | .40 | .60 | .25 | 1.00 |

Conduct an exploratory factor analysis on the correlation matrix. How many factors should be retained? Explain how you decided. Interpret the factors.

5. The Computer Anxiety Scale (CAS) is a measure of student anxiety in different situations related to computers. A study by Marcoulides (1989a), involving 225 college students, examined the reliability and factorial validity of the CAS. A sample correlation matrix is presented below.

|  | C1 | C2 | C3 | C4 | C5 | C6 |
|---|---|---|---|---|---|---|
| C1 | 1.000 | | | | | |
| C2 | 0.270 | 1.000 | | | | |
| C3 | 0.516 | 0.373 | 1.000 | | | |
| C4 | 0.194 | 0.320 | 0.403 | 1.000 | | |
| C5 | 0.325 | 0.271 | 0.273 | 0.425 | 1.000 | |
| C6 | 0.425 | 0.215 | 0.399 | 0.341 | 0.500 | 1.000 |
| C7 | 0.243 | 0.242 | 0.236 | 0.247 | 0.367 | 0.401 |
| C8 | 0.471 | 0.275 | 0.489 | 0.325 | 0.285 | 0.165 |
| C9 | 0.253 | 0.365 | 0.215 | 0.245 | 0.456 | 0.781 |
| C10 | 0.478 | 0.156 | 0.155 | 0.254 | 0.358 | 0.281 |
| C11 | 0.528 | 0.482 | 0.568 | 0.482 | 0.298 | 0.381 |
| C12 | 0.422 | 0.482 | 0.712 | 0.601 | 0.357 | 0.268 |
| C13 | 0.258 | 0.504 | 0.568 | 0.489 | 0.315 | 0.332 |
| C14 | 0.568 | 0.665 | 0.436 | 0.300 | 0.324 | 0.151 |
| C15 | 0.368 | 0.367 | 0.438 | 0.155 | 0.071 | 0.236 |
| C16 | 0.236 | 0.472 | 0.318 | 0.529 | 0.565 | 0.440 |
| C17 | 0.458 | 0.336 | 0.489 | 0.201 | 0.431 | 0.208 |
| C18 | 0.571 | 0.404 | 0.431 | 0.592 | 0.294 | 0.189 |
| C19 | 0.125 | 0.242 | 0.373 | 0.368 | 0.331 | 0.537 |
| C20 | 0.239 | 0.477 | 0.402 | 0.254 | 0.395 | 0.468 |

|  | C8 | C9 | C10 | C11 | C12 | C13 |
|---|---|---|---|---|---|---|
| C8 | 1.000 | | | | | |
| C9 | 0.256 | 1.000 | | | | |
| C10 | 0.582 | 0.459 | 1.000 | | | |
| C11 | 0.368 | 0.458 | 0.259 | 1.000 | | |
| C12 | 0.587 | 0.489 | 0.336 | 0.506 | 1.000 | |
| C13 | 0.414 | 0.432 | 0.431 | 0.358 | 0.404 | 1.000 |
| C14 | 0.428 | 0.477 | 0.360 | 0.291 | 0.389 | 0.529 |
| C15 | 0.324 | 0.150 | 0.381 | 0.468 | 0.590 | 0.540 |
| C16 | 0.408 | 0.360 | 0.254 | 0.395 | 0.422 | 0.443 |
| C17 | 0.320 | 0.340 | 0.402 | 0.477 | 0.422 | 0.456 |
| C18 | 0.458 | 0.442 | 0.181 | 0.341 | 0.459 | 0.504 |
| C19 | 0.389 | 0.529 | 0.249 | 0.448 | 0.443 | 0.477 |
| C20 | 0.537 | 0.324 | 0.151 | 0.367 | 0.319 | 0.479 |

*(Continued)*

|      | C15   | C16   | C17   | C18   | C19   | C20   |
|------|-------|-------|-------|-------|-------|-------|
| C15  | 1.000 |       |       |       |       |       |
| C16  | 0.593 | 1.000 |       |       |       |       |
| C17  | 0.485 | 0.589 | 1.000 |       |       |       |
| C18  | 0.125 | 0.251 | 1.256 | 1.000 |       |       |
| C19  | 0.258 | 0.521 | 0.328 | 0.125 | 1.000 |       |
| C20  | 0.394 | 0.258 | 0.125 | 0.258 | 0.456 | 1.000 |

Conduct an exploratory factor analysis on the above correlation matrix. How many factors should be retained? Explain how you decided. Interpret the factors.

6. Here is a correlation matrix for eight indicators of four variables, for a sample of 200 students. SES is an indicator of socioeconomic status; MA1 and MA2 are indicators of mental ability (MA); ASP1 and ASP2 are indicators of aspirations (ASP); and AA1, AA2, and AA3 are indicators of academic achievement (AA). Assume that SES affects ASP, and MA and ASP affect AA.

|      | SES   | MA1   | MA2   | ASP1  | ASP2  | AA1   | AA2   | AA3   |
|------|-------|-------|-------|-------|-------|-------|-------|-------|
| SES  | 1.00  |       |       |       |       |       |       |       |
| MA1  | 0.33  | 1.00  |       |       |       |       |       |       |
| MA2  | 0.30  | 0.71  | 1.00  |       |       |       |       |       |
| ASP1 | 0.34  | 0.10  | 0.18  | 1.00  |       |       |       |       |
| ASP2 | 0.29  | 0.22  | 0.10  | 0.50  | 1.00  |       |       |       |
| AA1  | 0.32  | 0.40  | 0.37  | 0.41  | 0.30  | 1.00  |       |       |
| AA2  | 0.28  | 0.40  | 0.36  | 0.39  | 0.20  | 0.69  | 1.00  |       |
| AA3  | 0.34  | 0.32  | 0.33  | 0.31  | 0.20  | 0.63  | 0.55  | 1.00  |
| Standard deviations |       |       |       |       |       |       |       |       |
|      | 1.09  | 60.12 | 58.63 | 5.62  | 10.36 | 7.84  | 6.59  | 7.18  |

Conduct an exploratory factor analysis on this correlation matrix. How many factors should be retained? Explain how you decided. Interpret the factors.

7. The following correlation matrix is presented for a sample of 75 students and includes five indicators for two dimensions of achievement. The Xs are indicators of mental achievement and the Ys are indicators of physical achievement.

|     | X1    | X2    | Y1    | Y2    | Y3    |
|-----|-------|-------|-------|-------|-------|
| X1  | 1.000 |       |       |       |       |
| X2  | 0.852 | 1.000 |       |       |       |
| Y1  | 0.321 | 0.325 | 1.000 |       |       |
| Y2  | 0.215 | 0.125 | 0.599 | 1.000 |       |
| Y3  | 0.012 | 0.085 | 0.875 | 0.458 | 1.000 |

Conduct an exploratory factor analysis on this correlation matrix. How many factors should be retained? Explain how you decided. Interpret the factors.

CHAPTER EIGHT

# Confirmatory Factor Analysis and Structural Equation Modeling

In the previous chapter the factor-analytic model was presented as an exploratory technique for discovering the factors underlying the interrelationships between a set of observed variables. In this chapter the factor-analytic model is examined as a confirmatory technique. In general terms, confirmatory factor analysis (CFA) is not concerned with discovering a factor (or latent variable) structure but with confirming the existence of a specific factor structure. Of course, in order for one to confirm the existence of a specific structure, one must have definite ideas about the composition of the underlying factor structure. In contrast, the approach taken in exploratory factor analysis is one of raw empiricism, where little is known about the factor structure (except, perhaps, what measures might prove useful in detecting it). As Thurstone noted at the early stages of the development of factor analysis (1947, p. 56):

> The exploratory nature of factor analysis is often not understood. Factor analysis has its principal usefulness at the borderline of science. It is naturally superseded by rational formulations in terms of the science involved. Factor analysis is useful, especially in those domains where basic and fruitful concepts are essentially lacking and where crucial experiments have been difficult to conceive. The new methods have a humble role. They enable us to make only the crudest first map of a new domain.

In fact, even as early as 1935 Thurstone had written about the exploratory use of factor analysis (1935, p. xi):

> No one would think of investigating the fundamental laws of classical mechanics by correlational methods or by factor methods, because the laws of

208

classical mechanics are already well known. If nothing were known about the law of falling bodies, it would be sensible to analyze, factorially, a great many attributes of objects that are dropped or thrown from an elevated point. It would then be discovered that one factor is heavily loaded with time of fall and with the distance fallen but that this factor has a zero loading in the weight of the object.

The characterization of exploratory factor analysis (EFA) given by Thurstone suggests an inductivist technique most useful when little is known about a domain. In contrast, confirmatory factor analysis (CFA) is a "hypotheticist procedure designed to test a hypothesis about the relationship of certain hypothetical common factor variables, whose number and interpretation *are given in advance*, to the observed variables" (Mulaik, 1988, p. 265—emphasis added). In fact, in CFA, knowledge of the domain should ideally be as advanced as to allow one to specify exact numerical values for the loadings in the factor pattern matrix.

## A SHORT HISTORY OF CONFIRMATORY FACTOR ANALYSIS

It is a great irony that factor analysis was developed at the turn of the century as an exclusively confirmatory technique. Spearman (1904) was interested in confirming his two-factor theory of intelligence, which held that a general factor $g$ of intelligence was the primary force responsible for the moderate to high correlations found among cognitively oriented tests. If, in fact, only one factor was responsible for the correlations among a set of cognitive variables, Spearman proposed that the correlations must form *tetrad equalities*. Tetrad equalities involve products of correlations among a set of four observed variables. For example, consider the following $X_1$, $X_2$, $X_3$, and $X_4$ as a set of four observed variables. Six correlations may be computed from these four observed variables:

$$r_{X_1,X_2} \quad r_{X_1,X_3} \quad r_{X_1,X_4} \quad r_{X_2,X_3} \quad r_{X_2,X_4} \quad r_{X_3,X_4}$$

In order to form tetrads, the correlations are taken, two at a time, and multiplied, in order to form three unique products of correlations:

$$r_{X_1,X_2} \times r_{X_3,X_4}$$
$$r_{X_1,X_3} \times r_{X_2,X_4}$$
$$r_{X_2,X_3} \times r_{X_1,X_4}$$

From these correlation products, three independent tetrad equalities can be formed:

$$r_{x_1,x_2} \times r_{x_3,x_4} = r_{x_1,x_3} \times r_{x_2,x_4}$$

$$r_{x_1,x_2} \times r_{x_3,x_4} = r_{x_2,x_3} \times r_{x_1,x_4}$$

$$r_{x_1,x_3} \times r_{x_2,x_4} = r_{x_2,x_3} \times r_{x_1,x_4}$$

If these equalities hold, then the existence of one factor is confirmed. This is because if the conditions imposed by tetrad equalities hold, the proportionality among the correlations implies a matrix of unit rank or a matrix having one nonzero eigenvalue as well as one nonzero determinant (for further discussion on matrices see chapter 2). A great deal of effort was expended by Spearman and his colleagues (Holzinger, 1930; Holzinger & Spearman, 1926) to develop a statistical test for when the equalities implied by the tetrads did not hold (i.e., whether the difference $r_{x_1,x_2} \times r_{x_3,x_4} - r_{x_1,x_3} \times r_{x_2,x_4}$ [termed a *vanishing tetrad*] did not equal zero). Kenny (1974) much later actually proposed a test for a vanishing tetrad in which the second canonical correlation extracted from the matrix must equal zero (which must be the case if only one eigenvalue is nonzero).

Spearman was not the only investigator to pursue this line of factor confirmation. Kelly (1935) also proposed the pentad criterion (for five variables) for confirming the existence of two factors. Mathematically, this criterion is written as:

$$r_{12}r_{23}r_{34}r_{45}r_{51} - r_{12}r_{23}r_{35}r_{41}r_{54} - r_{12}r_{24}r_{35}r_{43}r_{51} +$$

$$r_{12}r_{24}r_{31}r_{45}r_{53} - r_{12}r_{25}r_{34}r_{41}r_{53} - r_{12}r_{25}r_{31}r_{43}r_{54} -$$

$$r_{13}r_{24}r_{35}r_{41}r_{52} - r_{13}r_{25}r_{34}r_{42}r_{51} - r_{14}r_{23}r_{31}r_{45}r_{52} -$$

$$r_{14}r_{25}r_{32}r_{43}r_{51} - r_{15}r_{23}r_{31}r_{42}r_{54} - r_{15}r_{24}r_{32}r_{41}r_{53} = 0$$

If the pentad criterion held, then the correlation matrix among the five variables had two nonzero determinants or two nonzero eigenvalues. What made the tetrad and pentad criteria truly confirmatory approaches to factor analysis was that these criteria would hold regardless of the specific values of the correlations (it was their pattern of proportionality that was important).

Nevertheless, despite the original confirmatory nature of factor analysis, as the technique developed and was used in a broader spectrum of substantive areas, it quickly lost its confirmatory characteristic. One of the few attempts to reinstate confirmation in factor analysis was Cattell's Procrustes factor analysis (Hurley & Cattell, 1962)—although Mosier (1939) had proposed an almost identical method previously. Procrustes factor analysis, named after Procrustes from Greek mythology (the highwayman who either stretched or cut his victims to fit his bed), involves rotating a factor pattern matrix as forcefully as possible to a target matrix. The rotation can either be orthogonal or oblique. Recall from chapter 7 that the target matrix is a matrix of hypothesized factor loadings. Thus, the main goal of Procrustes factor analysis is to see how closely it is possible to rotate the original factor pattern into the target pattern. Unfortunately, because no significance test can be used to test the closeness between the initial and target patterns, the technique did not receive widespread support.

And yet, Procrustes factor analysis might have eventually become the preferred method of confirmatory factor analysis if not for the development of the maximum likelihood factor analysis (Jöreskog, 1967, 1969, 1979). Maximum likelihood quickly became popular because it provided a chi-square goodness-of-fit test between the original correlation (or covariance) matrix and the reproduced correlation (or covariance) matrix. Presumably, if the appropriate number of factors were extracted, the reproduced matrix did not differ significantly from the original matrix. Jöreskog (1969) also went on to propose the application of maximum likelihood factor analysis to situations in which only a subset of loadings in a factor pattern matrix was estimated (i.e., a subset consistent with a researcher's prior hypotheses). In fact, from this early work Jöreskog and Sörbom (1981) developed the first structural equation modeling program that incorporated confirmatory factor analysis. The program eventually became known as LISREL (for linear structural relations; Jöreskog & Sörbom, 1981). Since then, the LISREL program has become so popular that many researchers (incorrectly) refer to structural equation modeling by the name of the program.

## WHY THE TERM "STRUCTURAL EQUATION MODEL"?

Confirmatory factor analysis is one among many models that may be labeled generally as *structural equation models*. In its broadest sense the term *structural equation model* (SEM) is concerned with testing complex models for structure of functional relationships between observed variables and latent variables (Marcoulides, 1995). The functional relationships are described by parameters that indicate the magnitude of the effect (direct or indirect) that independent variables have on dependent variables. Thus, a structural equation model can be considered just like a series of linear regression equations relating dependent variables to independent and other dependent variables (Marcoulides, 1996). The coefficients determining the relations are usually the parameters we are interested in solving. For example, consider these two structural equations:

$$X = aA + bB + 2cC$$
$$Y = xX + 3bB$$

In these two equations, the variable $X$ is determined by $A$, $B$, and $C$, with unknown coefficients $a$, $b$, and $c$ to be solved, and the variable $Y$ is determined by $X$ and $B$, with unknown coefficients $x$ and $b$ to be solved. However, in SEM interest usually centers on explaining the correlation or covariance structure among the variables $X$ and $Y$. Thus, although the structural equations for $X$ and $Y$ are as just given, their correlation can be written as a function of:

$$r_{xy} = a + 3b^2 + c^2$$

As such, given enough information, one can easily solve for the unknown parameters $a$, $b$, and $c$ in terms of the correlations.

The term *structural equation modeling* was first coined by econometricians and is probably the most appropriate name for the process we have just briefly sketched. Yet other terms tend to be used interchangeably with it. For example, *path analysis*, developed by Sewall Wright (1920), is an early form of structural equation modeling that is restricted to observed variables. The observed variables are assumed to have been measured without error and are unidirectionally related to one another. As it turns out, the confusion with terms exists because the rules of path analysis are still used to identify the structural equations underlying the models. Using the path analysis approach, models are presented in the form of a drawing (often called a path diagram), and the structural equations of the model are inferred by reading this diagram correctly. Heise (1975), Kenny (1979), Li (1975), and Loehlin (1992) provide thorough discussions of path-analytic rules. In general, we feel that the term *path analysis* implies too many restrictions on the form of the model. Structural equation modeling, on the other hand, has grown to incorporate latent and observed variables that can be measured with and without error and have bidirectional relations among variables.

Another term used frequently is *causal analysis*. Unfortunately, this is also a misleading term. Although structural equation modeling may appear to imply causality, the structural equations are not causal relations but functional relations (e.g., $Y$ is a function of $A$, $B$, and $C$). Speculations about the nature of causality abound and there really is no single definition that is routinely employed in structural models. Bollen (1989) provides an excellent definition of causality oriented toward structural models. Bollen's (1989) definition includes isolation, association, and direction of causality as the three conditions that should always be used to establish a causal relation. However, because it is quite difficult to be certain that a cause and an effect are isolated from all other influences, one must regard all structural models as approximations to reality. Thus, one can never really prove a model or the causal relations within it. Nevertheless, one can at least disconfirm a proposed model.

*Covariance structure modeling* is another popular term that is used mostly by psychologists. Unfortunately, it too is restrictive. Although the covariance or correlation structure of observed data is the most commonly modeled, structural equation modeling can be used to model any moments of the data.[1] For example, mean structures are occasionally modeled, and facilities are provided for this in a number of structural equation modeling programs (for an excellent overview see Browne & Arminger, 1995). Modeling the third (skew) and fourth (kurtosis) moments of the data has also recently been discussed (see Bentler, 1983), but has received relatively little attention in the literature. This lack of attention is probably due to skepticism concerning how much unique information higher moments actually contribute beyond the mean and variance.

## USING DIAGRAMS TO REPRESENT MODELS

One of the clearest methods for communicating any mathematical model is to draw a picture of it (Marcoulides & Papadopoulos, 1993). As it turns out, a significant contribution of path analysis was the development of path diagrams for the representation of an SEM model. Figure 8.1 presents the most commonly used symbols for the representation of SEM models in path diagrams. Figure 8.2a presents a path diagram that can be used to represent the common factor model (see also Fig. 7.1), and Fig. 8.2b presents a path diagram to represent the principal components model.

In comparing the path diagrams of the common factor and principal components models, the reader should note an important difference as to how the observed variables are related to the latent variable (or factor). In the common factor model, the arrows point from the latent variable to the observed variable, whereas in the principal components (PCA) model, the arrows' direction is reversed. This is because factor analysis attempts to explain the intercorrelations among the observed variables, whereas PCA attempts to account for the variability in the observed variables. Another way to think of this issue is to consider the common factor model as a model in which the observed variables and their correlations are "affected" by the latent variables; on the other hand, in the principal components model, the observed variables "cause" variance in the components. Bollen and Lennox (1991) labeled the observed variables in common factor models as *effect indicators* because they indicate the effects of latent variables, and

---

[1]The discussion of SEM in this book is restricted to analyzing correlation/covariance structures. Of course, between the two, covariances are preferable for two reasons: (a) the statistical theory underlying estimation techniques used in SEM are based on the behavior of covariances and it is questionable to what extent the statistical theory can be generalized to correlations, and (b) correlations remove information concerning variability, which may be of great interest in comparing groups or in examining the behavior of a variable across time. Cudeck (1989) and Lee (1985) provided additional reasons for caution in using correlations.

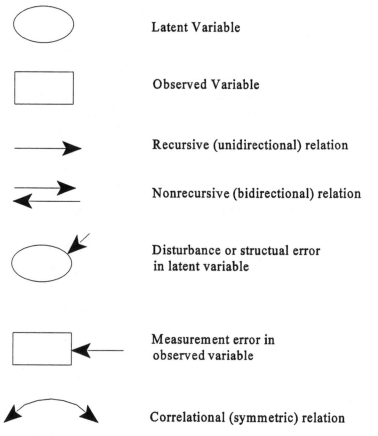

Latent Variable

Observed Variable

Recursive (unidirectional) relation

Nonrecursive (bidirectional) relation

Disturbance or structual error
in latent variable

Measurement error in
observed variable

Correlational (symmetric) relation

FIG. 8.1. Commonly used symbols for SEM models in path diagrams.

labeled the observed variables in principal component models as *causal indicators* because principal components are constructed from (caused by) a linear combination of the variables. With this in mind, one should think of the confirmatory factor model as just another common factor model with a priori defined structure. Confirmatory factor models are also commonly referred to as *measurement models* because they specify what observed variables are needed to measure latent variables.

## MATHEMATICAL REPRESENTATION
## OF STRUCTURAL EQUATION MODELS

Structural equation models are mathematical representations of hypothesized relations among a set of variables. As it turns out, each of the techniques discussed so far in this book can be conceived as representing a very specific

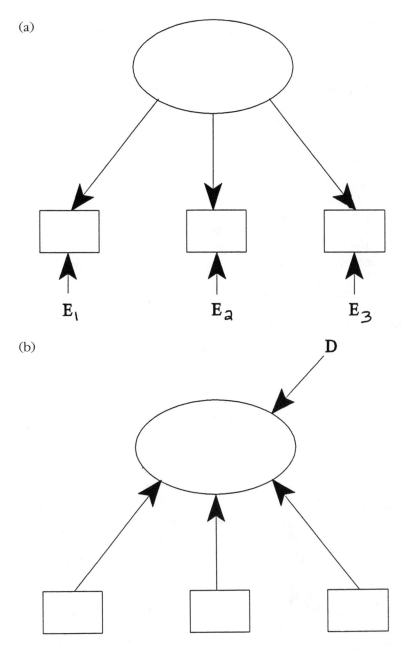

(a)

$E_1$          $E_2$          $E_3$

(b)                                    D

FIG. 8.2. (a) Path diagram to represent the common factor model. (b) Path diagram to represent the principal components model.

type of model of the data. For example, the canonical correlation model presented in chapter 6 simply posits relations between pairs of latent variables. An example path diagram of a canonical correlation model with two pairs of latent variables is given in Fig. 8.3. As can be seen from Fig. 8.3, the hypothesized relations within the canonical correlation model are represented by parameters, and it is these parameters that need to be determined. Of course, it should be clear that in order to determine the parameters of any model, there must be a mathematical model for representing these parameters as functions of the data.

To date, several mathematical models for SEM have been proposed in the literature. Although these mathematical models can translate data equally well into the model parameters, they differ in how parsimoniously this translation process is conducted. Perhaps the most well known of these mathematical models, the Keesling–Wiley–Jöreskog model, can require up to nine different parameters in order to represent a model. In contrast, the COSAN model can generally represent the same model using only two parameters. The mathematical model presented in this chapter is the Bentler–Weeks model (known also by the name of its computer implementation, EQS; Bentler, 1989, 1995). The Bentler–Weeks model can represent any model using only four parameters. Mathematically, the Bentler–Weeks model is represented by:

$$\eta = \beta\eta + \gamma\xi$$

where $\beta$ and $\gamma$ are coefficient matrices, and $\eta$ and $\xi$ are vectors of random variables. The random variables within $\eta$ are endogenous variables and the

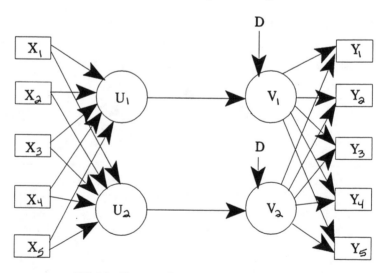

FIG. 8.3. Diagram of a canonical correlation model.

variables within $\xi$ are exogenous variables. Endogenous variables are dependent variables; exogenous variables are independent variables. Endogenous and exogenous variables can be either latent or observed. The matrix $\beta$ consists of coefficients (parameters) that describe the relations among the endogenous variables. The matrix $\xi$ consists of coefficients (parameters) that describe the relations between independent and dependent variables.

Two types of errors exist in the Bentler–Weeks model, one associated with latent variables, the other associated with observed variables. Latent variable errors are termed *structural errors* or *disturbances* because they represent variance in the latent variable unexplained by the model. Observed variable errors are measurement errors, and as measurement error, represent variance that is unique to the observed variable. Errors are considered to be a type of independent variable, and are placed within the exogenous variable vector $\xi$. Thus, in the Bentler–Weeks model, one is provided with a mathematical system for describing the interrelations among a set of latent and observed variables. It should also be noted that all variables (both latent and observed, exogenous and endogenous) are assumed to be in deviation score form (i.e., have a zero mean). Therefore, none of the structural models discussed in this chapter will have an intercept term. For those interested in incorporating intercepts into structural equations, a method for modeling mean differences between groups is presented in Byrne (1994) and Schumacker and Lomax (1996).

It is important to note that the primary interest in SEM modeling centers in describing the network of relations among the variables (implying that one is generally interested in the correlational or covariance structure among the variables). Although the SEM model is written in terms of equations linking the variables, the data used to solve the model parameters are actually covariances or correlations. In fact, this approach is no different from how the other models described in this book proceeded. For example, multiple regression uses a series of equation that link dependent to independent variables, but it is the correlational structure of the data that is used to solve for the regression coefficients. Similarly, in the Bentler–Weeks system, the sample covariance structure ($C$) among a set of variables $x,y$ is defined as

$$C = (x + y)(x + y)' = J(I - B)^{-1}\Gamma\Phi\Gamma'(I - B)^{-1}J'$$

where $\Gamma$ is a matrix of coefficients linking exogenous ($\xi$) with endogenous variables, $B$ is a matrix of coefficients linking endogenous variables ($\eta$), and $\Phi$ represents the covariances among the exogenous variables. The $J$ matrix serves as a "filter" for selecting the observed variables from the total number of variables to be included in the model.

Fortunately, in order to use the Bentler–Weeks model one rarely has to worry about explicitly communicating the covariance structure among the variables in the hypothesized model. What is required is the definition of a series of equations describing the hypothesized relations within the model. To simplify matters even more, one can write the equations using a simple four-letter system. This four-letter system consists of the following letters: $F$ (representing latent variables), $V$ (representing observed variables), $D$ (representing disturbances or unexplained effects in the latent variables), and $E$ (representing measurement error or unexplained effects in the observed variables). As discussed in the next section, every model suitable for evaluation by the Bentler–Weeks system (which includes most linear models) can be represented by equations using these four letters.

## THE CONFIRMATORY FACTOR ANALYSIS MODEL

In chapter 7, the exploratory factor model (EFA) was given as

$$X = \lambda f + e$$

and the correlation structure $(R)$ implied by the model was shown to be

$$R = (Z)(Z)' = (\lambda f + \varepsilon)(\lambda f + \varepsilon)' = \Lambda \Phi \Lambda' + \Theta$$

where $\Phi$ represents the correlations among the common factors and $\Theta$ refers to the errors or uniqueness. Recall that this expression implies that there is no correlation between the factors and errors of measurement or between the errors of measurement [i.e., $R(F,E) = R(E,E) = 0$].

As traditionally given, the confirmatory factor model in matrix notation is

$$Y = \Lambda \xi + e$$

where $Y$ are the scores on the observed variables, $\Lambda$ is a factor pattern loading matrix, $\xi$ are the common factors, and $e$ are the measurement errors in the observed variables. It should be clear to the reader that this notation for the CFA model is comparable to the common factor model.

As such, the covariance structure implied by the confirmatory factor model is defined as

$$C = \Lambda \Phi \Lambda' + \Psi$$

where $C$ is the sample variance–covariance matrix, $\Phi$ is a matrix of the factor variances–covariances, and $\Psi$ is a variance–covariance matrix among the measurement errors. Although the exploratory factor model and the common factor model share many similarities, there are several notable differences. First, note that the correlation matrix $(R)$ is modeled by EFA and the covariance matrix ($C$) is modeled by CFA. Thus, EFA generally involves an eigenanalysis of the observed correlation matrix, while CFA involves a covariance matrix (although both techniques can employ either a correlation or covariance matrix). Second, the assumption that the errors are uncorrelated [i.e., $R(E,E)$ = 0] is relaxed in CFA. In fact, errors across different variables can correlate in CFA, resulting in a nondiagonal $\Psi$ matrix. Finally, in EFA, if the factor pattern consists of $p$ rows and $r$ columns, a value for every cell of the $p \times r$ matrix is determined (i.e., all possible factor pattern loadings will be obtained). This is almost never true in CFA because the number of elements in the factor structure that are determined is generally less than $p \times r$ (i.e., the defined factor structure specifies a particular pattern of loadings).

In the Bentler–Weeks representation, the confirmatory factor model is generally expressed as

$$\eta = \beta\eta + \gamma\xi \ \text{ with } \ \beta = 0$$

and the covariance structure implied by the model is given as before as

$$\text{COV}(\eta\eta') = (0\eta + \gamma\xi)(0\eta + \gamma\xi)' = \Gamma\Phi\Gamma'$$

where the asymmetric relations in the model (the effects of the common and error factors on the observed variables) are in $\Gamma$ and the symmetric relations (the factor and error variances and covariances) are in $\Phi$. Note that for the confirmatory factor model the matrix $\beta$ is dropped from the complete set of equations in the Bentler–Weeks system presented earlier because in CFA there are no regression relations between endogenous variables.

So how does one communicate a hypothesized model in the Bentler–Weeks system? To answer this question let us consider an example study in which a researcher would like to confirm that intelligence consists of the two factors of crystallized and fluid intelligence (F1 and F2, respectively). The researcher also believes that these two factors are correlated with each other. In order to measure the crystallized intelligence factor three observed test scores (V1, V2, and V3) were recorded. In addition, three observed test scores of the fluid intelligence factor (V4, V5, and V6) were recorded. Each of these tests is known to have unique sources of variance (E1 through E6), but the researcher has no reason to believe they are correlated across tests. Based on this study description, the following six structural equations can be written:

$$V1 = F1 + E1$$
$$V2 = F1 + E2$$
$$V3 = F1 + E3$$
$$V4 = F2 + E4$$
$$V5 = F2 + E5$$
$$V6 = F2 + E6$$

with the side conditions that F1,F2 are correlated, and E1 through E6 are uncorrelated.

It is important to note that if one were taking an exploratory approach to this study, the structural equation model would differ in the following manner:

$$V1 = F1 + F2 + E1$$
$$V2 = F1 + F2 + E2$$
$$V3 = F1 + F2 + E3$$
$$V4 = F1 + F2 + E4$$
$$V5 = F1 + F2 + E5$$
$$V6 = F1 + F2 + E6$$

It should be clear that if, in fact, the confirmatory factor model "fits" the data, one will have confirmed that the two-factor theory of intelligence is consistent with the data. On the other hand, the EFA approach does not provide such a test. Although two factors are proposed, no distinction is made with respect to which tests should define the different factors.

## MODEL ESTIMATION

In general, confirmation of a model relies upon the retention of the following null hypothesis:

$H_0$: The data are consistent with the model.

Rejection of this null hypothesis is a rejection of one's model. Of course, the problem arises as to how "consistency" is to be determined. Thus, how consistent is consistent? Recall that the ultimate goal of CFA is to model the covariance or correlation structure of some data of interest. Unfortunately, all one has is an observed correlation or covariance matrix. Of course, there is great hope that this correlation or covariance matrix is a good approximation to the true correlation or covariance matrix in the population. Obviously, one would always prefer to test a proposed model using the population matrix instead of the sample matrix. Nevertheless, the sample matrix is all that is available. As such, it is the sample correlation or covariance matrix that is used to solve for the model parameters. Essentially, the structural equations are rewritten so that each of the parameters of the equations are a function of the elements of the sample matrix. Subsequently, after obtaining values for the parameters, if one were to substitute these values back into the expression for the correlation structure implied by the model, the resulting sample matrix (e.g., $S$) can be represented as $\hat{S}$. Clearly, $\hat{S}$

should be very close to $S$ because it was the elements of $S$ that assisted in solving for the model parameters. Thus, if one considers $S$ to represent the population matrix $\Sigma$, then the difference between $\hat{S}$ and $S$ (i.e., $S - \hat{S}$) should be small if the model is consistent with the data.

For example, assume that the following observed variance–covariance matrix $S$ is observed:

$$\begin{bmatrix} 0.80 & 0.60 \\ 0.60 & 0.70 \end{bmatrix}$$

and that the following model underlying the variances and covariances is proposed:

$$\begin{bmatrix} a^2 + b^2 & a^2 \\ a^2 & b^2 \end{bmatrix}$$

Writing the model parameters in terms of the variances and covariances and solving for $a$ and $b$ provides the following equations:

$$0.80 = a^2 + b^2$$
$$0.70 = b^2$$
$$0.60 = a^2$$

which yield $b = 0.84$ and $a = 0.77$. Substituting the values of 0.84 and 0.77 back into the variance–covariance equations provides the model-implied covariance matrix $(\hat{S})$:

$$\begin{bmatrix} 1.50 & 0.60 \\ 0.60 & 0.70 \end{bmatrix}$$

Finally, the difference between $S$ and $\hat{S}$,

$$\begin{bmatrix} 0.80 & 0.60 \\ 0.60 & 0.70 \end{bmatrix} - \begin{bmatrix} 1.50 & 0.60 \\ 0.60 & 0.70 \end{bmatrix} = \left| \begin{bmatrix} 0.70 & 0.00 \\ 0.00 & 0.00 \end{bmatrix} \right|$$

is evaluated for how large the difference is. The larger the difference is, the more the model misfits the data.

As it turns out, the evaluation of the difference between $\hat{S}$ and $S$ depends on the estimation method used to solve for the model parameters. The most commonly used estimation methods for solving the parameters of the confirmatory factor model (and other structural equation models) are unweighted least squares (ULS), generalized (weighted) least squares (GLS), and maximum likelihood (ML). With each estimation method, the structural equations of the model are solved iteratively or in steps, until optimal estimates of the parameters are obtained. Optimal parameter values are values that imply covariances ($\hat{S}$) close to the observed covariances ($S$) (i.e., are not too discrepant from the elements of $S$). As the reader may have surmised, this is another problem in partial differentiation. The difference between $\hat{S}$ and $S$ is known as a discrepancy function ($F$). In order to minimize this discrepancy function, the partial derivatives of the discrepancy function are taken with respect to the elements of $S - \hat{S}$. The form of the discrepancy function varies across the different estimation methods. However, the general form of this discrepancy function is

$$F = \sum_{ij} (S - \hat{S})' W(S - \hat{S})$$

in which a weighted sum of differences between the $I$ elements of $S$ and $\hat{S}$ is calculated. As $S$ and $\hat{S}$ become more different, the discrepancy function becomes larger, implying less correspondence between the model-implied covariances and the observed covariances.

Most currently available computer software packages capable of conducting CFA (and for that matter all SEM models) include ULS, GLS, and ML as standard estimation methods. For example, in SAS the CALIS (covariance analysis of linear structural equations) procedure estimates parameters and tests the appropriateness of a model using any one of these estimation procedures. It is important to note that only the ULS fit function as implemented in SAS does not assume that the variables are multivariate normally distributed (although, as becomes clear later, an evalution of the chi-square goodness-of-fit index does require multivariate normality). In contrast, the GLS and the ML estimation procedures both assume multivariate normality. If multivariate normality is present, a chi-square goodness-of-fit test for the model is available using the sample size and the value of the discrepancy function:

$$\chi^2 = (N - 1)(F)$$

with df = (the number of unique elements of $S$) – (the number of parameters solved, $q$). If chi-square is not significant, then no significant discrepancy exists between the implied and observed covariance/correlations. As such, the model fits the data and is confirmed. If chi-square is significant, there exists a significant discrepancy between the implied and observed matrices: The model does not fit the data and is rejected. Thus, this is one of those rare occassions in which a researcher is more interested in retaining the null hypothesis.

## THE PROBLEM OF IDENTIFICATION

Ironically, although SEM may appear very new to readers, most have really had more experience conducting SEM than multivariate analysis of variance (MANOVA), discriminant analysis, or any other multivariate procedure presented in this book. The reader may recall from high school algebra being given the tedious task of solving a set of ($k$) simultaneous equations in $k$ unknowns. For example, consider the following two equations with two unknowns,

$$20 = 2x + 3y$$
$$10 = x - 4y$$

One very popular method for obtaining the values of $x$ and $y$ is the Gaussian elimination approach. To some extent, obtaining the values of $x$ and $y$ using the Gaussian elimination approach is similar to conducting a simple form of SEM. For example, if we think of the two equations just given as structural equations (which they can be), 20 and 10 as two covariances, $x$ and $y$ as two parameters (which they certainly are), and Gaussian elimination as a method for obtaining optimal values for $x$ and $y$, then we are conducting SEM. Of course, SEM as practiced is more complex than these two simultaneous equations, but the logic is precisely the same: A series of equations is solved for which one is interested in obtaining optimal values.

Unfortunately, in order to obtain optimal values for structural equations, the solutions are determined iteratively. In contrast, the solution to the preceding simultaneous equations can be obtained in one step (i.e., the first values of $x$ and $y$ obtained are the best values possible). In fact, not only are they the best values possible, they are the *only* values possible. Why the only possible values? Because there are as many equations as there are unknowns. When such a situation arises, the model is said to be *just-identified*. In SEM, just-identification occurs when the number of data elements (e.g., correlations, covariances, etc.) equals the number of parameters. If the model is just-identified a solution can always be found for the parameter estimates that will result in perfect fit—a discrepancy function equal to zero. This implies that the both the chi-square goodness of fit test and its associated degrees of freedom will be zero. As such, there is no way one can really test/confirm the verisimilitude of a just-identified model. Thus, just-identified models will always fit the data perfectly. Of course, this is not necessarily bad. In fact, all of the statistical models discussed so far in this book are (at best) just-identified. For example, in a multiple regression model with five predictors and one criterion there are 21 unique elements in the variance–covariance matrix. Thus, 21 is exactly the number of parameters that are determined in the multiple regression problem (6 regression coefficients +

1 intercept + 14 covariances among the predictors = 21). Although the situation is not bad, it is rather limiting. Because one of the primary reasons for doing SEM is to test explicitly the fit of a proposed model, at a minimum, this requires positive degrees of freedom (i.e., more unique elements in the data matrix than parameters to be estimated).

To illustrate the identification problem further, consider the simple equation

$$20 = x + y$$

and attempt to find a solution for $x$ and $y$. Clearly, one solution could be $x = 5$, $y = 15$ and another could be $x = 1$, $y = 19$. Obviously, there is no unique solution for $x$ and $y$. In fact, there is an infinite number of solutions because there are more unknown parameters (two) than summary statistics (one). The model represented by this equation is said to be *underidentified* and any estimates obtained for parameters are meaningless.[2]

The most desirable condition to encounter in SEM is *overidentification*, where the number of available variances–covariances is always more than what is needed to obtain a unique solution. In fact, with an overidentified model, the degrees of freedom are always positive so that model fit can be explicitly tested. An overidentified model also implies that, for at least one of the model parameters, there is more than one equation that the solution to the parameter must satisfy. The number of additional equations the solution must satisfy is generally referred to as the number of *overidentifying constraints*. A simple way to think of overidentifying constraints is like hurdles in a horse race: The more hurdles (equations) the parameter solution satisfies (jumps over), the more confidence we can have in the accuracy of the solution. Not surprisingly, the greater the number of overidentifying constraints is, the more the degrees of freedom. Thus, before conducting any SEM analysis, one must verify the identification status of the proposed model. Unfortunately, herein lies the difficulty! Having positive degrees of freedom is a necessary but not a sufficient condition for identification. There are situations where the number of degrees of freedom for a model is quite high and yet some or all the parameters are underidentified.

---

[2]To some extent, even the discriminant analysis, canonical correlation, and common factor models would be underidentified if it were not for certain restrictions placed on the models. For example, the common factor model solves for $p \times r$ factor loadings in addition to $p$ uniquenesses. In the most extreme case, if one takes $r = p$, then one is solving for $p(1 + p)$ parameters when there are only half as many variance–covariance elements [$p(1 + p)/2$] available. Placing the restrictions that the common factors are extracted in order of magnitude, extracted to be orthogonal, and imposing the unit norm condition on the eigenvectors removes the identification problem to a degree (the pattern coefficients become identified up to a proportionality constant).

Model identification is an enormously complex issue, and generally much more complex than the simple example just given would suggest. For structural equation models in general, the most frequently invoked identification rules are the $t$-rule, and the order and rank conditions (Bollen, 1989). The $t$-rule is a simple rule to apply, but is only a necessary not a sufficient condition of identification. Basically, the $t$-rule for identification is that the number of nonredundant elements in the correlation/covariance matrix of the observed variables $(p)$ must be greater than or equal to the number of unknown parameters in the proposed model. Thus, if $t \leq p(p + 1)/2$ the necessary condition of identification is met. Unfortunately, although the $t$-rule is simple to apply, it is only good for determining underidentified models.

The order condition requires that for a model to be identified, the number of $p$ variables excluded from each structural equation must equal to $p - 1$. Unfortunately, the order condition is also a necessary but not sufficient criterion for model identification. Only the rank condition is both a necessary and sufficient condition for identification; however, it is not easy to apply. In general terms, the rank condition requires that the rank of any model matrices (e.g., $\Phi$, $B$, $\Gamma$) be of at least rank $p - 1$ for all submatrices formed by removing the parameter of interest. For example, with a confirmatory factor model the model is identified if:

1. $\Phi$ (the variance–covariance matrix for the factors) is a symmetric, positive definite matrix with unities in the diagonal.
2. $\Psi$ (the variance–covariance matrix for the measurement errors) is diagonal.
3. $\Lambda$ (the factor pattern matrix) has at least $r - 1$ fixed zeros in each column, $r$ equal to the number of factors.
4. $\Lambda^k$ has rank $r - 1$, where $\Lambda^k$ is the submatrix of $\Lambda$ consisting of the rows of $\Lambda$ that have zero elements in the $k$th column.

And yet, the usefulness of these criteria is doubtful because a failure to meet them does not necessarily mean the model is underidentified.

As it turns out, the only sure way to assess the identification status of a model prior to model fitting is to show through algebraic manipulation that each of the model parameters can be solved in terms of the variances and covariances. For most models, this is an ardouous task indeed! However, an efficient method for verifying the identification status of a model through algebraic manipulation of the variance–covariance equations does exist if one has access to a mathematics program such as Mathematica (Wolfram, 1991). The following example illustrates how Mathematica could be used for this task:

Consider the following three equations:

$$l = b^2 + c^2 + 2d$$
$$m = b^2 + 2d$$
$$n = b^2 + 2d$$

These can easily be analyzed in Mathematica using the Reduce command:

*Reduce[ {b^2+c^2+2\*d==1,b^2+2\*d==m,b^2+2\*d==n},{b,c,d}]*

which would provide the following results as output:

$$n{=}{=}m \ \&\& \ (c{=}{=}Sqrt[1{-}m] \ || \ c \ == \ -Sqrt[1{-}m]) \ \&\& \ d \ == \ \frac{-b^2 + m}{2}$$

where == means that two quantities are equal, && means that the equations must be satisfied simultaneously, and || means these are alternative solutions, only one of which need be satisfied. Properly interpreted, parameter *c* is overidentified (but trivially so), and parameters *b* and *d* are underidentified (i.e., they are linearly dependent).

Fortunately, if one attempts to conduct an SEM analysis in SAS PROC CALIS with an underidentified model, the procedure will detect this by reporting that certain parameters are linearly dependent on others (an example of which is shown later). However, not even PROC CALIS is perfect, and occasionally an identified parameter is labeled incorrectly as unidentified by the program. In general, this problem occurs in SAS PROC CALIS primarily when the data are simply not good enough to resolve the parameter's value. This situation is also commonly referred to as *empirical underidentification*; more technical discussions of the identification problem can be found in Fisher (1966), Long (1983), and Bekker, Merckens, and Wansbeek (1994).

## A NUMERICAL CFA EXAMPLE

In a recent study, Hull, Lehn, and Tedlie (1991) attempted to confirm a simple three-factor measurement model. The three-factor model proposed by Hull et al. (1991) is presented as a path diagram in Fig. 8.4. As can be seen from Fig. 8.4, each of the three factors included in this study was measured using three observed variables. The first factor, self-criticism, was measured with three observed variables: self-criticism 1 (SC1), self-criticism 2 (SC2), and self-criticism 3 (SC3). The second factor, high standards, was measured with three observed variables: high standards 1 (HS1), high standards 2 (HS2), and high standards 3 (HS3). The third factor, overgeneralization, was also measured with three observed variables: overgeneralization 1 (OG1), overgeneralization 2 (OG2), and overgeneralization 3 (OG3). For

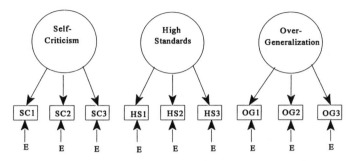

FIG. 8.4. Three-factor model proposed by Hull et al. (1991).

purposes of discussion, we first assume that there is no correlation among these three factors (later this assumption is changed).

## PROC CALIS Setup

The PROC CALIS procedure in SAS is an all-purpose structural equation modeling program that permits the user to work within the framework of several different SEM systems. Table 8.1 presents the PROC CALIS instructions for evaluating the Hull et al. model displayed in Fig. 8.4. These instructions are based on the Bentler–Weeks model discussed earlier in this chapter (an appendix is provided for readers who may be interested in seeing the same analysis conducted using the LISREL model). It is important to note that, by default, PROC CALIS analyzes the correlations among a set of observed variables. However, in deference to our preference to analyze variance–covariance matrices, the COV option in the procedure statement is selected in order to accomodate this preference (of course, in order to obtain a solution based on the COV option, one must provide both the correlation matrix among the nine variables and their standard deviations). Another default in PROC CALIS is the maximum likelihood method of estimation. If a different estimation method is preferred (e.g., ULS or GLS), a user may specify this method on the PROC CALIS line using the METHOD= statement. Finally, the specification of the number of observations (NOBS=255) is provided along with the maximum number of iterations (MAXITER=1000) that should be conducted to obtain a solution.

As noted earlier, PROC CALIS permits the user to define any one of several mathematical systems in order to solve the structural equations. Selecting the LINEQS command in the SAS program permits one to specify that the model is based on the Bentler–Weeks representation. Thus, the equations following the LINEQS command represent the Bentler–Weeks representation of the path diagram presented in Fig. 8.4. As can be seen by examining these equations, there are nine equations, one equation for each dependent variable in the model. Recall that terms beginning with V (as in

```
DATA ONE(TYPE=CORR);
_TYPE_='CORR';
INPUT_TYPE_$_NAME_$ V1-V9;
LABEL V1='SCI' V2='SC2' V3='SC3' V4='HS1' V5='HS2' V6='HS3'
      V7='OG1' V8='OG2' V9='OG3';
CARDS;
STD . 1.41 1.02 0.91 2.11 1.85 1.78 2.45 2.05 1.92
CORR V1    1.00 . . . . . . . .
CORR V2    0.40 1.00 . . . . . . .
CORR V3    0.45 0.37 1.00 . . . . . .
CORR V4    0.52 0.49 0.36 1.00 . . . . .
CORR V5    0.41 0.44 0.28 0.63 1.00 . . . .
CORR V6    0.35 0.43 0.20 0.74 0.60 1.00 . . .
CORR V7    0.34 0.20 0.51 0.35 0.15 0.20 1.00 . .
CORR V8    0.19 0.01 0.39 0.19 0.07 0.07 0.69 1.00 .
CORR V9    0.27 0.14 0.42 0.31 0.18 0.21 0.71 0.71 1.00
;
RUN;
PROC CALIS DATA=ONE ALL COV MAXITER=1000 NOBS=255;
LINEQS
V1 = V1F1 F1 + 0 F2 + 0 F3 + E1,
V2 = V2F1 F1 + 0 F2 + 0 F3 + E2,
V3 = V3F1 F1 + 0 F2 + 0 F3 + E3,
V4 = 0 F1 + V4F2 F2 + 0 F3 + E4,
V5 = 0 F1 + V5F2 F2 + 0 F3 + E5,
V6 = 0 F1 + V6F2 F2 + 0 F3 + E6,
V7 = 0 F1 + 0 F2 + V7F3 F3 + E7,
V8 = 0 F1 + 0 F2 + V8F3 F3 + E8,
V9 = 0 F1 + 0 F2 + V9F3 F3 + E9;

STD
E1-E9 = THE1-THE9,
F1-F3 = 1.00;

RUN;
```

V1 through V9) represent observed variables; terms beginning with E (as in E1 through E9) represent measurement errors (uniqueness) in the observed variables; and terms beginning with F (as in F1 through F3) represent latent variables. The first equation, for V1, stipulates that V1 (self-criticism 1) should load significantly on the first factor, with error measurement E1. In order to indicate that CALIS should solve for a parameter, a name for the parameter must be created and placed in front of the appropriate term in the equation. The name representing the loading of V1 on F1 in this example program is V1F1. Any time a term is included in an equation and no name is placed in front of it, the PROC CALIS procedure will assign to it as a coefficient a constant of 1. In fact, any number placed in front of an equation term will force that number to be the term's coefficient in the model. In this example (although it is not necessary), we have assigned 0 in front of F2 and F3 to emphasize that V1 does not load on either of these two factors. Implicitly, a 1 precedes E1. Accordingly, V1 through V3 are assigned to F1; V4 through V6, to F2; and V7 through V9, to F3.

The final command in the PROC CALIS program is the STD command. This command is optional, and is generally used to define the variances of the independent variables (either latent or observed), never the variances of the dependent variables. As can be seen by examining Table 8.1, by assigning names to the error variances (in this case, THE1–THE9), we have indicated that we want to obtain values for the error variance of each observed variable (remember that the errors are always considered to be independent variables). It is important to note that we could have alternatively assigned a value of 1 to E1 through E9 in the STD command, and solved for the standard deviation of the errors by giving them a name in the structural equations (e.g., THE1E1 in the first equation would have resulted in a solution for the standard deviation of E1).

The final specification in the STD command is of critical importance. As presented in Table 8.1, the variability of the factors has been set to unity. As it turns out, the decision to assign a value of 1 to the factor variances is based entirely on measurement issues. Because the factors (F1 through F3) have never been measured (all we have are indicators of those factors) and we wish to work with them numerically, they must have some unit of measurement assigned to them. Placing a constant value of 1 in the diagonal of the $\Phi$ matrix standardizes, and thus assigns, a metric to the latent variables. If instead we wished to explicitly solve for the factor variances, we could have done so with the requirement that we assign the unit of measurement to the factors in the structural equations. This is usually done by arbitrarily choosing one of the equations involving the latent variable, and assigning a coefficient of 1 to the latent variable in the equation. For example, we could have replaced the coefficient V1F1 with a 1. This does not mean V1 would be a perfect indicator of F1; instead, assigning a coefficient of 1 would assign V1's standard deviation to the factor, thus identifying it. Any constant value placed in front of F1 would have accomplished the same purpose; the only difference is the variance of F1 becomes the square of whatever the coefficient happens to be.

## Overview of the Proposed Model

Table 8.2 shows the first page of the SAS PROC CALIS output. This page provides a summary of the number of endogenous and exogenous variables defined in the proposed three-factor model. A description of the dimensions of the Bentler–Weeks model matrices is also provided. The _SEL_ matrix is

TABLE 8.2
First Page of SAS PROC CALIS Output,
Covariance Structure Analysis: Pattern and Initial Values

```
        LINEQS Model Statement
- - - - - - - - - - - - - - - - - - - - - - - - -

             Matrix        Rows & Cols         Matrix Type
  TERM       1- - - - - - - - - - - - - - - - - - - - - - - - - - - - - - - - - - - -
    1        _SEL_            9      21      SELECTION
    2        _BETA_          21      21      EQSBETA          IMINUSINV
    3        _GAMMA_         21      12      EQSGAMMA
    4        _PHI_           12      12      SYMMETRIC


              Number of endogenous variables = 9

  Manifest:      V1          V2          V3          V4          V5
                 V6          V7          V8          V9


              Number of exogenous variables = 12

  Latent:        F1          F2          F3
  Error:         E1          E2          E3          E4          E5
                 E6          E7          E8          E9
```

used to actively select out those observed variables the analyst wishes to use for a particular model. The dimensions of this matrix are always equal to the number of observed variables (here 9) by the total number of endogenous and exogenous variables in the model (here 21 = 9 endogenous observed variables + 9 exogenous errors of measurement + 3 exogenous latent variables). The _BETA_ matrix, which is a $21 \times 21$ matrix, represents the relations among the endogenous variables, and is by definition not used in the confirmatory factor model. The _GAMMA_ matrix, which has 21 (endogenous + exogenous) rows and 12 (exogenous) columns, consists of the regressions of the endogenous variables onto the exogenous variables. The _PHI_ matrix, which represents the variance–covariance matrix of the exogenous variables, is a $12 \times 12$ matrix. The summary of the model is continued in Table 8.3, with a recapitulation of the structural equations specified by the user. Finally, in terms of the overview of the model, Table 8.4 provides a listing of the covariance matrix to be used in the analysis of the proposed model along with some summary statistics for the observed variables.

**Details of the Proposed Model Estimation**

All iterative programs require starting values for the parameters in order to get the iterative process going. PROC CALIS provides its own starting values (the user does not have to but could, if desired, override the program's

TABLE 8.3
Summary of Model from SAS PROC CALIS Output,
Covariance Structure Analysis: Pattern and Initial Values

```
                        Manifest Variable Equations
                           Initial Estimates

        V1      =       .       *F1    + 1.0000 E1
                                L1V1
        V2      =       .       *F1    + 1.0000 E2
                                L1V2
        V3      =       .       *F1    + 1.0000 E3
                                L1V3
        V4      =       .       *F2    + 1.0000 E4
                                L2V4
        V5      =       .       *F2    + 1.0000 E5
                                L2V5
        V6      =       .       *F2    + 1.0000 E6
                                L2V6
        V7      =       .       *F3    + 1.0000 E7
                                L3V7
        V8      =       .       *F3    + 1.0000 E8
                                L3V8
        V9      =       .       *F3    + 1.0000 E9
                                L3V9


              Variances of Exogenous Variables
   - - - - - - - - - - - - - - - - - - - - - - - - - - - - - -
    Variable            Parameter                Estimate
   - - - - - - - - - - - - - - - - - - - - - - - - - - - - - -
     F1                                          1.000000
     F2                                          1.000000
     F3                                          1.000000
     E1                 THE1                         .
     E2                 THE2                         .
     E3                 THE3                         .
     E4                 THE4                         .
     E5                 THE5                         .
     E6                 THE6                         .
     E7                 THE7                         .
     E8                 THE8                         .
     E9                 THE9                         .
```

starting values) calculated through the method of instrumental variables. These starting values are given in Table 8.5. Jöreskog and Sörbom (1989) provided details on how the method of instrumental variables can be used to calculate starting values. In general, the closer the starting values are to the final parameter estimates, the smoother and faster the model's convergence. Frequently, the instrumental variables method provides starting values that are quite close to the final parameter values. The iterative history of the

TABLE 8.4
SAS System Covariance Structure Analysis:
Maximum Likelihood Estimation for Use with Proposed Model

| | | | |
|---|---|---|---|
| 255 Observations | | Model Terms | 1 |
| 9 Variables | | Model Matrices | 4 |
| 45 Informations | | Parameters | 18 |

| VARIABLE | Mean | Std Dev | |
|---|---|---|---|
| V1 | 0 | 1.410000000 | SC1 |
| V2 | 0 | 1.020000000 | SC2 |
| V3 | 0 | 0.910000000 | SC3 |
| V4 | 0 | 2.110000000 | HS1 |
| V5 | 0 | 1.850000000 | HS2 |
| V6 | 0 | 1.780000000 | HS3 |
| V7 | 0 | 2.450000000 | OG1 |
| V8 | 0 | 2.050000000 | OG2 |
| V9 | 0 | 1.920000000 | OG3 |

Covariances

| | V1 | V2 | V3 | V4 | V5 | |
|---|---|---|---|---|---|---|
| V1 | 1.988100 | 0.575280 | 0.577395 | 1.547052 | 1.069485 | SC1 |
| V2 | 0.575280 | 1.040400 | 0.343434 | 1.054578 | 0.830280 | SC2 |
| V3 | 0.577395 | 0.343434 | 0.828100 | 0.691236 | 0.471380 | SC3 |
| V4 | 1.547052 | 1.054578 | 0.691236 | 4.452100 | 2.459205 | HS1 |
| V5 | 1.069485 | 0.830280 | 0.471380 | 2.459205 | 3.422500 | HS2 |
| V6 | 0.878430 | 0.780708 | 0.323960 | 2.779292 | 1.975800 | HS3 |
| V7 | 1.174530 | 0.499800 | 1.137045 | 1.809325 | 0.679875 | OG1 |
| V8 | 0.549195 | 0.020910 | 0.727545 | 0.821845 | 0.265475 | OG2 |
| V9 | 0.730944 | 0.274176 | 0.733824 | 1.255872 | 0.639360 | OG3 |

| | V6 | V7 | V8 | V9 | |
|---|---|---|---|---|---|
| V1 | 0.878430 | 1.174530 | 0.549195 | 0.730944 | SC1 |
| V2 | 0.780708 | 0.499800 | 0.020910 | 0.274176 | SC2 |
| V3 | 0.323960 | 1.137045 | 0.727545 | 0.733824 | SC3 |
| V4 | 2.779292 | 1.809325 | 0.821845 | 1.255872 | HS1 |
| V5 | 1.975800 | 0.679875 | 0.265475 | 0.639360 | HS2 |
| V6 | 3.168400 | 0.872200 | 0.255430 | 0.717696 | HS3 |
| V7 | 0.872200 | 6.002500 | 3.465525 | 3.339840 | OG1 |
| V8 | 0.255430 | 3.465525 | 4.202500 | 2.794560 | OG2 |
| V9 | 0.717696 | 3.339840 | 2.794560 | 3.686400 | OG3 |

Determinant = 96.51 (Ln = 4.570)

*Note.* Some initial estimates computed by instrumental variable method.

TABLE 8.5
Starting Values for SAS System Covariance Structure Analysis:
Maximum Likelihood Estimation

Vector of Initial Estimates

| L1V1 | 1 | 0.93338 | Matrix Entry: _GAMMA_[1:1] |
| L1V2 | 2 | 0.58817 | Matrix Entry: _GAMMA_[2:1] |
| L1V3 | 3 | 0.62764 | Matrix Entry: _GAMMA_[3:1] |
| L2V4 | 4 | 1.99620 | Matrix Entry: _GAMMA_[4:2] |
| L2V5 | 5 | 1.34260 | Matrix Entry: _GAMMA_[5:2] |
| L2V6 | 6 | 1.34070 | Matrix Entry: _GAMMA_[6:2] |
| L3V7 | 7 | 2.16724 | Matrix Entry: _GAMMA_[7:3] |
| L3V8 | 8 | 1.56495 | Matrix Entry: _GAMMA_[8:3] |
| L3V9 | 9 | 1.64622 | Matrix Entry: _GAMMA_[9:3] |
| THE1 | 10 | 1.11691 | Matrix Entry: _PHI_[4:4] |
| THE2 | 11 | 0.69446 | Matrix Entry: _PHI_[5:5] |
| THE3 | 12 | 0.43416 | Matrix Entry: _PHI_[6:6] |
| THE4 | 13 | 0.46729 | Matrix Entry: _PHI_[7:7] |
| THE5 | 14 | 1.61992 | Matrix Entry: _PHI_[8:8] |
| THE6 | 15 | 1.37092 | Matrix Entry: _PHI_[9:9] |
| THE7 | 16 | 1.30558 | Matrix Entry: _PHI_[10:10] |
| THE8 | 17 | 1.75344 | Matrix Entry: _PHI_[11:11] |
| THE9 | 18 | 0.97636 | Matrix Entry: _PHI_[12:12] |

Predetermined Elements of the Predicted Moment Matrix

|    | V1 | V2 | V3 | V4 | V5 |     |
|----|----|----|----|----|----|-----|
| V1 | .  | .  | .  | 0  | 0  | SC1 |
| V2 | .  | .  | .  | 0  | 0  | SC2 |
| V3 | .  | .  | .  | 0  | 0  | SC3 |
| V4 | 0  | 0  | 0  | .  | .  | HS1 |
| V5 | 0  | 0  | 0  | .  | .  | HS2 |
| V6 | 0  | 0  | 0  | .  | .  | HS3 |
| V7 | 0  | 0  | 0  | 0  | 0  | OG1 |
| V8 | 0  | 0  | 0  | 0  | 0  | OG2 |
| V9 | 0  | 0  | 0  | 0  | 0  | OG3 |

|    | V6 | V7 | V8 | V9 |     |
|----|----|----|----|----|-----|
| V1 | 0  | 0  | 0  | 0  | SC1 |
| V2 | 0  | 0  | 0  | 0  | SC2 |
| V3 | 0  | 0  | 0  | 0  | SC3 |
| V4 | .  | 0  | 0  | 0  | HS1 |
| V5 | .  | 0  | 0  | 0  | HS2 |
| V6 | .  | 0  | 0  | 0  | HS3 |
| V7 | 0  | .  | .  | .  | OG1 |
| V8 | 0  | .  | .  | .  | OG2 |
| V9 | 0  | .  | .  | .  | OG3 |

WARNING:  The predicted moment matrix has 27 constant elements
          whose values differ from those of the observed moment matrix.
          The sum of squared differences is 20.235137516.

TABLE 8.6
Iterative History of PROC CALIS Run for Covariance Structure Analysis:
Maximum Likelihood Estimation

Levenberg-Marquardt Minimization
Algorithm for Hessian= 11
Maximum Iterations= 1000
Maximum Function Calls= 125
Maximum Absolute Gradient Criterion= 0.001
Number of Estimates= 18 Lower Bounds= 0 Upper Bounds= 0
Minimization Start: Active Constraints= 0 Criterion= 1.089
Maximum Gradient Element= 0.078 Radius= 0.196

| Iter | nfun | act | mincrit | maxgrad | difcrit | lambda | rhoratio |
|------|------|-----|---------|---------|---------|--------|----------|
| 1 | 2 | 0 | 1.03946 | 0.0559 | 0.0497 | 0.9790 | 0.9645 |
| 2 | 3 | 0 | 1.00202 | 0.0204 | 0.0374 | 0.0510 | 1.0243 |
| 3 | 4 | 0 | 1.00104 | 0.000390 | 0.000984 | 0 | 1.0195 |

Minimization Results: Iterations= 3 Function Calls= 4
Derivative Calls= 4
Active Constraints= 0 Criterion= 1
Maximum Gradient Element= 0.00039 Radius= 0.000963

NOTE: Convergence criterion satisfied.

PROC CALIS run for the example study is provided in Table 8.6. As can be seen, PROC CALIS needed to iterate three times before converging on optimal estimates for the parameters. We highly recommend that this page always be checked carefully to determine if there are convergence problems. If no convergence problems are encountered, the PROC CALIS provides the statement "NOTE: Convergence criterion satisfied."

**Assessing Model Fit**

The next part of the PROC CALIS output is devoted to describing how well the model fit the data. Among other details, the output includes information about the model-implied covariance matrix (i.e., $\hat{S}$ or the predicted model matrix, displayed in Table 8.7), approximately two dozen measures of goodness of fit (displayed in Table 8.8), the residual matrix ($S-\hat{S}$, also displayed in Table 8.8), and descriptive statistics concerning the residuals (Tables 8.9 and 8.10).

As can be seen in Table 8.8, there appears to be a bewildering array of measures for assessing the goodness of fit of a model. Because many of them define goodness of fit in different ways, we provide a brief description of each measure later in this chapter. For now, it is enough to note that the chi-square = 254.26, with df = 27, $p < .0001$, indicates a rejection of the fit of the model to the data.

TABLE 8.7
Predicted Model Matrix

|    | V1 | V2 | V3 | V4 | V5 | |
|----|----|----|----|----|----|----|
| V1 | 1.988155 | 0.575276 | 0.5773633 | 0.000000 | 0.000000 | SC1 |
| V2 | 0.575276 | 1.040400 | 0.3434363 | 0.000000 | 0.000000 | SC2 |
| V3 | 0.577363 | 0.343436 | 0.8281182 | 0.000000 | 0.000000 | SC3 |
| V4 | 0.000000 | 0.000000 | 0.0000000 | 4.452490 | 2.459121 | HS1 |
| V5 | 0.000000 | 0.000000 | 0.0000000 | 2.459121 | 3.422518 | HS2 |
| V6 | 0.000000 | 0.000000 | 0.0000000 | 2.778949 | 1.975874 | HS3 |
| V7 | 0.000000 | 0.000000 | 0.0000000 | 0.000000 | 0.000000 | OG1 |
| V8 | 0.000000 | 0.000000 | 0.0000000 | 0.000000 | 0.000000 | OG2 |
| V9 | 0.000000 | 0.000000 | 0.0000000 | 0.000000 | 0.000000 | OG3 |

|    | V6 | V7 | V8 | V9 | |
|----|----|----|----|----|----|
| V1 | 0.000000 | 0.000000 | 0.000000 | 0.000000 | SC1 |
| V2 | 0.000000 | 0.000000 | 0.000000 | 0.000000 | SC2 |
| V3 | 0.000000 | 0.000000 | 0.000000 | 0.000000 | SC3 |
| V4 | 2.778949 | 0.000000 | 0.000000 | 0.000000 | HS1 |
| V5 | 1.975874 | 0.000000 | 0.000000 | 0.000000 | HS2 |
| V6 | 3.168702 | 0.000000 | 0.000000 | 0.000000 | HS3 |
| V7 | 0.000000 | 6.002570 | 3.465463 | 3.339842 | OG1 |
| V8 | 0.000000 | 3.465463 | 4.202555 | 2.794558 | OG2 |
| V9 | 0.000000 | 3.339842 | 2.794558 | 3.686400 | OG3 |

Determinant = 262.9 (Ln = 5.572)

Tables 8.9 and 8.10 provide detailed information concerning the proposed model residuals (i.e., $S - \hat{S}$). Examination of the residual information can be useful in two respects. First, elements of the residual matrix that are unusually high indicate variables whose relations are not well explained by the model. If the model is rejected, then these unusually high residuals may indicate which parts of the model are most in need of revision. Generally, because of scale differences among the variables, examination of the asymptotically standardized residuals is most useful. A second useful purpose for examining the residuals concerns how well the data meet the assumption of multivariate normality. Clearly, the skewness of the histogram shown in Table 8.10 makes the assumption of multivariate normality for this data rather suspect.

**Parameter Values**

Tables 8.11 and 8.12 present both the unstandardized and standardized solutions along with the parameters' standard errors and $t$ values. If the data are multivariate normally distributed, the $t$ values may be interpreted as standard normal deviates (remember that a $t$ value above 1.96 indicates a parameter significant at $p < .05$). As can be seen by examining these tables,

TABLE 8.8
Measures for Assessing Goodness of Fit

```
Fit criterion.............................................   1.0010
Goodness of Fit Index (GFI).................................   0.8277
GFI Adjusted for Degrees of Freedom (AGFI)..................   0.7129
Root Mean Square Residual (RMR).............................   0.6706
Parsimonious GFI (Mulaik, 1989).............................   0.6208
Chi-square = 254.2636          df = 27          Prob>chi**2 = 0.0001
Null Model Chi-square:         df = 36                       1112.0091
RMSEA Estimate ..................   0.1820     90%C.I.[0.1620, 0.2028]
Probability of Close Fit....................................   0.0000
ECVI Estimate ...................   1.1486     90%C.I.[0.9589, 1.3689]
Bentler's Comparative Fit Index.............................   0.7888
Normal Theory Reweighted LS Chi-square......................  237.8681
Akaike's Information Criterion..............................  200.2636
Bozdogan's (1987) CAIC......................................   77.6495
Schwarz's Bayesian Criterion................................  104.6495
McDonald's (1989) Centrality................................   0.6404
Bentler & Bonett's (1980) Non-normed Index..................   0.7184
Bentler & Bonett's (1980) NFI...............................   0.7713
James, Mulaik, & Brett (1982) Parsimonious NFI..............   0.5785
Z-Test of Wilson & Hilferty (1931).........................   12.3451
Bollen (1986) Normed Index Rho1.............................   0.6951
Bollen (1988) Non-normed Index Delta2.......................   0.7905
Hoelter's (1983) Critical N.................................      42
```

Residual Matrix

|     | V1 | V2 | V3 | V4 | V5 | |
|-----|------|------|------|------|------|-----|
| V1  | -0.000055 | 0.000004 | 0.000032 | 1.547052 | 1.069485 | SC1 |
| V2  | 0.000004 | -0.000000 | -0.000002 | 1.054578 | 0.830280 | SC2 |
| V3  | 0.000032 | -0.000002 | -0.000018 | 0.691236 | 0.471380 | SC3 |
| V4  | 1.547052 | 1.054578 | 0.691236 | -0.000390 | 0.000084 | HS1 |
| V5  | 1.069485 | 0.830280 | 0.471380 | 0.000084 | -0.000018 | HS2 |
| V6  | 0.878430 | 0.780708 | 0.323960 | 0.000343 | -0.000074 | HS3 |
| V7  | 1.174530 | 0.499800 | 1.137045 | 1.809325 | 0.679875 | OG1 |
| V8  | 0.549195 | 0.020910 | 0.727545 | 0.821845 | 0.265475 | OG2 |
| V9  | 0.730944 | 0.274176 | 0.733824 | 1.255872 | 0.639360 | OG3 |

|     | V6 | V7 | V8 | V9 | |
|-----|------|------|------|------|-----|
| V1  | 0.878430 | 1.174530 | 0.549195 | 0.730944 | SC1 |
| V2  | 0.780708 | 0.499800 | 0.020910 | 0.274176 | SC2 |
| V3  | 0.323960 | 1.137045 | 0.727545 | 0.733824 | SC3 |
| V4  | 0.000343 | 1.809325 | 0.821845 | 1.255872 | HS1 |
| V5  | -0.000074 | 0.679875 | 0.265475 | 0.639360 | HS2 |
| V6  | -0.000302 | 0.872200 | 0.255430 | 0.717696 | HS3 |
| V7  | 0.872200 | -0.000070 | 0.000062 | -0.000002 | OG1 |
| V8  | 0.255430 | 0.000062 | -0.000055 | 0.000002 | OG2 |
| V9  | 0.717696 | -0.000002 | 0.000002 | -0.000000 | OG3 |

Average Absolute Residual = 0.4625
Average Off-diagonal Absolute Residual = 0.5781

## TABLE 8.9
### Proposed Model Residuals, Part I

Rank Order of 10 Largest Residuals

| V7,V4 | V4,V1 | V9,V4 | V7,V1 | V7,V3 | V5,V1 |
|-------|-------|-------|-------|-------|-------|
| 1.8093 | 1.5471 | 1.2559 | 1.1745 | 1.1370 | 1.0695 |

| V4,V2 | V6,V1 | V7,V6 | V5,V2 |
|-------|-------|-------|-------|
| 1.0546 | 0.8784 | 0.8722 | 0.8303 |

Asymptotically Standardized Residual Matrix

|     | V1 | V2 | V3 | V4 | V5 |     |
|-----|----|----|----|----|----|-----|
| V1 | 0.000000 | 0.000000 | 0.000000 | 8.286959 | 6.534217 | SC1 |
| V2 | 0.000000 | 0.000000 | 0.000000 | 7.808972 | 7.012427 | SC2 |
| V3 | 0.000000 | 0.000000 | 0.000000 | 5.737142 | 4.462405 | SC3 |
| V4 | 8.286959 | 7.808972 | 5.737142 | 0.000000 | 0.000000 | HS1 |
| V5 | 6.534217 | 7.012427 | 4.462405 | 0.000000 | 0.000000 | HS2 |
| V6 | 5.577739 | 6.852745 | 3.187289 | 0.000000 | 0.000000 | HS3 |
| V7 | 5.418602 | 3.187456 | 8.127926 | 5.577806 | 2.390586 | OG1 |
| V8 | 3.028040 | 0.159373 | 6.215468 | 3.027949 | 1.115606 | OG2 |
| V9 | 4.303032 | 2.231233 | 6.693625 | 4.940371 | 2.868720 | OG3 |

|     | V6 | V7 | V8 | V9 |     |
|-----|----|----|----|----|-----|
| V1 | 5.577739 | 5.418602 | 3.028040 | 4.303032 | SC1 |
| V2 | 6.852745 | 3.187456 | 0.159373 | 2.231233 | SC2 |
| V3 | 3.187289 | 8.127926 | 6.215468 | 6.693625 | SC3 |
| V4 | 0.000000 | 5.577806 | 3.027949 | 4.940371 | HS1 |
| V5 | 0.000000 | 2.390586 | 1.115606 | 2.868720 | HS2 |
| V6 | 0.000000 | 3.187305 | 1.115556 | 3.346690 | HS3 |
| V7 | 3.187305 | 0.000000 | 0.000000 | 0.000000 | OG1 |
| V8 | 1.115556 | 0.000000 | 0.000000 | 0.000000 | OG2 |
| V9 | 3.346690 | 0.000000 | 0.000000 | 0.000000 | OG3 |

Average Standardized Residual = 2.72

Average Off-diagonal Standardized Residual = 3.4

Rank Order of 10 Largest Asymptotically Standardized Residuals

| V4,V1 | V7,V3 | V4,V2 | V5,V2 | V6,V2 | V9,V3 |
|-------|-------|-------|-------|-------|-------|
| 8.2870 | 8.1279 | 7.8090 | 7.0124 | 6.8527 | 6.6936 |

| V5,V1 | V8,V3 | V4,V3 | V7,V4 |
|-------|-------|-------|-------|
| 6.5342 | 6.2155 | 5.7371 | 5.5778 |

TABLE 8.10
Proposed Model Residuals, Part II

Distribution of Asymptotically Standardized Residuals
(Each * represents 1 residuals)

| | | | | | |
|---|---|---|---|---|---|
| -0.10652 | - | 0.14348 | 18 | 40.00% | ****************** |
| 0.14348 | - | 0.39348 | 1 | 2.22% | * |
| 0.39348 | - | 0.64348 | 0 | 0.00% | |
| 0.64348 | - | 0.89348 | 0 | 0.00% | |
| 0.89348 | - | 1.14348 | 2 | 4.44% | ** |
| 1.14348 | - | 1.39348 | 0 | 0.00% | |
| 1.39348 | - | 1.64348 | 0 | 0.00% | |
| 1.64348 | - | 1.89348 | 0 | 0.00% | |
| 1.89348 | - | 2.14348 | 0 | 0.00% | |
| 2.14348 | - | 2.39348 | 2 | 4.44% | ** |
| 2.39348 | - | 2.64348 | 0 | 0.00% | |
| 2.64348 | - | 2.89348 | 1 | 2.22% | * |
| 2.89348 | - | 3.14348 | 2 | 4.44% | ** |
| 3.14348 | - | 3.39348 | 4 | 8.89% | **** |
| 3.39348 | - | 3.64348 | 0 | 0.00% | |
| 3.64348 | - | 3.89348 | 0 | 0.00% | |
| 3.89348 | - | 4.14348 | 0 | 0.00% | |
| 4.14348 | - | 4.39348 | 1 | 2.22% | * |
| 4.39348 | - | 4.64348 | 1 | 2.22% | * |
| 4.64348 | - | 4.89348 | 0 | 0.00% | |
| 4.89348 | - | 5.14348 | 1 | 2.22% | * |
| 5.14348 | - | 5.39348 | 0 | 0.00% | |
| 5.39348 | - | 5.64348 | 3 | 6.67% | *** |
| 5.64348 | - | 5.89348 | 1 | 2.22% | * |
| 5.89348 | - | 6.14348 | 0 | 0.00% | |
| 6.14348 | - | 6.39348 | 1 | 2.22% | * |
| 6.39348 | - | 6.64348 | 1 | 2.22% | * |
| 6.64348 | - | 6.89348 | 2 | 4.44% | ** |
| 6.89348 | - | 7.14348 | 1 | 2.22% | * |
| 7.14348 | - | 7.39348 | 0 | 0.00% | |
| 7.39348 | - | 7.64348 | 0 | 0.00% | |
| 7.64348 | - | 7.89348 | 1 | 2.22% | * |
| 7.89348 | - | 8.14348 | 1 | 2.22% | * |
| 8.14348 | - | 8.39348 | 1 | 2.22% | * |

all of the parameters are highly significant. However, because the model as a whole has been rejected, interpretation of the model in this example study is not justified. If the model as a whole was found to fit the data well, then one could proceed with an interpretation of the individual parameters in the model. Another set of information that can be used to assist with this interpretation is the squared multiple correlations presented in Table 8.12. The squared multiple correlations indicate the amount of variance explained for each observed variable by the latent variables. As can be seen in Table

## TABLE 8.11
### Parameter Values for Solutions, Part I

Manifest Variable Equations

| | | | | |
|---|---|---|---|---|
| V1 | = | 0.9834*F1 | + | 1.0000 E1 |
| Std Err | | 0.1100 V1F1 | | |
| t Value | | 8.9374 | | |

| | | | | |
|---|---|---|---|---|
| V2 | = | 0.5850*F1 | + | 1.0000 E2 |
| Std Err | | 0.0749 V2F1 | | |
| t Value | | 7.8097 | | |

| | | | | |
|---|---|---|---|---|
| V3 | = | 0.5871*F1 | + | 1.0000 E3 |
| Std Err | | 0.0692 V3F1 | | |
| t Value | | 8.4866 | | |

| | | | | |
|---|---|---|---|---|
| V4 | = | 1.8597*F2 | + | 1.0000 E4 |
| Std Err | | 0.1159 V4F2 | | |
| t Value | | 16.0454 | | |

| | | | | |
|---|---|---|---|---|
| V5 | = | 1.3223*F2 | + | 1.0000 E5 |
| Std Err | | 0.1068 V5F2 | | |
| t Value | | 12.3823 | | |

| | | | | |
|---|---|---|---|---|
| V6 | = | 1.4943*F2 | + | 1.0000 E6 |
| Std Err | | 0.0992 V6F2 | | |
| t Value | | 15.0706 | | |

| | | | | |
|---|---|---|---|---|
| V7 | = | 2.0351*F3 | + | 1.0000 E7 |
| Std Err | | 0.1329 V7F3 | | |
| t Value | | 15.3171 | | |

| | | | | |
|---|---|---|---|---|
| V8 | = | 1.7028*F3 | + | 1.0000 E8 |
| Std Err | | 0.1112 V8F3 | | |
| t Value | | 15.3171 | | |

| | | | | |
|---|---|---|---|---|
| V9 | = | 1.6411*F3 | + | 1.0000 E9 |
| Std Err | | 0.1031 V9F3 | | |
| t Value | | 15.9209 | | |

Variances of Exogenous Variables

| Variable | Parameter | Estimate | Standard Error | t Value |
|---|---|---|---|---|
| F1 | | 1.000000 | 0 | 0.000 |
| F2 | | 1.000000 | 0 | 0.000 |
| F3 | | 1.000000 | 0 | 0.000 |
| E1 | THE1 | 1.021038 | 0.179253 | 5.696 |
| E2 | THE2 | 0.698206 | 0.082665 | 8.446 |
| E3 | THE3 | 0.483436 | 0.069850 | 6.921 |
| E4 | THE4 | 0.993883 | 0.212851 | 4.669 |
| E5 | THE5 | 1.674046 | 0.177925 | 9.409 |
| E6 | THE6 | 0.935850 | 0.150125 | 6.234 |
| E7 | THE7 | 1.860914 | 0.251556 | 7.398 |
| E8 | THE8 | 1.302885 | 0.176121 | 7.398 |
| E9 | THE9 | 0.993143 | 0.151641 | 6.549 |

TABLE 8.12
Parameter Values for Solutions, Part II

Equations with Standardized Coefficients

| V1 | = | 0.6975*F1 | + | 0.7166 E1 |
| | | V1F1 | | |

| V2 | = | 0.5735*F1 | + | 0.8192 E2 |
| | | V2F1 | | |

| V3 | = | 0.6452*F1 | + | 0.7641 E3 |
| | | V3F1 | | |

| V4 | = | 0.8814*F2 | + | 0.4725 E4 |
| | | V4F2 | | |

| V5 | = | 0.7148*F2 | + | 0.6994 E5 |
| | | V5F2 | | |

| V6 | = | 0.8394*F2 | + | 0.5435 E6 |
| | | V6F2 | | |

| V7 | = | 0.8307*F3 | + | 0.5568 E7 |
| | | V7F3 | | |

| V8 | = | 0.8306*F3 | + | 0.5568 E8 |
| | | V8F3 | | |

| V9 | = | 0.8547*F3 | + | 0.5190 E9 |
| | | V9F3 | | |

Squared Multiple Correlations

| Variable | | Error Variance | Total Variance | R-squared |
|---|---|---|---|---|
| 1 | V1 | 1.021038 | 1.988155 | 0.486440 |
| 2 | V2 | 0.698206 | 1.040400 | 0.328907 |
| 3 | V3 | 0.483436 | 0.828118 | 0.416224 |
| 4 | V4 | 0.993883 | 4.452490 | 0.776780 |
| 5 | V5 | 1.674046 | 3.422518 | 0.510873 |
| 6 | V6 | 0.935850 | 3.168702 | 0.704658 |
| 7 | V7 | 1.860914 | 6.002570 | 0.689981 |
| 8 | V8 | 1.302885 | 4.202555 | 0.689978 |
| 9 | V9 | 0.993143 | 3.686400 | 0.730593 |

Predicted Moments of Latent Variables

| | F1 | F2 | F3 |
|---|---|---|---|
| F1 | 1.000000000 | 0.000000000 | 0.000000000 |
| F2 | 0.000000000 | 1.000000000 | 0.000000000 |
| F3 | 0.000000000 | 0.000000000 | 1.000000000 |

8.12, consistent with the highly significant parameter results, most of the squared multiple correlations are quite high.

## Model Modification

In general, if an initial proposed model has been rejected, the process of confirmation is over unless new data are collected. Nevertheless, using the same data, one can attempt to revise the model so that it fits the data better. Obviously, any attempt to revise a proposed model is more in the realm of exploration than confirmation. Although exploratory data analyses with the main goal of improving fit (more commonly termed *specification searches*) are not necessarily bad, the final result (in terms of an adequately fitting model) should not be taken too seriously unless the model is validated against a new sample. As such, all the cautions voiced in chapter 6 concerning the conduct of stepwise regression apply in this situation with equal, if not more, force.

To assist with model modification, PROC CALIS provides two different types of diagnostic statistics indicating which parameters, when revised, would increase the overall fit of the model. A selected part of the model modification information that is provided by PROC CALIS as part of its standard output is displayed in Table 8.13. There are two statistics represented here: the Lagrange multiplier statistic (which indicates which parameters should be added to the model), and the Wald index (which indicates which parameters should be dropped from the model). In summary, the value of the Lagrange multiplier statistic indicates how much the model's chi-square would decrease if a particular parameter were freed, whereas the value of the Wald index indicates how much the model's chi-square would increase if the parameter were fixed to zero (i.e., if it were deleted from the model). As such, the status of every possible model parameter is provided with respect to the Lagrange multiplier statistic (if the parameter is currently not in the model) or with respect to the Wald index (if the parameter is currently in the model). For example, the value of 2.077 represents the Lagrange multiplier statistic of the relationship between E1 and E7 (i.e., the error terms of the variables, which is currently fixed to zero in the model), and indicates that permitting this relationship to exist in the model will decrease the chi-square by 2.077. In addition to the Lagrange multiplier statistic, the actual estimated value of the relationship between E1 and E7 is provided (i.e., $r = .175$) along with its corresponding significance level ($p = .150$). As can be seen, not much of an improvement in the model fit would be gained if the two error terms were permitted to be correlated. On the other hand, the Lagrange multiplier statistic suggests that there is a relationship between the self-criticism (F1) factor and the high-standards (F2) factor (i.e., F2:F1 = 80.7687).

TABLE 8.13
PROC CALIS Diagnostic Statistics

Lagrange Multiplier and Wald Test Indices _PHI_[12:12]
Diagonal Matrix
Univariate Tests for Constant Constraints (Contd.)

| | E7 | | E8 | | E9 | |
|------|--------|--------|--------|--------|--------|--------|
| F1 | 23.143 | | 2.624 | | 2.228 | |
| | 0.000 | 0.584 | 0.105 | −0.165 | 0.136 | 0.138 |
| F2 | 7.885 | | 5.141 | | 6.149 | |
| | 0.005 | 0.302 | 0.023 | −0.204 | 0.013 | 0.203 |
| F3 | SING | | SING | | SING | |
| | . | . | . | . | . | . |
| E1 | 2.077 | | 0.602 | | 0.299 | |
| | 0.150 | 0.175 | 0.438 | −0.079 | 0.585 | 0.051 |
| E2 | 1.745 | | 9.624 | | 0.618 | |
| | 0.187 | 0.121 | 0.002 | −0.237 | 0.432 | 0.055 |
| E3 | 6.906 | | 3.012 | | 0.145 | |
| | 0.009 | 0.208 | 0.083 | 0.115 | 0.704 | 0.023 |
| E4 | 10.271 | | 0.025 | | 0.015 | |
| | 0.001 | 0.428 | 0.875 | −0.018 | 0.903 | 0.012 |
| E5 | 1.527 | | 0.023 | | 0.689 | |
| | 0.217 | −0.172 | 0.879 | −0.018 | 0.407 | 0.088 |
| E6 | 0.272 | | 1.849 | | 1.105 | |
| | 0.602 | −0.061 | 0.174 | −0.132 | 0.293 | 0.093 |
| E7 | 54.725 | [THE7] | SING | | SING | |
| | | | . | . | . | . |
| E8 | SING | | 54.726 | [THE8] | SING | |
| | . | . | | | . | . |
| E9 | SING | | SING | | 42.893 | [THE9] |
| | . | . | . | . | | |

Rank order of 10 largest Lagrange multipliers in _PHI_

| F2 | : | F1 | | F3 | : | F1 | | E3 | : | F3 |
|---------|---|-------|--|---------|---|-------|--|---------|---|-------|
| 80.7687 | : | 0.000 | | 1.9675 | : | 0.000 | | 41.8976 | : | 0.000 |
| E4 | : | F1 | | E2 | : | F2 | | E7 | : | F1 |
| 29.5160 | : | 0.000 | | 27.2891 | : | 0.000 | | 23.1433 | : | 0.000 |
| E1 | : | F2 | | E4 | : | F3 | | F3 | : | F2 |
| 20.0396 | : | 0.000 | | 18.6238 | : | 0.000 | | 18.4600 | : | 0.000 |
| | | | | E4 | : | E1 | | | | |
| | | | | 11.0448 | : | 0.001 | | | | |

The values presented next to the bracket notation (e.g., 54.725 [THE7] ) represent the Wald statistic for each parameter already specified in the model. For example, if the THE7 parameter were deleted from the model, the model chi-square would increase by a magnitude of 54.725 points. But how can deleting parameters based on the Wald statistic improve model fit if the chi-square will increase with the parameter's deletion? Remember that the significance of a chi-square is based on its degrees of freedom. Given the same chi-square value for two models, the model with the greater degrees of freedom will fit better than the other. If a parameter does not contribute significantly to the fit of the model (it reduces chi-square by a trivial amount) one degree of freedom has been wasted. Thus, as a general strategy, a proposed model should be revised by first adding parameters based on the Lagrange multiplier statistic, and then once an adequate model fit has been achieved, by removing trivial parameters. Only one parameter should be freed or fixed to zero at a time; the chi-square value given by the Lagrange and Wald tests is only applicable under those circumstances.

Finally, another very helpful bit of information presented in Table 8.13 relates to which parameters, if freed, would result in a noinidentified solution. These are indicated in the output using the term SING.

## GOODNESS-OF-FIT INDICES

A great many indices of fit have been proposed for confirmatory factor and other SEM models. In this section, an overview is given of the indices provided by PROC CALIS (although there is considerable overlap between what other SEM programs provide). One naturally wonders why there have been so many measures of goodness of fit developed. There are several reasons. First, not all the indices define goodness of fit in the same way. Although most define goodness of fit in terms of the discrepancy between the observed and model implied covariance matrices, others combine this criterion with a parsimony criterion (and, in general, more parsimonious models are to be preferred—are better fitting—than less parsimonious models). Still other criteria (not represented among the CALIS indices) define fit in terms of model *complexity* (Bozdogan, 1990). Model complexity combines the triple criteria of small discrepancy, parsimony, and small sampling variances of the estimated parameters. Still others (e.g., Hershberger, 1995) define fit in terms of the number of models equivalent (statistically identical to in fit) to a preferred model, with preferred models more favorably evaluated if they have fewer equivalent models.

A second reason there are so many fit indices has to do with dissatisfaction with many of the indices' reliance on sample size, either explicitly (e.g., the chi-square goodness-of-fit test) or implicitly (e.g., the goodness-of-fit index,

GFI). In theory, the fit of a model should be independent of the size of the sample used to test it. A third reason concerns the unlikely assumption made for those indices that use the central chi-square distribution; this assumption implies that the model is perfectly true. Unfortunately, the "perfectly" part of this assumption is probably never met in practice. A more reasonable assumption is that the model is "approximately" true, which suggests the use of the noncentral chi-square distribution in the fit index.

There are many ways in which these various fit indices can be categorized. In this section they are placed them into one of three categories: (a) tests of significance, (b) descriptive, and (c) information. Good sources for additional information concerning these indices are Bollen (1989), and the volumes edited by Bollen and Long (1993) and Marcoulides and Schumacker (1996).

### Category I: Tests of Significance

These indices either provide explicit tests of significance for the fit of the model or contribute to such tests.

1. *Fit criterion.* This number is literally the minimum value of the discrepancy function $F$ discussed earlier and is equal to

$$\frac{\chi^2}{N-1}$$

2. $\chi^2$. This is the chi-square goodness-of-fit discussed earlier, which if the model fits should be nonsignificant (generally, $p > .05$).

3. *Normal theory reweighted least squares chi-square.* In case the data depart from multivariate normality, a more robust estimator is GLS; this value of chi-square is obtained from the minimum of the fit function using GLS. Note in the example that the GLS estimate (237.87) differs from the ML estimate (254.26).

4. *Z-test of Wilson and Hilferty.* Wilson and Hilferty (1931) proposed the following Z-test:

$$Z = \sqrt{\frac{\sqrt[3]{\frac{x^2}{df}} - \left(1 - \frac{2}{9df}\right)}{\sqrt{\frac{2}{9\,df}}}}$$

assuming multivariate normality. In this example, the $z$ value of 12.35 is highly significant, again suggesting rejection of the model.

5. *Probability of close fit.* Browne and Cudeck (1992) argued that the chi-square goodness-of-fit test, an "exact" test of model fit, was unrealistic because the model tested is almost certainly never exactly true. A test of "close" fit was proposed in which the null hypothesis of the exact test (no difference between $S$ and $\hat{S}$) was replaced by a null hypothesis permitting a range of difference or discrepancy between $S$ and $\hat{S}$. The root-mean-square error of approximation (RMSEA, $\varepsilon_a$) = $\sqrt{\frac{F}{df}}$, is this measure of model discrepancy ($F$), and measures amount of model discrepancy per degree of freedom (Steiger, 1990; see RMSEA, discussed later). The chi-square goodness of fit test is still invoked but the region of rejection is now made smaller by a factor of 0.05. In this model, with RMSEA = .182, the null hypothesis is rejected ($p < .00001$ under "Probability of Close Fit").

**Category II: Descriptive Indices**

There are no significance tests for the descriptive indices because the sampling distribution of these indices is unknown. Nevertheless, experience with the indices has lead to specific suggestions concerning how large the index must be in order to confirm the model.

1. *Goodness-of-fit index (GFI).* The GFI ( Jöreskog & Sörbom, 1993) measures the relative amount of the variances and covariances in $S$ that are predictable from $\hat{S}$. The form of GFI differs, based on the method of estimation used. For ML it is:

$$1 - \frac{\text{tr}[(\hat{S}^{-1}S - I)^2]}{\text{tr}[(\hat{S}^{-1}S)^2]}$$

GFI is scaled to range between 0 and 1, with larger values indicating better fit. Values above .90 are considered generally to indicate well-fitting models.

2. *Adjusted goodness-of-fit index (AGFI).* As noted before, the more parameters ($q$) that are defined for a model, the better it will fit. The AGFI ( Jöreskog & Sörbom, 1993) adjusts the GFI downward based on the number of parameters. For ML it is:

$$1 - \left[ \frac{q(q+1)}{2df} \right] (1 - \text{GFI})$$

The AGFI is also scaled to range between 0 and 1, with values above .90 suggesting good model fit.

3. *Null model chi-square.* The null model chi-square is used as a baseline model to compare other models, and is part of several of the descriptive indices described later. The null model is defined as a complete absence of

covariance structure; in the variance–covariance matrix, all off-diagonal elements are hypothesized to equal zero. This is a worst-case scenario, and invariably, the analyst's proposed model will fit better. The idea is to examine how much better the proposed model does relative to the null model.

4. *Bentler and Bonett's (1980) normed fit index (NFI)*. The NFI measures the relative improvement in fit obtained by a proposed $m$ model compared against the null model $b$:

$$NFI = \frac{\chi_b^2 - \chi_m^2}{\chi_b^2}$$

The NFI is also scaled to range between 0 and 1, with higher values (above .90 typically) indicating good model fit.

5. *James, Mulaik, and Brett's (1982) parsimonious NFI*. Similar to the function of the AGFI, the PNFI corrects the NFI for overparameterization:

$$PNFI = \frac{df_m}{df_b} \frac{\chi_b^2 - \chi_m^2}{\chi_b^2}$$

The more parameters the proposed model $m$ contains, the smaller the parsimony ratio $\frac{df_m}{df_b}$, and hence, the smaller the NFI.

6. *Bollen's (1986) normed index $\rho_1$*. Bollen (1986) also criticized the NFI for not rewarding parsimoniously parameterized models and suggested the following index as an alternative:

$$\rho_1 = \frac{\dfrac{\chi_b^2}{df_b} - \dfrac{\chi_m^2}{df_m}}{\dfrac{\chi_b^2}{df_b}}$$

Values of $\rho_1$ also range between 0 and 1, with higher values (above .90) indicating good model fit.

7. *Bentler and Bonett's (1986) nonnormed fit index (NNFI)*. The NNFI is defined as:

$$NNFI = \frac{\dfrac{\chi_b^2}{df_b} - \dfrac{\chi_m^2}{df_m}}{\left(\dfrac{\chi_b^2}{df_b}\right) - 1}$$

The NNFI, like Bollen's $\rho_1$, corrects for the number of parameters in the model. However, it is not scaled to range between 0 and 1.

8. *Bollen's (1988) nonnormed index* $\Delta_2$. Another fit index that corrects the Bentler and Bonett NFI for lack of parsimony is $\Delta_2$:

$$\Delta_2 = \frac{\chi_b^2 - \chi_m^2}{\chi_b^2 - df_m}$$

$\Delta_2$ is not scaled to fall between 0 and 1.

9. *Bentler's (1990) comparative fit index (CFI)*. Primarily developed to compensate for the tendency of the NFI to indicate falsely a lack of fit in small samples, it too compares a preferred model to a baseline or null model:

$$CFI = \left| \frac{(\chi_b^2 - df_b) - (\chi_m^2 - df_m)}{(\chi_b^2 - df_b)} \right|$$

The CFI is also scaled to range between 0 and 1.

10. *Root-mean-square residual (RMR; Jöreskog & Sörbom, 1993)*. Defined as

$$RMR = \sqrt{\sum_{i=1}^{q} \sum_{j=1}^{i} \frac{(s_{ij} - \hat{s}_{ij})^2}{q(q+1)}}$$

RMR provides an overall summary of the magnitude of the residuals. The magnitude of the RMR depends also on the scaling of the variables (if using covariances), but in most applications RMRs .05 or below tend to indicate good fitting models.

11. *Hoelter's (1983) critical N (CN)*. Hoelter proposed the following statistic

$$CN = \frac{Critical \ \chi^2}{F} + 1$$

where the critical $\chi^2$ is a critical value from the chi-square distribution with degrees of freedom equal to the model's degrees of freedom, and alpha is equal to the model-specified alpha. Hoelter suggested that CN $\geq$ 200 indicates a good fitting model.

12. *McDonald's (1989) centrality index (CENT)*. Given that a proposed model is never strictly true, the use of the central chi-square distribution for evaluating model fit (which assumes the model is correct) has been criticized. If the model is not too misspecified, the noncentral chi-square distribution provides a more realistic basis for evaluating a model. The centrality index,

$$\text{CENT} = \exp\left[-\frac{(\chi^2 - df)}{2N}\right]$$

scales the noncentrality parameter (mean of the noncentral chi-square) to range between 0 and 1. In the CENT index, the noncentrality parameter = $\chi^2 - df$.

13. *Root-mean-squared error of approximation (RMSEA; Steiger, 1989)*. As noted earlier, the RMSEA,

$$\varepsilon_a = \sqrt{\frac{F}{df}}$$

measures the discrepancy per degree of freedom for the model. At a minimum, RMSEA = 0 (when the model fits perfectly) but has no upper bound. Browne and Cudeck (1992) recommended values of .05 or less as indicating acceptable model fit.

14. *Expected cross-validation index (ECVI; Browne & Cudeck, 1992)*. Although important for model validation, cross-validating a model can be expensive in terms of sample size if an initial sample is split into a training sample and a test sample. Browne and Cudeck (1992) showed that the degree of cross-validation expected for a model on additional samples can be obtained from the full sample using the expected cross-validation index (ECVI):

$$\text{ECVI} = F + \frac{q}{2N}$$

A model is preferred if it minimizes the value of ECVI relative to other models. It can be shown that the ECVI is a rescaled version of Akaike's information criterion (see later discussion).

**Category III: Information Indices**

Information indices consider simultaneously both the goodness of fit of the model and the number of parameters used to obtain the fit. A penalty function is applied to the fit function as the number of parameters increases relative to the degree of fit.

1. *Akaike's information criterion (AIC; Akaike, 1987)*. The AIC, which is used in many model selection contexts including time-series modeling, suggests the best number of parameters to include in a model. The model with the number of parameters that produces the smallest AIC is considered best:

$$AIC = \chi^2 - 2\ df$$

2. *Consistent Akaike's information criterion (CAIC; Bozdogan, 1987).* Bozdogan suggested a correction to the AIC,

$$CAIC = \chi^2 - [\ln(N) + 1]df$$

due to the apparent tendency of the AIC to not penalize severely enough overparameterized models. Again, models that minimize CAIC are preferred.

3. *Schwartz's Bayesian criterion (SBC; Sclove, 1987).* Schwartz's Bayesian criterion,

$$SBC = \chi^2 - \ln(N)df$$

was developed explicitly from Bayesian theory, whereas the AIC and CAIC were developed from Kullback–Leibler information theory (for a comparison of these two approaches, see Sclove, 1987). The SBC tends to favor models with fewer parameters than either the AIC or CAIC.

### Issue of Fit with Nested Models

In addition to assessing the fit of a particular model, many of the fit indices just described also provide a means by which models can be compared in fit. When two or more models are *nested* within each other, the fit of the models can always be compared. A model is nested within another if the nested model can be obtained from the first model by fixing one or more of the first model's parameters to zero or some other constant. The nested model will have a greater number of degrees on freedom; the difference between the chi-squares of the two models provides a test of whether fixing one or more parameters results in a significant decrement in fit. This difference chi-square ($\chi^2_{\text{diff}}$) is referred to as a chi-square distribution with degrees of freedom equal to the difference between the degrees of freedom of the models being compared. Of course, model comparison can procede in the other direction: models can have parameters sequentially freed, and the successive models can be compared to assess whether the freeing of a particular parameter resulted in a significant increment in fit. Figure 8.5 illustrates the process of comparing nested models. In this simple two-factor example, model 3 (Fig. 8.5c) is nested in models 2 (Fig. 8.5b) and 1 (Fig. 8.5a); model 2 is nested within model 1. Therefore chi-square goodness-of-fit indices can be compared across nested models. In addition, there is no necessity to compare a hypothesized model with the null model in the CFI, NNFI, NFI, PNFI, $\rho_1$, and $\Delta_2$ indices; any pair of nested models may be compared in these *incremental fit indices*. On the other hand, the information indices

(a)

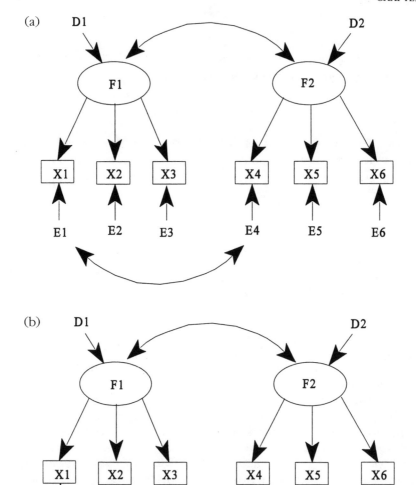

(b)

FIG. 8.5. *(Continued)*

(AIC, CAIC, and SBIC) do not require two models to be nested in order to be compared in a pair of models; whichever has the lower index value is considered the preferred model. It is also possible to transform the information indices into incremental fit measures that resemble the CFI and others:

$$AIC = \frac{AIC_m - AIC_{null}}{df_{null} - \chi^2_{null}}$$

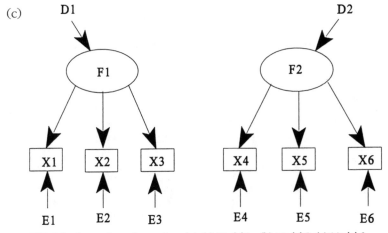

FIG. 8.5. Comparison of nested models.(a) Model 1. (b) Model 2. (c) Model 3.

$$CAIC = \frac{CAIC_m - CAIC_{null}}{df_{null} - [\ln(N) + 1] - \chi^2_{null}}$$

$$SBIC = \frac{SBIC_m - SBIC_{null}}{df_{null} - \ln(N) - \chi^2_{null}}$$

where the null model in these indices can be any model one wishes to compare the hypothesized model against.

## Modification of the Initial Model

Based on all the fit criteria (see Table 8.8), the initial model proposed by Hull et al. (1991) does not fit the data. In order to find out whether a good-fitting model actually exists, we successively free additional parameters from within the original proposed model. As criteria to examine improvement in the fit of the model, we use one significance test (chi-square difference test), one descriptive index (CFI), and one information index (AIC). It is important to note that, for purposes of illustration, the specification search presented next is completely statistically driven (i.e., freeing of a particular parameter does not imply that we believe the parameter to make any theoretical sense). Obviously, selecting parameters to free in an actual empirical study would require extra attention to pertinent substantive issues.

The results of an example specification search are presented in Table 8.14. As can be seen in Table 8.14, the Initial fit indices correspond to those obtained from the first analysis reported in Table 8.8. The other fit indices provided in Table 8.14 correspond to the results obtained by freeing one parameter at a time in the original model. For example, freeing the parameter

TABLE 8.14
Results of a Specification Search

| Model | $\chi^2_{(df)}$ | $p$ | $\chi^2_{\text{diff}(df)}$ | $p$ | CFI | AIC |
|---|---|---|---|---|---|---|
| Initial | $254.26_{(27)}$ | .0001 | — | — | .79 | 200.26 |
| V3F3 | $203.25_{(26)}$ | .0001 | $51.01_{(1)}$ | .0001 | .84 | 151.25 |
| V2F2 | $160.96_{(25)}$ | .0001 | $42.29_{(1)}$ | .0001 | .87 | 110.96 |
| V1F1 and V1F3 | $90.48_{(23)}$ | .0001 | $70.48_{(2)}$ | .0001 | .94 | 44.48 |
| V4F3 | $68.42_{(22)}$ | .0001 | $22.06_{(1)}$ | .0001 | .96 | 24.42 |
| V3F2 | $54.94_{(21)}$ | .0001 | $13.48_{(1)}$ | .0001 | .97 | 12.94 |
| V7F1 and V8F2 | $37.07_{(19)}$ | .0078 | $17.87_{(2)}$ | .0001 | .98 | −0.93 |
| V6F1 | $27.87_{(18)}$ | .0641 | $9.20_{(1)}$ | .0001 | .99 | −8.13 |
| V6F3 | $22.80_{(17)}$ | .0721 | $5.07_{(1)}$ | .0243 | .99 | −11.20 |
| V2F3 | $21.50_{(16)}$ | .1602 | $1.30_{(1)}$ | .2542 | .99 | −10.50 |

V6F1 in the original model provides a chi-square of the model that is non-significant ($p$ = .0641), and the other indices of fit (CFI = .99 and AIC = −8.13) also show improvement. An examination of the next chi-square difference measure (5.07, $df$ = 1) suggests that if the modification continues, and parameter V6F3 is freed, there is still a significant improvement in the model fit. However, further freeing parameter V2F3 does not result in a significant improvement in model fit; the chi-square difference is nonsignificant and the AIC actually worsens (becomes more positive). Thus model modification ceases with the freeing of parameter V6F3.

It is important to note that the freeing of parameters in the original model can also cause some serious problems in the SEM analysis. For example, if during the specification search the parameter V1E1 was freed, a warning note would appear as presented in Table 8.15. Basically, this note serves as a mechanism for detecting freed parameters that are not identified in the model. Of course with V1E1 this was expected, because one cannot estimate both the variance of a measurement error and its path (standard deviation) with its variable.

## MULTISAMPLE ANALYSIS

A question that frequently comes up in SEM modeling is whether a confirmatory factor model is the same across multiple groups. For example, in the Hull et al. (1991) study the researchers could have wondered whether the three-factor structure is the same in males and females. Most currently available structural equation modeling programs (e.g., AMOS, EQS, LISREL, Mx, SEPATH) are capable of dealing with the assessment of the equivalence of confirmatory factor models directly. Unfortunately, the current version of PROC CALIS does not permit the comparison of multiple groups simultane-

TABLE 8.15
Specification Search with Parameter V1E1 Freed

| | |
|---|---|
| Fit criterion | 0.0846 |
| Goodness of Fit Index (GFI) | 0.9816 |
| GFI Adjusted for Degrees of Freedom (AGFI) | 0.9448 |
| Root Mean Square Residual (RMR) | 0.2581 |
| Parsimonious GFI (Mulaik, 1989) | 0.4090 |

| Chi-square = 21.4962 | df = 15 | Prob>chi**2 = 0.1217 |
|---|---|---|
| Null Model Chi-square: | df = 36 | 1112.0091 |

| | | |
|---|---|---|
| RMSEA Estimate | 0.0413 | 90%C.I.[., 0.0776] |
| Probability of Close Fit | | 0.6086 |
| ECVI Estimate | 0.3305 | 90%C.I.[., 0.3965] |
| Bentler's Comparative Fit Index | | 0.9940 |
| Normal Theory Reweighted LS Chi-square | | 21.4128 |
| Akaike's Information Criterion | | −8.5038 |
| Bozdogan's (1987) CAIC | | −76.6227 |
| Schwarz's Bayesian Criterion | | −61.6227 |
| McDonald's (1989) Centrality | | 0.9873 |
| Bentler & Bonett's (1980) Non-normed Index | | 0.9855 |
| Bentler & Bonett's (1980) NFI | | 0.9807 |
| James, Mulaik, & Brett (1982) Parsimonious NFI | | 0.4086 |
| Z-Test of Wilson & Hilferty (1931) | | 1.1687 |
| Bollen (1986) Normed Index Rho1 | | 0.9536 |
| Bollen (1988) Non-normed Index Delta2 | | 0.9941 |
| Hoelter's (1983) Critical N | | 297 |

NOTE: Covariance matrix is not full rank.
      Not all parameters are identified.
NOTE: Some parameter estimates are linearly related to other
      parameter estimates as shown in the following equations:

$$V1E1 = 7.5463 * THE1 - 17.9997$$

ously, but the user can "trick" the program into analyzing multiple groups if each group has the same sample size. Table 8.16 presents an example PROC CALIS program for comparing a particular factor structure with six observed variables (representing various facial characteristic measurements) on a group of male and female children (the data are adapted from Flury & Riedwyl, 1988).

In order for PROC CALIS to analyze the two groups simultaneously, their covariance matrices must be stacked block diagonally, as shown in Table 8.16. By doing this, PROC CALIS is actually assuming that it is analyzing a single group with 12 variables instead of two groups with 6 variables. A sample size of 58 exists for each group; therefore a combined sample size of 116 is indicated. GLS estimation is invoked for this model. According to the proposed measurement model, a single-factor structure appears in males,

## TABLE 8.16
### Sample PROC CALIS Program to Compare
### Factor Structure and Observed Variables

```
DATA ONE(TYPE=COV);
_TYPE_='COV';
INPUT_TYPE_$ NAME_$ V1-V12;
LABEL V1='M_MFB' V2='M_BAM' V3='M_TFH' V4='M_LGAN' V5='M_LTN' V6='M_LTG'
      V7='F_MFB' V8='F_BAM' V9='F_TFH' V10='F_LGAN' V11='F_LTN' V12='F_LTG'
;
CARDS;
COV V1     26.90 ...........
COV V2     12.62 27.25 ..........
COV V3     5.38 2.88 35.23 .........
COV V4     2.93 2.06 10.37 17.85 ........
COV V5     8.18  7.13 6.03 2.92 15.37 .......
COV V6     12.11 11.44 7.97 4.99 14.52 31.84 ......
COV V7     0.00 0.00 0.00 0.00 0.00 0.00 63.20 .....
COV V8     0.00 0.00 0.00 0.00 0.00 0.00 13.15 35.89 ....
COV V9     0.00 0.00 0.00 0.00 0.00 0.00 4.39 -.69 47.81 ...
COV V10    0.00 0.00 0.00 0.00 0.00 0.00 -16.12 -1.75 5.73 19.39 ..
COV V11    0.00 0.00 0.00 0.00 0.00 0.00 0.04 8.35 9.57 6.72 26.06 .
COV V12    0.00 0.00 0.00 0.00 0.00 0.00 0.47 5.00 5.00 3.85 12.89 37.20
;
*NOTE:
*      M_REFERS TO MALES; F_REFERS TO FEMALES
*
*      MFB = 'MINIMAL FRONTAL BREADTH'
*      BAM= 'BREADTH OF ANGULUS MANDIBULAE'
*      TFH= 'TRUE FACIAL HEIGHT'
*      LGAN='LENGTH FROM GLABELLA TO APEX NASI'
*      LTN='LENGTH FROM TRAGION TO NASION'
*      LTG='LENGTH FROM TRAGION TO GNATHION';
RUN;
PROC CALIS DATA=ONE ALL COV MAXITER=1000 NOBS=116 METHOD=GSL;
LINEQS
V1  = V1F1 F1 + E1,
V2  = V2F1 F1 + E2,
V3  = V3F1 F1 + E3,
V4  = V4F1 F1 + E4,
V5  = V5F1 F1 + E5,
V6  = V6F1 F1 + E6,
V7  = V1F2 F2 + E1,
V8  = V2F2 F2 + E2,
V9  = V3F2 F2 + E3,
V10 = V4F3 F3 + E4,
V11 = V5F3 F3 + E5,
V12 = V6F3 F3 + E6;

STD
F1 F2 F3 = 1.00,
E1-E6 = THE1-THE6;

RUN;
```

254

TABLE 8.17
Goodness of Fit for Covariance Structure Analysis:
Generalized Least-Squares Estimation

| | | |
|---|---|---|
| Fit criterion | | 2.4813 |
| Goodness of Fit Index (GFI) | | 0.5864 |
| GFI Adjusted for Degrees of Freedom (AGFI) | | 0.4624 |
| Root Mean Square Residual (RMR) | | 8.1370 |
| Parsimonious GFI (Mulaik, 1989) | | 0.5331 |
| Chi-square = 285.3509 | df = 60 | Prob>chi**2 = 0.0001 |
| Null Model Chi-square: | df = 66 | 123.4453 |
| RMSEA Estimate        0.1807 | 90%C.I.[0.1599, 0.2021] | |
| Probability of Close Fit | | 0.0000 |
| ECVI Estimate        2.7944 | 90%C.I.[2.3696, 3.2847] | |
| Bentler's Comparative Fit Index | | -2.9229 |
| Akaike's Information Criterion | | 165.3509 |
| Bozdogan's (1987) CAIC | | -59.8645 |
| Schwarz's Bayesian Criterion | | 0.1355 |
| McDonald's (1989) Centrality | | 0.3786 |
| Bentler & Bonett's (1980) Non-normed Index | | -3.3152 |
| Bentler & Bonett's (1980) NFI | | -1.3116 |
| James, Mulaik, & Brett (1982) Parsimonious NFI | | -1.1923 |
| Z-Test of Wilson & Hilferty (1931) | | 11.2620 |
| Bollen (1986) Normed Index Rho1 | | -1.5427 |
| Bollen (1988) Non-normed Index Delta2 | | -2.5519 |
| Hoelter's (1983) Critical N | | 33 |

whereas a two-factor structure appears in females. As such, the first three variable loadings on the male facial characteristic factor (V1F1, V2F1, and V3F1) are believed to be equal to the loadings on the first female factor (V1F2, V2F2, and V3F2) and the second three variable loadings on the male factor (V4F1, V5F1, and V6F1) are believed to be equal to the loadings on the second female factor. Note how parameters are equated in PROC CALIS: The same name is assigned.

The assessment of the fit of the multisample analysis is provided in Table 8.17. As can be seen from Table 8.17, all the goodness-of-fit criteria indicate that the proposed model should be rejected. As such, the attempt to confirm the proposed factor structure across male and female groups is abandoned and the analyst is again confronted with the issue of model modification.

## SECOND-ORDER FACTOR ANALYSIS

A second-order factor analysis is a simple extension to the confirmatory factor analysis model. A second-order factor analysis refers to a second factor analysis that is performed on a model with significant correlations among previously defined factors. In general, this second factor analysis is based on an underlying second-order factor structure. For example, in a previous section of this chapter a model that involved the factors of the self-criticism, high standards, and overgeneralization was examined. Based on the analysis conducted it was discovered that the three factors were highly correlated with each other. In order to examine the interrelationships among the three factors further, a second-order factor analysis must be conducted. Figure 8.6 presents a second-order factor model of the Hull et al. (1991) study. As can be seen in Fig. 8.6, the second-order factor has been labeled a "self-punitive attitudes" factor.

Table 8.18 provides the PROC CALIS instructions for analyzing the second-order factor model presented in Fig. 8.6. As can be seen in Table 8.18 (in comparison to the setup shown in Table 8.1), factors one, two, and three (F1, F2, F3) are now endogenous variables, whose correlations are to be explained by a new factor (F4, self-punitive attitudes), the second-order factor. Because F1, F2, and F3 are now endogenous variables, their variances can no longer be specified as free or fixed under the STD command. Instead, their disturbances (D1, D2, and D3) are specified to be solved. The inclusion of the factor disturbances is necessary for the same reason as including the measurement errors (E1 through E9). Even under the most optimistic circumstances, one would hardly expect the factors F1, F2, and F3 to be perfectly predicted by factor F4. Thus, a residual term must be specified in the regression equation predicting F1, F2, and F3. However, one is still confronted with the problem of specifying a metric for the four factor. In this model, this is accomplished this by fixing one indicator for each of F1, F2, and F3 to 1 and fixing F4's variance to 1 under STD.

**Results**

The results of analyzing the second-order factor model are presented in Tables 8.19 and 8.20. Not surprisingly, the chi-square statistics for the second-order factor model are no different from those presented for the CFA

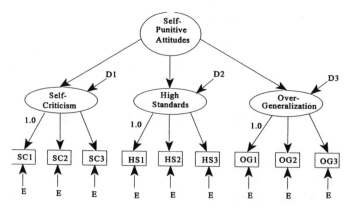

FIG. 8.6.  Second-order factor model of the study by Hull et al. (1991).

TABLE 8.18
PROC CALIS Instructions for Analyzing a Second-Order Factor Model

```
DATA ONE(TYPE=CORR);
_TYPE_='CORR';
INPUT_TYPE_$_NAME_$ V1-V9;
LABEL V1='SC1' V2='SC2' V3='SC3' V4='HS1' V5='HS2' V6='HS3'
      V7='OG1' V8='OG2' V9='OG3';
CARDS;
STD .  1.41 1.02 0.91 2.11 1.85 1.78 2.45 2.05 1.92
CORR V1    1.00 ........
CORR V2    0.40 1.00 .......
CORR V3    0.45 0.37 1.00 ......
CORR V4    0.52 0.49 0.36 1.00 .....
CORR V5    0.41 0.44 0.28 0.63 1.00 ....
CORR V6    0.35 0.43 0.20 0.74 0.60 1.00 ...
CORR V7    0.34 0.20 0.51 0.35 0.15 0.20 1.00 ..
CORR V8    0.19 0.01 0.39 0.19 0.07 0.07 0.69 1.00 .
CORR V9    0.27 0.14 0.42 0.31 0.18 0.21 0.71 0.71 1.00
;
RUN;
PROC CALIS DATA=ONE ALL COV MAXITER=1000 NOBS=255;
LINEQS
V1 =    1 F1 +     0 F2 +    0 F3 + E1,
V2 =  V2F1 F1 +    0 F2 +    0 F3 + E2,
V3 =  V3F1 F1 +    0 F2 +    0 F3 + E3,
V4 =    0 F1 +     1 F2 +    0 F3 + E4,
V5 =    0 F1 +  V5F2 F2 +    0 F3 + E5,
V6 =    0 F1 +  V6F2 F2 +    0 F3 + E6,
V7 =    0 F1 +     0 F2 +    1 F3 + E7,
V8 =    0 F1 +     0 F2 + V8F3 F3 + E8,
V9 =    0 F1 +     0 F2 + V9F3 F3 + E9,

F1 = BE13 F4 +      D1,
F2 = BE24 F4 +      D2,
F3 = BE34 F4 +      D3;

STD
E1-E9 = THE1-THE9,
F4 = 1.00,
D1-D3 = PSI1-PSI3;

RUN;
```

in the previous section (i.e., $\chi^2 = 104.097$, $df = 24$, $p < .00001$). This is because there is really no statistical difference between the two models. In other words, specifying that the factor correlations are due to a fourth (second-order) factor versus specifying that just three factors are correlated results in identical chi-square and degrees of freedom values. In fact, the two models are equivalent models: statistically indistinguishable yet representing differ-

# TABLE 8.19
## Results of Analyzing Second-Order Factor Model, Part I

```
Fit criterion .................................................   0.4098
Goodness of Fit Index (GFI) .....................................   0.9173
GFI Adjusted for Degrees of Freedom (AGFI) ......................   0.8450
Root Mean Square Residual (RMR) .................................   0.2112
Parsimonious GFI (Mulaik, 1989) .................................   0.6115
Chi-square = 104.0970          df = 24              Prob>chi**2 = 0.0001
Null Model Chi-square:         df = 36                       1112.0091
RMSEA Estimate ...................  0.1146     90%C.I.[0.0926, 0.1377]
Probability of Close Fit.........................................   0.0000
ECVI Estimate ...................  0.5820     90%C.I.[0.4706, 0.7243]
Bentler's Comparative Fit Index .................................   0.9256
Normal Theory Reweighted LS Chi-square...........................  103.0262
Akaike's Information Criterion ..................................   56.0970
Bozdogan's (1987) CAIC...........................................  -52.8933
Schwarz's Bayesian Criterion ....................................  -28.8933
McDonald's (1989) Centrality ....................................   0.8547
Bentler & Bonett's (1980) Non-normed Index ......................   0.8883
Bentler & Bonett's (1980) NFI ...................................   0.9064
James, Mulaik, & Brett (1982) Parsimonious NFI ..................   0.6043
Z-Test of Wilson & Hilferty (1931) ..............................   6.6520
Bollen (1986) Normed Index Rho1 .................................   0.8596
Bollen (1988) Non-normed Index Delta2 ...........................   0.9264
Hoelter's (1983) Critical N .....................................      90
```

```
WARNING: The central parameter matrix _PHI_ has probably 1
         negative eigenvalue(s).
```

### Residual Matrix

|    | V1 | V2 | V3 | V4 | V5 | |
|----|-----------|-----------|-----------|-----------|-----------|-----|
| V1 | -.0000004 | -.0190182 | 0.0180267 | 0.1104272 | 0.1145119 | SC1 |
| V2 | -.0190182 | -.0000006 | -.0031718 | 0.1643908 | 0.2385425 | SC2 |
| V3 | 0.0180267 | -.0031718 | -.0000002 | -.1466305 | -.0855782 | SC3 |
| V4 | 0.1104272 | 0.1643908 | -.1466305 | -.0000009 | -.0775432 | HS1 |
| V5 | 0.1145119 | 0.2385425 | -.0855782 | -.0775432 | -.0000003 | HS2 |
| V6 | -.1625516 | 0.1356763 | -.2831600 | 0.0140745 | 0.1376664 | HS3 |
| V7 | 0.10000024 | -.1660180 | 0.5103604 | 0.4498532 | -.2238119 | OG1 |
| V8 | -.3094314 | -.5111274 | 0.2267780 | -.2644727 | -.4566371 | OG2 |
| V9 | -.1116963 | -.2479559 | 0.2423804 | 0.1897795 | -.0693077 | OG3 |

|    | V6 | V7 | V8 | V9 | |
|----|-----------|-----------|-----------|-----------|-----|
| V1 | -.1625516 | 0.1000024 | -.3094314 | -.1116963 | SC1 |
| V2 | 0.1356763 | -.1660180 | -.5111274 | -.2479559 | SC2 |
| V3 | -.2831600 | 0.5103604 | 0.2267780 | 0.2423804 | SC3 |
| V4 | 0.0140745 | 0.4498532 | -.2644727 | 0.1897795 | HS1 |
| V5 | 0.1376664 | -.2238119 | -.4566371 | -.0693077 | HS2 |
| V6 | -.0000002 | -.1128764 | -.5317183 | -.0547971 | HS3 |
| V7 | -.1128764 | -.0000002 | -.0041541 | -.0652401 | OG1 |
| V8 | -.5317183 | -.0041541 | 0.0000000 | 0.0736513 | OG2 |
| V9 | -.0547971 | -.0652401 | 0.0736513 | 0.0000000 | OG3 |

Average Absolute Residual = 0.1474
Average Off-diagonal Absolute Residual = 0.1843

TABLE 8.20
Results of Analyzing Second-Order Factor Model, Part II

Manifest Variable Equations

| V1 | = | 1.0000 F1 | + | 1.0000 E1 |
|---|---|---|---|---|

| V2 | = | 0.6196*F1 | + | 1.0000 E2 |
|---|---|---|---|---|
| Std Err | | 0.0801 L1V2 | | |
| t Value | | 7.7354 | | |

| V3 | = | 0.5832*F1 | + | 1.0000 E3 |
|---|---|---|---|---|
| Std Err | | 0.0723 L1V3 | | |
| t Value | | 8.0694 | | |

| V4 | = | 1.0000 F2 | + | 1.0000 E4 |
|---|---|---|---|---|

| V5 | = | 0.6647*F2 | + | 1.0000 E5 |
|---|---|---|---|---|
| Std Err | | 0.0531 L2V5 | | |
| t Value | | 12.5082 | | |

| V6 | = | 0.7246*F2 | + | 1.0000 E6 |
|---|---|---|---|---|
| Std Err | | 0.0495 L2V6 | | |
| t Value | | 14.6503 | | |

| V7 | = | 1.0000 F3 | + | 1.0000 E7 |
|---|---|---|---|---|

| V8 | = | 0.7991*F3 | + | 1.0000 E8 |
|---|---|---|---|---|
| Std Err | | 0.0552 L3V8 | | |
| t Value | | 14.4861 | | |

| V9 | = | 0.7842*F3 | + | 1.0000 E9 |
|---|---|---|---|---|
| Std Err | | 0.0519 L3V9 | | |
| t Value | | 15.1076 | | |

Latent Variable Equations

| F1 | = | 1.0656*F4 | + | 1.0000 D1 |
|---|---|---|---|---|
| Std Err | | 0.1207 BE13 | | |
| t Value | | 8.8265 | | |

| F2 | = | 1.3482*F4 | + | 1.0000 D2 |
|---|---|---|---|---|
| Std Err | | 0.1591 BE24 | | |
| t Value | | 8.4723 | | |

| F3 | = | 1.0084*F4 | + | 1.0000 D3 |
|---|---|---|---|---|
| Std Err | | 0.1591 BE34 | | |
| t Value | | 6.3386 | | |

*(Continued)*

TABLE 8.20
*(Continued)*

Variances of Exogenous Variables

| Variable | Parameter | Estimate | Standard Error | t Value |
|----------|-----------|----------|----------------|---------|
| F4 |       | 1.000000 | 0 | 0.000 |
| E1 | THE1 | 1.028995 | 0.127259 | 8.086 |
| E2 | THE2 | 0.672151 | 0.070990 | 9.468 |
| E3 | THE3 | 0.501866 | 0.055082 | 9.111 |
| E4 | THE4 | 0.635914 | 0.179332 | 3.546 |
| E5 | THE5 | 1.736238 | 0.176319 | 9.847 |
| E6 | THE6 | 1.164717 | 0.139242 | 8.365 |
| E7 | THE7 | 1.660372 | 0.240819 | 6.895 |
| E8 | THE8 | 1.429972 | 0.176905 | 8.083 |
| E9 | THE9 | 1.016150 | 0.147814 | 6.875 |
| D1 | PSI1 | -0.176404 | 0.185650 | -0.950 |
| D2 | PSI2 | 1.998596 | 0.386264 | 5.174 |
| D3 | PSI3 | 3.325308 | 0.458867 | 7.247 |

ent models and, therefore, different hypotheses about the data (Hershberger, 1994). In order to specify that the factors can correlate freely, one can simply add the following statements to the program presented in Table 8.1:

```
COV
F1  F2  F3  =  PHI1-PHI3;
```

As can be seen from Table 8.19, the descriptive fit indices suggest that the two-factor model fits the data quite well (e.g., CFI = .925). Thus, given the improvement in fit for the second-order model, it is not surprising to see in Table 8.20 that the paths between F1, F2, and F3 with F4 are all significant (i.e., the $t$ value for each parameter is greater than 2).

## AN EXTENSION TO A SIMPLE
## STRUCTURAL EQUATION MODEL

The Hull et al. (1991) second-order factor model can also be extended to the "full" structural equation model presented in Fig. 8.7. As can be seen in Fig. 8.7, this model hypothesizes that self-punitive attitudes directly affect level of depression (factor F5) and that level of depression has a residual disturbance correlation with overgeneralization (F3). It is important to note that F4 (because it is the exogenous variable) does not have disturbance term.

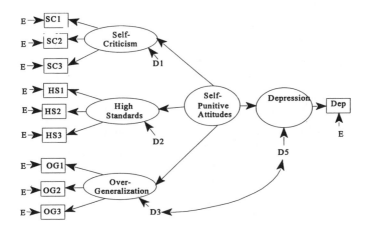

FIG. 8.7. Full structural equation model.

Table 8.21 presents the PROC CALIS setup with the depression variable added to program. Although the results of this analysis (not provided in tables) confirm that there is a significant path between depression and self-punitive attitudes (standardized coefficient F4F5 = .188, $p < .01$) and the F5,F3 disturbance correlation is significant as well (D3,D5 = .587, $p < .0000$), the whole model does not fit the data ($\chi^2 = 119.587$, $df = 33$). In fact, the presence of a significant correlated disturbance between F3 and F5 implies that there are sources of covariance between the two factors not included in the model. Of course, the fact that depression would have a unique relation with overgeneralization and not with self-criticism or high standards could have been predicted by examining the last row of the correlation matrix. The overgeneralization indicators form a cluster of variables significantly correlated with depression; this, with the exception of self-criticism 3, is not the case for the other indicators. Nonetheless, our criticism voiced earlier concerning the atheoretical, fit-maximizing character of correlated indicator measurement errors applies with equal, if not more, force for correlated factor disturbances. If at all possible, such correlations should be avoided, and certainly not included in a model if not initially predicted.

In preference to the correlated disturbances of F3 and F5 would be a directed relation between F3 and F5: either F3 → F5 or F5 → F3 based on prior theory or logic. But there is a problem statistically distinguishing the effects D3 ↔ D5, F3 → F5, and F5 → F3: Any one of the three effects will result in an equivalent model. There is one way in which one can at least rule out a model with the F3 → F5 effect. As it turns out, when an F3 → F5 model is actually run (not in tables), the parameter F4 → F5 is no longer found to be significant. Thus, if prior theory insists on a significant relation between depression and self-punitive attitudes, the model with F3 → F5

## TABLE 8.21
### PROC CALIS Setup with Depression Variable Added

```
DATA ONE(TYPE=CORR);
_TYPE_='CORR';
INPUT_TYPE_$_NAME_$ V1-V10;
LABEL V1='SC1' V2='SC2' V3='SC3' V4='HS1' V5='HS2' V6='HS3'
      V7='OG1' V8='OG2' V9='OG3' V10='DEPRESS';
CARDS;
STD . 1.41 1.02 0.91 2.11 1.85 1.78 2.45 2.05 1.92 5.40
CORR V1    1.00 .........
CORR V2    0.40 1.00 ........
CORR V3    0.45 0.37 1.00 .......
CORR V4    0.52 0.49 0.36 1.00 ......
CORR V5    0.41 0.44 0.28 0.63 1.00 .....
CORR V6    0.35 0.43 0.20 0.74 0.60 1.00 ....
CORR V7    0.34 0.20 0.51 0.35 0.15 0.20 1.00 ...
CORR V8    0.19 0.01 0.39 0.19 0.07 0.07 0.69 1.00 ..
CORR V9    0.27 0.14 0.42 0.31 0.18 0.21 0.71 0.71 1.00 .
CORR V10   0.16 -.01 0.29 0.08 -.03 0.04 0.50 0.54 0.44 1.00
;
RUN;
PROC CALIS DATA=ONE ALL COV MAXITER=1000 NOBS=255;
LINEQS


V1  =     1 F1 +         0 F2 +         0 F3 + E1,
V2  = V2F1 F1 +          0 F2 +         0 F3 + E2,
V3  = V3F1 F1 +          0 F2 +         0 F3 + E3,
V4  =     0 F1 +         1 F2 +         0 F3 + E4,
V5  =     0 F1 +      V5F2 F2 +         0 F3 + E5,
V6  =     0 F1 +      V6F2 F2 +         0 F3 + E6,
V7  =     0 F1 +         0 F2 +         1 F3 + E7,
V8  =     0 F1 +         0 F2 +      V8F3 F3 + E8,
V9  =     0 F1 +         0 F2 +      V9F3 F3 + E9,
V10 =     0 F1 +         0 F2 +         0 F3 + 0 F4 + 1 F5 + E10,
F1 = 1 F4 + D1,
F2 = BE24 F4 + D2,
F3 = BE34 F4 + D3,
F5 = BE54 F4 + D5;


STD
E1-E9 = THE1-THE9,
F4 = 1.00,
D2-D3 = PSI2-PSI3,
D5 = PSI5;


COV
D5 D3 = PSI53;


RUN;
```

would not be a desirable alternative. It is important to note that if one attempts to have it both ways by modeling F5 → F3 and F3 → F5, the following identification warning is provided on the output:

```
NOTE:  Covariance matrix is not full rank.
       Not all parameters are identified.
NOTE:  Some parameter estimates are linearly related to
       other parameter estimates as shown in the following
       equation:
BE53 = 0.0936 * BE35 - 0.0951*BE34 + 1.0397 *BE54
       - 0.1943 * PSI3 + 0.3228*PSI5 - 198.8
```

## A SEM WITH RECURSIVE AND NONRECURSIVE PATHS

One of the primary motivating factors for the development of structural equation modeling lies in the desirability of having a method for resolving nonrecursive relations (i.e., variables are reciprocally related in such a way that each affects and depends on the other). As noted earlier, as originally conceived by Wright, the path analysis approach only includes recursive relations between variables (a model is said to be recursive if all the linkages between variables run one-way). The reason for this restriction has to do with the use of ordinary least-squares regression to estimate the parameters of the model. However, with nonrecursive relations the disturbance terms of the structural equations involved in the relation become correlated, a fundamental violation of the ordinary least-squares regression assumption of independence of errors. For this reason, econometricians developed a number of estimation techniques as alternatives to ordinary least-squares regression to deal with the correlated disturbance problem, including weighted least squares, two-stage least squares, three-stage least squares, and generalized least squares, among others. Of course, correlated disturbances, as illustrated in the last example, are generally no problem in contemporary structural equation modeling. What still remains a problem, however, is the identification of nonrecursive relations in a model.

The identifiability of nonrecursive relations can be much more difficult than recursive relations. The identification difficulty, perhaps as well as limitations in theories that involve nonrecursive relations, probably accounts for the dearth of example models in the research literature. Essentially, in order for a nonrecursive path to be identified, at least one predictor for each of the variables involved in the nonrecursive relation must be unique. This unique predictor has been termed an *instrumental variable*. Instrumental variables may be a part of the original set of variables or may be added for the specific purpose of identification.

For example, if $X$ is an instrumental variable for $Y$ in the nonrecursive relation $Y \Leftrightarrow Z$, the following conditions must be met:

1. $X$ has no direct effect on $Z$.
2. $X$ does affect $Y$, either directly or indirectly, or is correlated with $Y$.
3. Neither $Y$ nor $Z$ directly or indirectly affects $X$.
4. The disturbances of $X$ are uncorrelated with the disturbances of $Z$.

Figure 8.8a provides an example of an instrumental variable F1 for F2. In contrast, in Fig. 8.8b, F1 is not an instrument for F2 because F2 affects F1. Therefore, the relation F2 $\rightleftharpoons$ F3 is not identified.

Although it may appear straightforward, the identification of nonrecursive relations can be tricky in two respects. Sometimes rules 1 through 4 may be more subtly violated. For example, in Fig. 8.8c, F1 is not an instrument for F2 because it is correlated with the disturbance for F3 via F4 and its disturbance. Yet another reason for difficulty in identifying nonrecursive relations has to do with the empirical, if not theoretical, necessity of $X$, if is correlated with $Y$, and $Y$ is correlated with $Z$, being correlated with $Z$. Even if such a variable could be found that was not correlated with $Z$ under these circumstances, theoretically minded researchers would be loath to add it to the model solely for reasons of identification.

Figure 8.9 presents a model that is used quite often in determining employment eligibility. The model includes a nonrecursive relation between

(a)

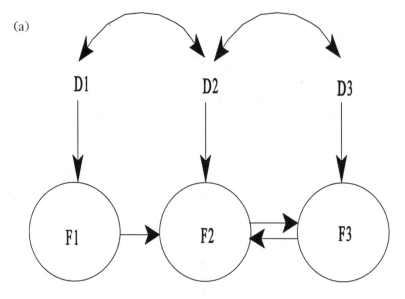

FIG. 8.8. *(Continued)*

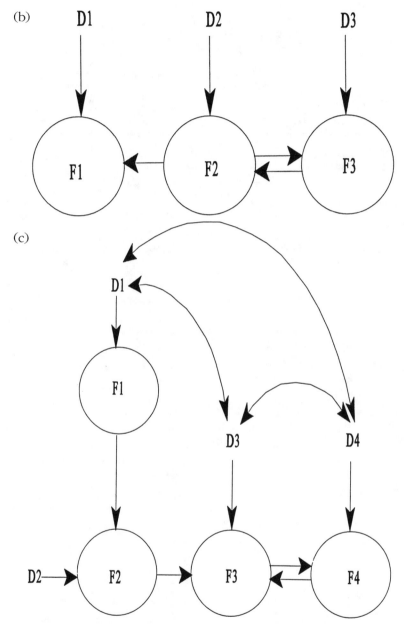

FIG. 8.8.  (a) Instrumental variables: F1 is an instrumental variable for F2. (b) F1 is not an instrument for F2. (c) Nonrecursive relations.

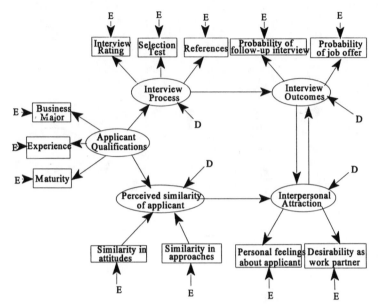

FIG. 8.9. Model for determining employment eligibility.

the factors called interview outcomes and interpersonal attraction. This model is identified because interview process is an instrument for interview outcomes and perceived similarity of applicant is an instrument for interpersonal attraction.

The PROC CALIS program for analyzing a correlation matrix obtained from 165 job applicants is provided in Table 8.22. Table 8.23 presents the results of the parameter estimates for the proposed model concerning employment eligibility. As can be seen from Table 8.23, the relation $F4 \rightarrow F5$ is significant, whereas the relation $F5 \rightarrow F4$, is not: The interview outcomes affect interpersonal attraction, whereas interpersonal attraction (fortunately) does not affect interview outcomes. Thus, one may decide that the $F5 \rightarrow F4$ parameter should be removed from the model and that the model should be reestimated as a recursive model.

TABLE 8.22
PROC CALIS Program for Analyzing a
Correlation Matrix from 165 Job Applicants

```
DATA ONE(TYPE=CORR);
_TYPE_='CORR';
INPUT_TYPE_$_NAME_$ V1-V12;
CARDS;
STD  .  1.04 0.73 0.59 0.81 1.01 0.91 0.85 5.33 0.98 0.50 0.65 0.28
CORR V1     1.00 ...........
CORR V2     0.64 1.00 ..........
CORR V3     0.53 0.72 1.00 .........
CORR V4    -.01 -.09 -.20 1.00 ........
CORR V5     0.06 0.13 0.08 0.17 1.00 .......
CORR V6     0.31 0.20 0.14 0.41 0.29 1.00 ......
CORR V7     0.17 0.21 0.14 0.18 0.21 0.44 1.00 .....
CORR V8    -.09 0.03 0.01 0.03 -.01 0.11 0.33 1.00 ....
CORR V9     0.23 0.11 0.13 0.07 0.09 0.11 -.05 -.46 1.00 ...
CORR V10    0.28 0.23 0.27 0.09 0.15 0.26 0.10 -.28 0.49 1.00 ..
CORR V11    0.25 0.28 0.21 -.09 0.06 -.05 -.26 -.42 0.47 0.26 1.00 .
CORR V12    0.06 0.00 0.03 -.20 0.01 -.34 -.46 -.46 0.27 0.14 0.48 1.00
;
RUN;
PROC CALIS DATA=ONE ALL COV NOBS=165 MAXIT=1000;


LINEQS

V1  =     1 F1 +   E1,
V2  =   V2F1 F1 +   E2,
V3  =   V3F1 F1 +   E3,
V4  =     1 F2 +   E4,
V5  =   V5F2 F2 +   E5,
V6  =   V6F2 F2 +   E6,
V7  =     1 F3 +   E7,
V8  =   V8F3 F3 +   E7,
V9  =     1 F4 +   E9,
V10 = V10F4 F4 +   E9,
V11 =     1 F5 +   E11,
V12 = V12F5 F5 +   E11,

F2 = F2F1 F1 +             D2,
F3 = F3F1 F1 + F3F2 F2 + D3,
F4 = F4F3 F3 + F4F5 F5 + D4,
F5 = F5F2 F2 + F5F4 F4 + D5;


STD
E1-E6 = THE1-THE6,
E7 E9 E11 = THE7 THE9 THE11,
F1 = PHI1,
D2-D5 = PSI2-PSI5;


RUN;
```

## TABLE 8.23
### Results of Proposed Model for Job Eligibility

Manifest Variable Equations

| | | | | |
|---|---|---|---|---|
| V1 | = | 1.0000 F1 | + | 1.0000 E1 |
| V2 | = | 0.9192*F1 | + | 1.0000 E2 |
| Std Err | | 0.1006 V2F1 | | |
| t Value | | 9.1335 | | |
| V3 | = | 0.6313*F1 | + | 1.0000 E3 |
| Std Err | | 0.0697 V3F1 | | |
| t Value | | 9.0563 | | |
| V4 | = | 1.0000 F2 | + | 1.0000 E4 |
| V5 | = | 0.8197*F2 | + | 1.0000 E5 |
| Std Err | | 0.2672 V5F2 | | |
| t Value | | 3.0675 | | |
| V6 | = | 3.6594*F2 | + | 1.0000 E6 |
| Std Err | | 2.0836 V6F2 | | |
| t Value | | 1.7563 | | |
| V7 | = | 1.0000 F3 | + | 1.0000 E7 |
| V8 | = | 34.8370*F3 | + | 1.0000 E7 |
| Std Err | | 14.6266 V8F3 | | |
| t Value | | 2.3818 | | |
| V9 | = | 1.0000 F4 | + | 1.0000 E9 |
| V10 | = | -0.0137*F4 | + | 1.0000 E9 |
| Std Err | | 0.0469 V10F4 | | |
| t Value | | -0.2931 | | |
| V11 | = | 1.0000 F5 | + | 1.0000 E11 |
| V12 | = | 0.0267*F5 | + | 1.0000 E11 |
| Std Err | | 0.0363 V12F5 | | |
| t Value | | 0.7373 | | |

Latent Variable Equations

| | | | | | | |
|---|---|---|---|---|---|---|
| F2 | = | 0.0846*F1 | + | 1.0000 D2 | | |
| Std Err | | 0.0582 F2F1 | | | | |
| t Value | | 1.4529 | | | | |
| F3 | = | 0.0255*F2 | − | 0.0079*F1 | + | 1.0000 D3 |
| Std Err | | 0.0371 F3F2 | | 0.0179 F3F1 | | |
| t Value | | 0.6874 | | -0.4391 | | |
| F4 | = | -2.6621*F3 | − | 0.6699*F5 | + | 1.0000 D4 |
| Std Err | | 1.3204 F4F3 | | 0.4191 F4F5 | | |
| t Value | | -2.0161 | | -1.5983 | | |
| F5 | = | 0.1976*F2 | + | 0.4788*F4 | + | 1.0000 D5 |
| Std Err | | 0.1406 F5F2 | | 0.1487 F5F4 | | |
| t Value | | 1.4056 | | 3.2200 | | |

*(Continued)*

268

TABLE 8.23
*(Continued)*

Variances of Exogenous Variables

| Variable | Parameter | Estimate | Standard Error | t Value |
|----------|-----------|----------|----------------|---------|
| F1 | PHI1 | 0.530445 | 0.110400 | 4.805 |
| E1 | THE1 | 0.551155 | 0.073012 | 7.549 |
| E2 | THE2 | 0.084686 | 0.034751 | 2.437 |
| E3 | THE3 | 0.136706 | 0.021918 | 6.237 |
| E4 | THE4 | 0.572043 | 0.076763 | 7.452 |
| E5 | THE5 | 0.963622 | 0.109799 | 8.776 |
| E6 | THE6 | -0.297530 | 0.593713 | -0.501 |
| E7 | THE7 | 0.699668 | 0.077265 | 9.055 |
| E9 | THE9 | 0.249866 | 0.027593 | 9.055 |
| E11 | THE11 | 0.078154 | 0.008631 | 9.055 |
| D2 | PSI2 | 0.080258 | 0.051806 | 1.549 |
| D3 | PSI3 | 0.022762 | 0.019839 | 1.147 |
| D4 | PSI4 | 0.921110 | 0.308624 | 2.985 |
| D5 | PSI5 | 0.348258 | 0.069334 | 5.023 |

## EXERCISES

1. McCloy, Campbell, and Cudeck (1994) proposed three factors of military performance: declarative knowledge, procedural knowledge, and motivation. Their study had eight observed variables: job knowledge test 1 (JK1); job knowledge test 2 (JK2); hands-on test 1 (HO1); hands-on test 2 (HO2); file measure 1 (FM1); file measure 2 (FM2); average task rating score by supervisor (TRS); and average task rating score by peers (TRP). The authors hypothesized that all eight variables would load on declarative knowledge; that all the variables except JK1 and JK2 would load on procedural knowledge; and that FM1, FM2, TRS, and TRP would load on motivation. The correlation matrix that follows is for 255 infantrymen. Do the data support this model of performance? If not, what modifications to the model would you need to specify in order to obtain an adequate fit to the data?

```
STD          11.80 14.71 9.33 11.32 2.35 0.97 0.64 0.59
CORR  JK1    1.00 . . . . . . .
CORR  JK2    0.59 1.00 . . . . . .
CORR  HO1    0.30 0.25 1.00 . . . . .
CORR  HO2    0.25 0.35 0.41 1.00 . . . .
CORR  FM1    0.08 0.12 0.09 0.11 1.00 . . .
CORR  FM2    -.08 -.15 -.08 -.05 -.08 1.00 . .
CORR  TRS    0.19 0.32 0.25 0.25 0.05 -.23 1.00 .
CORR  TRP    0.11 0.20 0.21 0.17 0.19 -.11 0.36 1.00
```

2. Fit a second-order factor model to the preceding data. To what extent does this model better explain the data than the first-order model?

3. Judge and Watanabe (1993) proposed, in a paper titled "Another look at the job satisfaction–life satisfaction relationship," the model here:

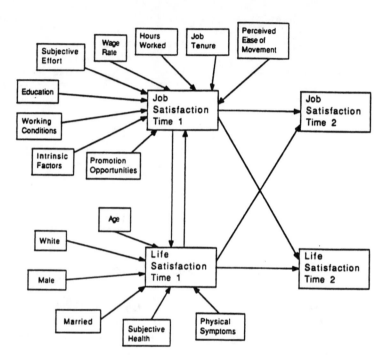

Evaluate this model with the data shown in the following table.

| Variable | M | SD | 1 | 2 | 3 | 4 | 5 | 6 | 7 | 8 | 9 | 10 | 11 | 12 | 13 | 14 | 15 | 16 | 17 | 18 | 19 |
|---|---|---|---|---|---|---|---|---|---|---|---|---|---|---|---|---|---|---|---|---|---|---|
| 1. Job satisfaction—Time 1 | 19.16 | 5.04 | — | | | | | | | | | | | | | | | | | | |
| 2. Life satisfaction—Time 1 | 63.04 | 11.65 | .41 | — | | | | | | | | | | | | | | | | | |
| 3. Job satisfaction—Time 2 | 18.62 | 4.43 | .38 | .35 | — | | | | | | | | | | | | | | | | |
| 4. Life satisfaction—Time 2 | 62.37 | 10.96 | .25 | .51 | .44 | — | | | | | | | | | | | | | | | |
| 5. Age | 35.97 | 11.19 | .14 | .06 | .07 | .15 | — | | | | | | | | | | | | | | |
| 6. Male | 0.73 | 0.44 | .03 | -.04 | .01 | -.03 | -.03 | — | | | | | | | | | | | | | |
| 7. White | 0.93 | 0.25 | .17 | .06 | .13 | -.02 | .03 | .23 | — | | | | | | | | | | | | |
| 8. Married | 0.79 | 0.41 | .05 | .13 | .03 | .01 | .08 | .32 | .18 | — | | | | | | | | | | | |
| 9. Subjective health | 6.09 | 0.96 | .14 | .25 | .04 | .11 | -.07 | .07 | .03 | .06 | — | | | | | | | | | | |
| 10. Education | 4.58 | 1.33 | .17 | .07 | .12 | .08 | -.03 | .06 | .10 | -.01 | .09 | — | | | | | | | | | |
| 11. Job tenure | 4.51 | 1.60 | .06 | -.04 | -.01 | .01 | .37 | .03 | -.03 | .04 | .04 | -.02 | — | | | | | | | | |
| 12. Hours worked per week | 35.05 | 14.13 | .09 | .02 | .04 | -.02 | .05 | .37 | .06 | .25 | .06 | .04 | .06 | — | | | | | | | |
| 13. Wage rate | 4.96 | 3.36 | .12 | .02 | .07 | .08 | .20 | .40 | .16 | .26 | .12 | .30 | .23 | .34 | — | | | | | | |
| 14. Ease of movement | 2.93 | 1.57 | .15 | .12 | .05 | -.01 | -.13 | .04 | .00 | -.02 | .13 | -.02 | -.11 | .08 | -.08 | — | | | | | |
| 15. Subjective effort | 2.32 | 0.57 | -.07 | -.08 | -.09 | -.15 | -.02 | .02 | -.04 | .04 | .06 | .06 | .01 | .23 | .15 | .02 | — | | | | |
| 16. Working conditions | 3.12 | 0.95 | .33 | .28 | .20 | .15 | .10 | -.10 | .02 | -.03 | .11 | .08 | .05 | -.03 | .02 | .08 | -.10 | — | | | |
| 17. Intrinsic factors | 39.22 | 7.73 | .51 | .30 | .21 | .20 | .18 | .12 | .15 | .14 | .07 | .24 | .19 | .21 | .28 | .08 | .03 | .30 | — | | |
| 18. Promotion opportunities | 0.85 | 1.48 | .03 | -.09 | .02 | .03 | .05 | .08 | .01 | .00 | -.05 | .04 | .01 | .06 | .16 | -.14 | -.03 | -.08 | .05 | — | |
| 19. Physical symptoms | 33.03 | 5.33 | .22 | .32 | .15 | .14 | -.06 | .17 | .02 | -.02 | .41 | .12 | .01 | .02 | .08 | .06 | -.07 | .19 | .14 | .05 | — |

*Note.* Correlations greater than .08 are significant at the .05 level (two-tailed). Variables 5–19 were assessed only at Time 1.

4. A correlation matrix for a sample of 100 graduate students is presented here for six indicators of two dimensions of self-concept. The Xs are indicators of social self-concept and the Ys are indicators of academic self-concept, as shown in the following figure.

|     | Y1     | Y2     | Y3     | X1    | X2    | X3    |
|-----|--------|--------|--------|-------|-------|-------|
| Y1  | 1.000  |        |        |       |       |       |
| Y2  | 0.512  | 1.000  |        |       |       |       |
| Y3  | 0.628  | 0.545  | 1.000  |       |       |       |
| X1  | 0.032  | 0.054  | 0.312  | 1.000 |       |       |
| X2  | −0.012 | 0.005  | −0.003 | 0.412 | 1.000 |       |
| X3  | −0.031 | −0.009 | −0.036 | 0.623 | 0.399 | 1.000 |

(a) Run a confirmatory factor analysis on the matrix using the proposed model. (b) Interpret your results, indicating goodness-of-fit measures.

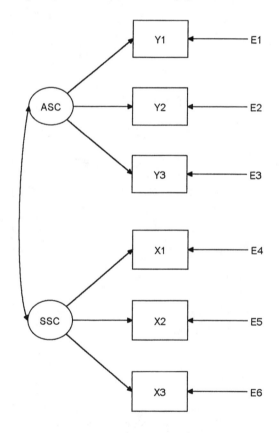

5. The figure here represents a theoretical model of achievement in medical school proposed by Sheehan and Sanford (1989). The model presents two outcome variables: learning about medicine, that is, acquisition of medical knowledge, and learning to do medicine, that is, performing as a clinician. The two latent variables that influence the acquisition of medical knowledge are college achievement and science aptitude. The two variables that influence clinical performance are attitudes and values, and medical knowledge.

Test the fit of the data presented here to this model and discuss the results. The data are based on a sample of 428 medical students, and the variables used in the study are described in the list of variables.

A Structural Model of Medical Student Performance

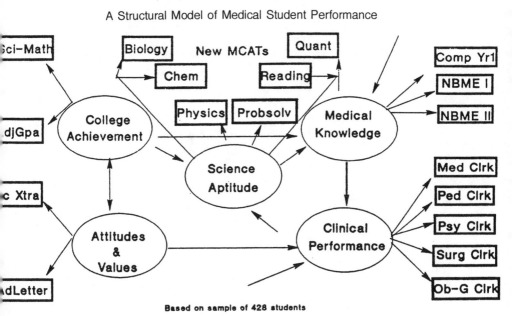

Based on sample of 428 students

Correlations Among 18 Variables with Means, Standard Deviations, and Reliabilities

| | COMP1 | NBMEI | NBMEII | MED | OB | PEDS | PSYCH | SURG |
|---|---|---|---|---|---|---|---|---|
| 1 COMP1 | 1.000 | | | | | | | |
| 2 NBMEI | .733 | 1.000 | | | | | | |
| 3 NBMEII | .620 | .751 | 1.000 | | | | | |
| 4 MED | .145 | .155 | .155 | 1.000 | | | | |
| 5 OB | .081 | .124 | .089 | .217 | 1.000 | | | |
| 6 PEDS | .103 | .093 | .058 | .315 | .194 | 1.000 | | |
| 7 PSYCH | .040 | .060 | .073 | .333 | .251 | .160 | 1.000 | |
| 8 SURG | .225 | .252 | .229 | .400 | .216 | .152 | .199 | 1.000 |
| 9 BIOMC | .237 | .321 | .319 | -.075 | .013 | -.013 | -.110 | .032 |
| 10 CHEMMC | .419 | .463 | .419 | .013 | .012 | .044 | -.028 | .058 |
| 11 PHYSMC | .250 | .342 | .317 | -.069 | -.106 | -.041 | -.010 | -.029 |
| 12 PROBMC | .357 | .426 | .371 | -.056 | .015 | -.024 | -.114 | .010 |
| 13 READMC | .159 | .167 | .311 | .058 | .106 | .051 | .086 | -.014 |
| 14 QUANTMC | .203 | .317 | .340 | -.081 | -.041 | -.048 | .001 | .020 |
| 15 SCIMGPA | .377 | .313 | .231 | .017 | .047 | .075 | .002 | .043 |
| 16 ADJSTGPA | .335 | .214 | .224 | .081 | .012 | .062 | .047 | .071 |
| 17 ACXTRA | .101 | .052 | -.053 | .089 | .116 | .108 | .142 | .114 |
| 18 LETTER | .109 | .066 | -.002 | .137 | .062 | .108 | .087 | .137 |
| MEAN | 70.363 | 521.752 | 508.563 | 3.694 | 3.776 | 3.537 | 3.813 | 3.484 |
| S.D. | 7.268 | 90.081 | 94.680 | .644 | .781 | .675 | .693 | .658 |
| RELIABILITY | .620 | .867 | .646 | .488 | .148 | .163 | .202 | .301 |

| | BIOMC | CHEMMC | PHYSMC | PROBMC | READMC | QUANTMC |
|---|---|---|---|---|---|---|
| BIOMC | 1.000 | | | | | |
| CHEMMC | .412 | 1.000 | | | | |
| PHYSMC | .378 | .596 | 1.000 | | | |
| PROBMC | .580 | .727 | .685 | 1.000 | | |
| READMC | .303 | .269 | .193 | .298 | 1.000 | |
| QUANTMC | .268 | .437 | .425 | .483 | .378 | 1.000 |
| SCIMGPA | .133 | .242 | .163 | .213 | .060 | .035 |
| ADJSTGPA | .066 | .196 | .156 | .141 | .028 | .000 |
| ACXTRA | .024 | -.016 | -.074 | -.003 | -.006 | -.008 |
| LETTER | -.054 | -.021 | -.069 | -.082 | -.059 | -.078 |
| MEAN | 10.322 | 10.526 | 10.451 | 10.528 | 9.661 | 9.942 |
| S.D. | 1.599 | 1.633 | 2.000 | 1.687 | 1.491 | 1.784 |
| RELIABILITY | .351 | .634 | .542 | .841 | .122 | .296 |

| | SCIMGPA | ADJSTGPA | ACXTRA | LETTER |
|---|---|---|---|---|
| SCIMGPA | 1.000 | | | |
| ADJSTGPA | .675 | 1.000 | | |
| ACXTRA | .041 | -.085 | 1.000 | |
| LETTER | .110 | .110 | .356 | 1.000 |
| MEAN | 3.499 | 3.313 | 19.283 | 27.206 |
| S.D. | .332 | 2.772 | 9.346 | 5.412 |
| RELIABILITY | .893 | .511 | .269 | .474 |

*Note.* 1, compl, first year comprehensive exam; 2, nbmei, National Board of Medical Examiners, Part 1; 3, nbmeii, National Board of Medical Examiners, Part II; 4, med, grade in medicine clerkship; 5, ob, grade in OB/Gyn clerkship; 6, peds, grade in pediatrics clerkship; 7, psych, grade in psychiatry clerkship; 8, surg, grade in surgery clerkship; 9, biomc, biology MCAT; 10, chemmc, chemistry MCAT; 11, physmc, physics MCAT; 12, probmc, problem-solving MCAT; 13, readmc, reading MCAT; 14, quantmc, quantitative skills MCAT; 15, scimgpa, average college grades for biology, chemistry, physics, and math; 16, adjstgpa, adjusted GPA, overall GPA adjusted for college selectivity; 17, acxtra, academic extras, scored 0 to 40; 18, letter, letter of recommendation, scored 0 to 40.

6. The figure that follows represents a theoretical model of achievement proposed by Lomax (1993). The model proposes that home background and ability influence aspirations, which in turn influences achievement. Test the fit of the data presented next to this proposed model and discuss the results. The data are in the form of a covariance matrix obtained from a sample of 200 students.

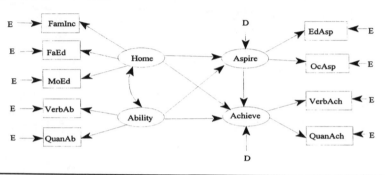

```
1.024
 .792   1.077
1.027    .919   1.844
 .756    .697   1.244   1.286
 .567    .537    .876    .632    .852
 .445    .424    .677    .526    .518    .670
 .434    .389    .635    .498    .475    .545    .716
 .580    .564    .893    .716    .546    .422    .373    .851
 .491    .499    .888    .646    .508    .389    .339    .629    .871
```

7. Organizational structure is believed to be a prominent, but not completely understood, contextual variable that may impact on decision making practices of administrators. The figures shown next present a theoretical and structural model for personnel allocation decisions that were developed by Heck, Marcoulides, and Glasman (1989) and validated by Heck and Marcoulides (1989). A sample correlation matrix is also presented for 74 small school districts and 62 large school districts.

TSM, teacher–student matching
    TEAS, teacher skills
    EVTP, teacher performance
    TEAA, teacher attitudes toward students, parents, and curriculum
    NOS, perceived needs of students
    STR, ratings of students

OC, organizational concerns
    DOC, distribution of children across grade levels
    OSN, overall staffing needs in the school
    DIP, district policies and union agreements

IPC, internal political concerns
  TPRE, teacher preferences for assignment
  COOT, characteristics of other teachers
  TEAA, teacher attitudes
  CFT, correspondence data from the teacher
  IM, indirect measures about teacher suitability
PI, parent input
  PPRE, importance principal attaches to parent preferences
  CFP, personal correspondence with parents
  STR, student ratings
DS, data sources
  PCO, planned classroom observations
  UO, unplanned observations and recollections
  SAI, student achievement informations
  TER, teacher ratings
  GUT, intuitive feelings
AD, allocation decisions
  Q5, teacher issues
  Q6, schoolwide issues
  Q7, teacher demands and bargaining
  Q8, parent wishes
  Q9, organizational policies and union agreements

(a) Use the structural model to test if the same model structure is present across two sizes of school districts: small and large. (b) If it is not, determine to what degree it differs, and how.

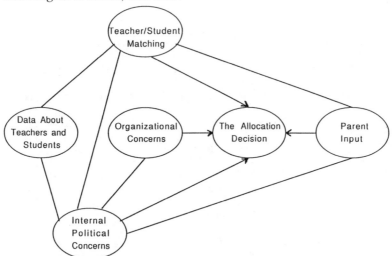

Predictive model of variables influencing personnel allocation decisions.

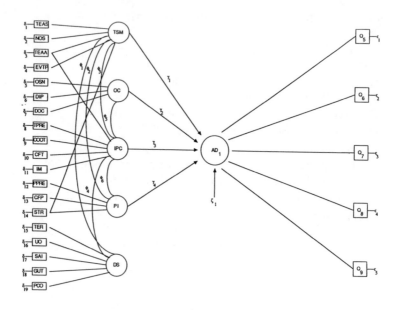

LISREL representation of model for personnel allocation decisions.

Large School Districts

| TPRE | PPRE | TEAS | OSC | DIP | NOS | EVTP | COOT | TEAA | DOC | CFP | STR |
|------|------|------|------|------|------|------|------|------|------|------|------|
| 1.00 | | | | | | | | | | | |
| 0.35 | 1.00 | | | | | | | | | | |
| 0.35 | 0.21 | 1.00 | | | | | | | | | |
| 0.46 | 0.20 | 0.01 | 1.00 | | | | | | | | |
| 0.05 | 0.04 | 0.16 | 0.13 | 1.00 | | | | | | | |
| 0.31 | −0.32 | 0.25 | 0.45 | 0.08 | 1.00 | | | | | | |
| 0.36 | −0.15 | 0.09 | 0.35 | 0.07 | −0.06 | 1.00 | | | | | |
| 0.22 | 0.24 | 0.16 | −0.15 | −0.12 | 0.32 | 0.48 | 1.00 | | | | |
| 0.10 | 0.27 | 0.15 | 0.12 | 0.04 | −0.05 | 0.08 | 0.29 | 1.00 | | | |
| 0.23 | 0.16 | 0.15 | 0.08 | −0.13 | −0.06 | 0.33 | 0.31 | 0.37 | 1.00 | | |
| 0.02 | 0.20 | 0.08 | 0.02 | 0.16 | 0.07 | 0.27 | 0.03 | 0.26 | 0.27 | 1.00 | |
| 0.01 | −0.04 | 0.41 | 0.08 | −0.19 | −0.04 | 0.04 | 0.23 | 0.26 | 0.15 | 0.07 | 1.00 |
| 0.14 | −0.12 | 0.10 | 0.11 | −0.17 | 0.10 | 0.17 | 0.03 | 0.15 | 0.19 | 0.09 | 0.32 |
| 0.11 | 0.12 | 0.01 | 0.04 | 0.04 | −0.04 | 0.05 | 0.15 | 0.15 | 0.30 | 0.19 | 0.12 |
| −0.09 | 0.45 | 0.00 | 0.10 | 0.30 | 0.08 | 0.04 | 0.18 | 0.02 | 0.10 | 0.23 | 0.18 |
| −0.05 | 0.22 | 0.20 | −0.15 | 0.02 | 0.28 | 0.15 | 0.23 | 0.39 | 0.47 | 0.11 | −0.05 |
| −0.01 | 0.10 | 0.02 | 0.06 | 0.13 | 0.04 | −0.04 | −0.04 | 0.06 | 0.02 | 0.02 | −0.11 |
| 0.00 | 0.07 | 0.05 | −0.11 | 0.06 | −0.32 | 0.38 | 0.15 | 0.29 | 0.10 | 0.15 | 0.11 |
| 0.18 | −0.13 | −0.22 | −0.08 | −0.09 | 0.14 | 0.03 | 0.24 | 0.10 | 0.04 | 0.23 | 0.19 |
| 0.54 | 0.13 | 0.27 | −0.05 | 0.04 | 0.23 | 0.04 | 0.20 | 0.22 | 0.14 | 0.04 | 0.02 |
| 0.12 | 0.23 | 0.01 | 0.11 | 0.30 | 0.40 | 0.07 | 0.21 | 0.44 | 0.09 | 0.06 | 0.09 |
| 0.18 | −0.09 | 0.07 | −0.13 | 0.11 | −0.09 | −0.11 | −0.18 | 0.07 | 0.02 | 0.08 | 0.30 |
| 0.38 | 0.12 | 0.01 | −0.26 | 0.13 | −1.07 | 0.11 | −0.10 | 0.06 | 0.02 | 0.28 | 0.22 |
| 0.47 | −0.03 | 0.11 | −0.20 | 0.09 | 0.03 | 0.10 | 0.09 | −0.01 | 0.08 | 0.24 | 0.10 |
| 0.52 | −0.10 | 0.08 | −0.21 | 0.16 | 0.17 | 0.03 | 0.10 | −0.11 | −0.01 | 0.21 | 0.20 |

| PCO | CFT | GUT | SAI | UO | TER | IM | Q5 | Q6 | Q7 | Q8 | Q9 |
|---|---|---|---|---|---|---|---|---|---|---|---|
| 1.00 | | | | | | | | | | | |
| 0.32 | 1.00 | | | | | | | | | | |
| 0.06 | 0.12 | 1.00 | | | | | | | | | |
| −0.33 | 0.01 | 0.13 | 1.00 | | | | | | | | |
| 0.03 | 0.04 | 0.32 | 0.21 | 1.00 | | | | | | | |
| 0.23 | 0.31 | 0.11 | 0.46 | 0.23 | 1.00 | | | | | | |
| 0.19 | 0.06 | 0.12 | 0.21 | 0.26 | 0.26 | 1.00 | | | | | |
| 0.17 | 0.15 | −0.05 | 0.35 | 0.40 | 0.16 | −0.04 | 1.00 | | | | |
| 0.21 | 0.06 | 0.10 | 0.02 | 0.05 | 0.17 | 0.08 | 0.18 | 1.00 | | | |
| 0.14 | 0.32 | 0.23 | 0.47 | 0.14 | −0.24 | 0.10 | −0.01 | 0.12 | 1.00 | | |
| −0.07 | 0.11 | 0.03 | 0.03 | 0.02 | 0.02 | −0.11 | −0.01 | 0.04 | 0.01 | 1.00 | |
| 0.28 | −0.11 | 0.23 | −0.14 | 0.07 | 0.22 | 0.03 | −0.08 | −0.01 | 0.08 | 0.01 | 1.00 |
| 0.00 | −0.19 | 0.01 | −0.10 | −0.12 | 0.03 | −0.14 | −0.03 | −0.05 | 0.14 | 0.51 | 0.89 |

Small School Districts

| TPRE | PPRE | TEAS | OSC | DIP | NOS | EVTP | COOT | TEAA | DOC | CFP | STR |
|---|---|---|---|---|---|---|---|---|---|---|---|
| 1.00 | | | | | | | | | | | |
| 0.17 | 1.00 | | | | | | | | | | |
| 0.27 | 0.10 | 1.00 | | | | | | | | | |
| 0.29 | 0.07 | 0.19 | 1.00 | | | | | | | | |
| 0.11 | 0.16 | 0.06 | 0.13 | 1.00 | | | | | | | |
| 0.13 | −0.07 | 0.25 | 0.45 | 0.08 | 1.00 | | | | | | |
| 0.06 | −0.02 | 0.09 | 0.35 | 0.07 | −0.06 | 1.00 | | | | | |
| 0.02 | 0.24 | 0.16 | −0.15 | −0.12 | 0.32 | 0.21 | 1.00 | | | | |
| −0.10 | 0.27 | 0.15 | 0.12 | 0.04 | −0.05 | 0.08 | 0.29 | 1.00 | | | |
| −0.23 | 0.16 | 0.15 | 0.08 | −0.13 | −0.06 | 0.33 | 0.01 | 0.37 | 1.00 | | |
| −0.02 | 0.20 | 0.08 | 0.02 | 0.16 | 0.07 | 0.27 | 0.03 | 0.26 | 0.27 | 1.00 | |
| −0.01 | −0.04 | 0.41 | 0.08 | −0.19 | −0.04 | 0.04 | 0.23 | 0.26 | 0.15 | 0.07 | 1.00 |
| 0.14 | −0.12 | 0.10 | 0.11 | −0.17 | 0.10 | 0.17 | 0.03 | 0.15 | 0.19 | 0.09 | 0.32 |
| 0.11 | 0.22 | 0.01 | 0.04 | 0.04 | −0.04 | 0.05 | 0.15 | 0.15 | 0.30 | 0.19 | 0.12 |
| 0.23 | 0.20 | 0.00 | 0.10 | 0.30 | 0.08 | 0.04 | 0.18 | 0.02 | 0.10 | 0.23 | 0.18 |
| −0.21 | 0.22 | 0.20 | −0.15 | 0.02 | 0.28 | 0.00 | 0.23 | 0.19 | 0.17 | 0.11 | −0.05 |
| −0.01 | 0.19 | 0.02 | 0.06 | 0.13 | 0.04 | −0.04 | −0.04 | 0.06 | 0.02 | 0.02 | −0.11 |
| 0.00 | 0.07 | 0.05 | −0.11 | 0.06 | −0.32 | 0.05 | 0.15 | 0.29 | 0.10 | 0.15 | 0.11 |
| 0.18 | −0.13 | −0.22 | −0.08 | −0.09 | 0.14 | 0.03 | 0.24 | 0.10 | 0.04 | 0.23 | 0.19 |
| 0.30 | 0.13 | 0.27 | −0.05 | 0.04 | 0.23 | 0.04 | 0.20 | 0.22 | 0.14 | 0.04 | 0.02 |
| 0.12 | 0.23 | 0.01 | 0.11 | 0.30 | 0.40 | 0.07 | 0.21 | 0.44 | 0.09 | 0.06 | 0.09 |
| 0.18 | −0.09 | 0.07 | −0.13 | 0.11 | −0.09 | −0.11 | −0.18 | 0.07 | 0.02 | 0.08 | 0.10 |
| 0.08 | 0.12 | −0.01 | −0.26 | 0.13 | −1.07 | 0.11 | −0.10 | 0.06 | 0.02 | 0.28 | 0.22 |
| 0.07 | −0.03 | 0.21 | −0.20 | 0.09 | 0.03 | 0.10 | 0.09 | −0.01 | 0.08 | 0.24 | 0.10 |
| 0.10 | −0.10 | 0.08 | −0.21 | 0.16 | 0.17 | 0.03 | 0.10 | −0.11 | −0.01 | 0.21 | 0.00 |

*(Continued)*

| PCO | CFT | GUT | SAI | UO | TER | IM | Q5 | Q6 | Q7 | Q8 | Q9 |
|---|---|---|---|---|---|---|---|---|---|---|---|
| 1.00 | | | | | | | | | | | |
| 0.05 | 1.00 | | | | | | | | | | |
| 0.18 | 0.32 | 1.00 | | | | | | | | | |
| −0.09 | 0.31 | 0.03 | 1.00 | | | | | | | | |
| 0.16 | 0.34 | 0.15 | 0.11 | 1.00 | | | | | | | |
| 0.03 | 0.31 | 0.01 | 0.16 | 0.13 | 1.00 | | | | | | |
| 0.19 | 0.36 | 0.12 | 0.01 | 0.16 | 0.26 | 1.00 | | | | | |
| 0.07 | 0.15 | −0.05 | 0.35 | 0.00 | 0.06 | −0.04 | 1.00 | | | | |
| 0.01 | 0.06 | 0.10 | 0.02 | 0.05 | 0.07 | 0.28 | 0.18 | 1.00 | | | |
| 0.04 | 0.00 | 0.03 | 0.17 | 0.14 | −0.04 | 0.10 | −0.01 | 0.22 | 1.00 | | |
| −0.07 | 0.11 | 0.03 | 0.03 | 0.02 | 0.02 | −0.11 | −0.01 | 0.04 | 0.14 | 1.00 | |
| 0.08 | −0.11 | 0.13 | −0.14 | 0.07 | 0.02 | 0.03 | −0.08 | −0.01 | 0.14 | 0.61 | 1.00 |
| 0.00 | −0.19 | 0.01 | −0.10 | −0.12 | 0.03 | −0.14 | −0.03 | −0.05 | 0.14 | 0.51 | 0.89 |

# Appendix A

TABLE A.1
Standard Normal Distribution

| $z$ | $Pr(Z{\leq}z)$ | $z$ | $Pr(Z{\leq}z)$ | $z$ | $Pr(Z{\leq}z)$ |
|------|-----------|------|-----------|------|-----------|
| 0.0  | 0.5000000 | 0.86 | 0.8051055 | 1.72 | 0.9572838 |
| 0.01 | 0.5039894 | 0.87 | 0.8078498 | 1.73 | 0.9581849 |
| 0.02 | 0.5079783 | 0.88 | 0.8105703 | 1.74 | 0.9590705 |
| 0.03 | 0.5119665 | 0.89 | 0.8132671 | 1.75 | 0.9599408 |
| 0.04 | 0.5159534 | 0.90 | 0.8159399 | 1.76 | 0.9607961 |
| 0.05 | 0.5199388 | 0.91 | 0.8185887 | 1.77 | 0.9616364 |
| 0.06 | 0.5239222 | 0.92 | 0.8212136 | 1.78 | 0.9624620 |
| 0.07 | 0.5279032 | 0.93 | 0.8238145 | 1.79 | 0.9632730 |
| 0.08 | 0.5318814 | 0.94 | 0.8263912 | 1.80 | 0.9640697 |
| 0.09 | 0.5358564 | 0.95 | 0.8289439 | 1.81 | 0.9648521 |
| 0.10 | 0.5398278 | 0.96 | 0.8314724 | 1.82 | 0.9656205 |
| 0.11 | 0.5437953 | 0.97 | 0.8339768 | 1.83 | 0.9663750 |
| 0.12 | 0.5477584 | 0.98 | 0.8364569 | 1.84 | 0.9671159 |
| 0.13 | 0.5517168 | 0.99 | 0.8389129 | 1.85 | 0.9678432 |
| 0.14 | 0.5556700 | 1.00 | 0.8413447 | 1.86 | 0.9685572 |
| 0.15 | 0.5596177 | 1.01 | 0.8437524 | 1.87 | 0.9692581 |
| 0.16 | 0.5635595 | 1.02 | 0.8461358 | 1.88 | 0.9699460 |
| 0.17 | 0.5674949 | 1.03 | 0.8484950 | 1.89 | 0.9706210 |
| 0.18 | 0.5714237 | 1.04 | 0.8508300 | 1.90 | 0.9712834 |
| 0.19 | 0.5753454 | 1.05 | 0.8531409 | 1.91 | 0.9719334 |
| 0.20 | 0.5792597 | 1.06 | 0.8554277 | 1.92 | 0.9725711 |
| 0.21 | 0.5831662 | 1.07 | 0.8576903 | 1.93 | 0.9731966 |

*(Continued)*

| z | Pr(Z≤z) | z | Pr(Z≤z) | z | Pr(Z≤z) |
|---|---------|---|---------|---|---------|
| 0.22 | 0.5870644 | 1.08 | 0.8599289 | 1.94 | 0.9738102 |
| 0.23 | 0.5909541 | 1.09 | 0.8621434 | 1.95 | 0.9744119 |
| 0.24 | 0.5948349 | 1.10 | 0.8643339 | 1.96 | 0.9750021 |
| 0.25 | 0.5987063 | 1.11 | 0.8665005 | 1.97 | 0.9755808 |
| 0.26 | 0.6025681 | 1.12 | 0.8686431 | 1.98 | 0.9761482 |
| 0.27 | 0.6064199 | 1.13 | 0.8707619 | 1.99 | 0.9767045 |
| 0.28 | 0.6102612 | 1.14 | 0.8728568 | 2.00 | 0.9772499 |
| 0.29 | 0.6140919 | 1.15 | 0.8749281 | 2.01 | 0.9777844 |
| 0.30 | 0.6179114 | 1.16 | 0.8769756 | 2.02 | 0.9783083 |
| 0.31 | 0.6217195 | 1.17 | 0.8789995 | 2.03 | 0.9788217 |
| 0.32 | 0.6255158 | 1.18 | 0.8809999 | 2.04 | 0.9793248 |
| 0.33 | 0.6293000 | 1.19 | 0.8829768 | 2.05 | 0.9798178 |
| 0.34 | 0.6330717 | 1.20 | 0.8849303 | 2.06 | 0.9803007 |
| 0.35 | 0.6368307 | 1.21 | 0.8868606 | 2.07 | 0.9807738 |
| 0.36 | 0.6405764 | 1.22 | 0.8887676 | 2.08 | 0.9812372 |
| 0.37 | 0.6443088 | 1.23 | 0.8906514 | 2.09 | 0.9816911 |
| 0.38 | 0.6480273 | 1.24 | 0.8925123 | 2.10 | 0.9821356 |
| 0.39 | 0.6517317 | 1.25 | 0.8943502 | 2.11 | 0.9825708 |
| 0.40 | 0.6554217 | 1.26 | 0.8961653 | 2.12 | 0.9829970 |
| 0.41 | 0.6590970 | 1.27 | 0.8979577 | 2.13 | 0.9834142 |
| 0.42 | 0.6627573 | 1.28 | 0.8997274 | 2.14 | 0.9838226 |
| 0.43 | 0.6664022 | 1.29 | 0.9014747 | 2.15 | 0.9842224 |
| 0.44 | 0.6700314 | 1.30 | 0.9031995 | 2.16 | 0.9846137 |
| 0.45 | 0.6736448 | 1.31 | 0.9049021 | 2.17 | 0.9849966 |
| 0.46 | 0.6772419 | 1.32 | 0.9065825 | 2.18 | 0.9853713 |
| 0.47 | 0.6808225 | 1.33 | 0.9082409 | 2.19 | 0.9857379 |
| 0.48 | 0.6843863 | 1.34 | 0.9098773 | 2.20 | 0.9860966 |
| 0.49 | 0.6879331 | 1.35 | 0.9114920 | 2.21 | 0.9864474 |
| 0.50 | 0.6914625 | 1.36 | 0.9130850 | 2.22 | 0.9867906 |
| 0.51 | 0.6949743 | 1.37 | 0.9146565 | 2.23 | 0.9871263 |
| 0.52 | 0.6984682 | 1.38 | 0.9162067 | 2.24 | 0.9874545 |
| 0.53 | 0.7019440 | 1.39 | 0.9177356 | 2.25 | 0.9877755 |

(Continued)

TABLE A.1
(Continued)

| z | Pr(Z≤z) | z | Pr(Z≤z) | z | Pr(Z≤z) |
|---|---------|---|---------|---|---------|
| 0.54 | 0.7054015 | 1.40 | 0.9192433 | 2.26 | 0.9880894 |
| 0.55 | 0.7088403 | 1.41 | 0.9207302 | 2.27 | 0.9883962 |
| 0.56 | 0.7122603 | 1.42 | 0.9221962 | 2.28 | 0.9886962 |
| 0.57 | 0.7156612 | 1.43 | 0.9236415 | 2.29 | 0.9889893 |
| 0.58 | 0.7190427 | 1.44 | 0.9250663 | 2.30 | 0.9892759 |
| 0.59 | 0.7224047 | 1.45 | 0.9264707 | 2.31 | 0.9895559 |
| 0.60 | 0.7257469 | 1.46 | 0.9278550 | 2.32 | 0.9898296 |
| 0.61 | 0.7290691 | 1.47 | 0.9292191 | 2.33 | 0.9900969 |
| 0.62 | 0.7323711 | 1.48 | 0.9305634 | 2.34 | 0.9903581 |
| 0.63 | 0.7356527 | 1.49 | 0.9318879 | 2.35 | 0.9906133 |
| 0.64 | 0.7389137 | 1.50 | 0.9331928 | 2.36 | 0.9908625 |
| 0.65 | 0.7421539 | 1.51 | 0.9344783 | 2.37 | 0.9911060 |
| 0.66 | 0.7453731 | 1.52 | 0.9357445 | 2.38 | 0.9913437 |
| 0.67 | 0.7485711 | 1.53 | 0.9369916 | 2.39 | 0.9915758 |
| 0.68 | 0.7517478 | 1.54 | 0.9382198 | 2.40 | 0.9918025 |
| 0.69 | 0.7549029 | 1.55 | 0.9394292 | 2.41 | 0.9920237 |
| 0.70 | 0.7580363 | 1.56 | 0.9406201 | 2.42 | 0.9922397 |
| 0.71 | 0.7611479 | 1.57 | 0.9417924 | 2.43 | 0.9924506 |
| 0.72 | 0.7642375 | 1.58 | 0.9429466 | 2.44 | 0.9926564 |
| 0.73 | 0.7673049 | 1.59 | 0.9440826 | 2.45 | 0.9928572 |
| 0.74 | 0.7703500 | 1.60 | 0.9452007 | 2.46 | 0.9930531 |
| 0.75 | 0.7733726 | 1.61 | 0.9463011 | 2.47 | 0.9932443 |
| 0.76 | 0.7763727 | 1.62 | 0.9473839 | 2.48 | 0.9934309 |
| 0.77 | 0.7793501 | 1.63 | 0.9484493 | 2.49 | 0.9936128 |
| 0.78 | 0.7823046 | 1.64 | 0.9494974 | 2.50 | 0.9937903 |
| 0.79 | 0.7852361 | 1.65 | 0.9505285 | 2.51 | 0.9939634 |
| 0.80 | 0.7881446 | 1.66 | 0.9515428 | 2.52 | 0.9941323 |
| 0.81 | 0.7910299 | 1.67 | 0.9525403 | 2.53 | 0.9942969 |
| 0.82 | 0.7938919 | 1.68 | 0.9535213 | 2.54 | 0.9944574 |
| 0.83 | 0.7967306 | 1.69 | 0.9544860 | 2.55 | 0.9946139 |
| 0.84 | 0.7995458 | 1.70 | 0.9554345 | 2.56 | 0.9947664 |
| 0.85 | 0.8023375 | 1.71 | 0.9563671 | 2.57 | 0.9949151 |

*Note.* This table was computed with IMSL subroutine MDNOR.

TABLE A.2
Critical Values of $t$ Distribution

| $v$ | $t_{.100}$ | $t_{.050}$ | $t_{.025}$ | $t_{.010}$ | $t_{.005}$ | $t_{.001}$ | $t_{.0005}$ |
|---|---|---|---|---|---|---|---|
| 1 | 3.078 | 6.314 | 12.706 | 31.821 | 63.657 | 318.310 | 636.620 |
| 2 | 1.886 | 2.920 | 4.303 | 6.965 | 9.925 | 22.326 | 31.598 |
| 3 | 1.638 | 2.353 | 3.182 | 4.541 | 5.841 | 10.213 | 12.924 |
| 4 | 1.533 | 2.132 | 2.776 | 3.747 | 4.604 | 7.173 | 8.610 |
| 5 | 1.476 | 2.015 | 2.571 | 3.365 | 4.032 | 5.893 | 6.869 |
| 6 | 1.440 | 1.943 | 2.447 | 3.143 | 3.707 | 5.208 | 5.959 |
| 7 | 1.415 | 1.895 | 2.365 | 2.998 | 3.499 | 4.785 | 5.408 |
| 8 | 1.397 | 1.860 | 2.306 | 2.896 | 3.355 | 4.501 | 5.041 |
| 9 | 1.383 | 1.833 | 2.262 | 2.821 | 3.250 | 4.297 | 4.781 |
| 10 | 1.372 | 1.812 | 2.228 | 2.764 | 3.169 | 4.144 | 4.587 |
| 11 | 1.363 | 1.796 | 2.201 | 2.718 | 3.106 | 4.025 | 4.437 |
| 12 | 1.356 | 1.782 | 2.179 | 2.681 | 3.055 | 3.930 | 4.318 |
| 13 | 1.350 | 1.771 | 2.160 | 2.650 | 3.012 | 3.852 | 4.221 |
| 14 | 1.345 | 1.761 | 2.145 | 2.624 | 2.977 | 3.787 | 4.140 |
| 15 | 1.341 | 1.753 | 2.131 | 2.602 | 2.947 | 3.733 | 4.073 |
| 16 | 1.337 | 1.746 | 2.120 | 2.583 | 2.921 | 3.686 | 4.015 |
| 17 | 1.333 | 1.740 | 2.110 | 2.567 | 2.898 | 3.646 | 3.965 |
| 18 | 1.330 | 1.732 | 2.101 | 2.552 | 2.878 | 3.610 | 3.922 |
| 19 | 1.328 | 1.729 | 2.093 | 2.539 | 2.861 | 3.579 | 3.883 |
| 20 | 1.325 | 1.725 | 2.086 | 2.528 | 2.845 | 3.552 | 3.850 |
| 21 | 1.323 | 1.721 | 2.080 | 2.518 | 2.831 | 3.527 | 3.819 |
| 22 | 1.321 | 1.717 | 2.074 | 2.508 | 2.819 | 3.505 | 3.792 |
| 23 | 1.319 | 1.714 | 2.069 | 2.500 | 2.807 | 3.485 | 3.767 |
| 24 | 1.318 | 1.711 | 2.064 | 2.492 | 2.797 | 3.467 | 3.745 |
| 25 | 1.316 | 1.708 | 2.060 | 2.485 | 2.787 | 3.450 | 3.725 |
| 26 | 1.315 | 1.706 | 2.056 | 2.479 | 2.779 | 3.435 | 3.707 |
| 27 | 1.314 | 1.703 | 2.052 | 2.473 | 2.771 | 3.421 | 3.690 |
| 28 | 1.313 | 1.701 | 2.048 | 2.467 | 2.763 | 3.408 | 3.674 |
| 29 | 1.311 | 1.699 | 2.045 | 2.462 | 2.756 | 3.396 | 3.659 |
| 30 | 1.310 | 1.697 | 2.042 | 2.457 | 2.750 | 3.385 | 3.646 |
| 40 | 1.303 | 1.684 | 2.021 | 2.423 | 2.704 | 3.307 | 3.551 |
| 60 | 1.296 | 1.671 | 2.000 | 2.390 | 2.660 | 3.232 | 3.460 |
| 120 | 1.289 | 1.658 | 1.980 | 2.358 | 2.617 | 3.160 | 3.373 |
| $\infty$ | 1.282 | 1.645 | 1.960 | 2.326 | 2.576 | 3.090 | 3.291 |

## TABLE A.3
### Values of $f$ Such That $\Pr(F \leq f) = .95$ Where $F$ has an $F$ Distribution With $v_1$ and $v_2$ Degrees of Freedom

| $v_1$ | | | | | $v_2$ | | | | | |
|---|---|---|---|---|---|---|---|---|---|---|
| | 1 | 2 | 3 | 4 | 5 | 6 | 7 | 8 | 9 | 10 |
| 5 | 6.61 | 5.79 | 5.41 | 5.19 | 5.05 | 4.95 | 4.88 | 4.82 | 4.77 | 4.73 |
| 6 | 5.99 | 5.14 | 4.76 | 4.53 | 4.39 | 4.28 | 4.21 | 4.15 | 4.10 | 4.06 |
| 7 | 5.59 | 4.74 | 4.35 | 4.12 | 3.97 | 3.87 | 3.79 | 3.73 | 3.68 | 3.64 |
| 8 | 5.32 | 4.46 | 4.07 | 3.84 | 3.69 | 3.58 | 3.50 | 3.44 | 3.39 | 3.35 |
| 9 | 5.12 | 4.26 | 3.86 | 3.63 | 3.48 | 3.37 | 3.29 | 3.23 | 3.18 | 3.14 |
| 10 | 4.96 | 4.10 | 3.71 | 3.48 | 3.33 | 3.22 | 3.14 | 3.07 | 3.02 | 2.98 |
| 11 | 4.84 | 3.98 | 3.59 | 3.36 | 3.20 | 3.09 | 3.01 | 2.95 | 2.90 | 2.85 |
| 12 | 4.75 | 3.89 | 3.49 | 3.26 | 3.11 | 3.00 | 2.91 | 2.85 | 2.80 | 2.75 |
| 13 | 4.67 | 3.81 | 3.41 | 3.18 | 3.03 | 2.92 | 2.83 | 2.77 | 2.71 | 2.67 |
| 14 | 4.60 | 3.74 | 3.34 | 3.11 | 2.96 | 2.85 | 2.76 | 2.70 | 2.65 | 2.60 |
| 15 | 4.54 | 3.68 | 3.29 | 3.06 | 2.90 | 2.79 | 2.71 | 2.64 | 2.59 | 2.54 |
| 16 | 4.49 | 3.63 | 3.24 | 3.01 | 2.85 | 2.74 | 2.66 | 2.59 | 2.54 | 2.49 |
| 17 | 4.45 | 3.59 | 3.20 | 2.96 | 2.81 | 2.70 | 2.61 | 2.55 | 2.49 | 2.45 |
| 18 | 4.41 | 3.55 | 3.16 | 2.93 | 2.77 | 2.66 | 2.58 | 2.51 | 2.46 | 2.41 |
| 19 | 4.38 | 3.52 | 3.13 | 2.90 | 2.74 | 2.63 | 2.54 | 2.48 | 2.42 | 2.38 |
| 20 | 4.35 | 3.49 | 3.10 | 2.87 | 2.71 | 2.60 | 2.51 | 2.45 | 2.39 | 2.35 |
| 21 | 4.32 | 3.47 | 3.07 | 2.84 | 2.68 | 2.57 | 2.49 | 2.42 | 2.37 | 2.32 |
| 22 | 4.30 | 3.44 | 3.05 | 2.82 | 2.66 | 2.55 | 2.46 | 2.40 | 2.34 | 2.30 |
| 23 | 4.28 | 3.42 | 3.03 | 2.80 | 2.64 | 2.53 | 2.44 | 2.37 | 2.32 | 2.27 |
| 24 | 4.26 | 3.40 | 3.01 | 2.78 | 2.62 | 2.51 | 2.42 | 2.36 | 2.30 | 2.25 |
| 25 | 4.24 | 3.39 | 2.99 | 2.76 | 2.60 | 2.49 | 2.40 | 2.34 | 2.28 | 2.24 |
| 26 | 4.23 | 3.37 | 2.98 | 2.74 | 2.59 | 2.47 | 2.39 | 2.32 | 2.27 | 2.22 |
| 27 | 4.21 | 3.35 | 2.96 | 2.73 | 2.57 | 2.46 | 2.37 | 2.31 | 2.25 | 2.20 |
| 28 | 4.20 | 3.34 | 2.95 | 2.71 | 2.56 | 2.45 | 2.36 | 2.29 | 2.24 | 2.19 |
| 29 | 4.18 | 3.33 | 2.93 | 2.70 | 2.55 | 2.43 | 2.35 | 2.28 | 2.22 | 2.18 |
| 30 | 4.17 | 3.32 | 2.92 | 2.69 | 2.53 | 2.42 | 2.33 | 2.27 | 2.21 | 2.16 |
| 40 | 4.08 | 3.23 | 2.84 | 2.61 | 2.45 | 2.34 | 2.25 | 2.18 | 2.12 | 2.08 |
| 60 | 4.00 | 3.15 | 2.76 | 2.53 | 2.37 | 2.25 | 2.17 | 2.10 | 2.04 | 1.99 |
| 120 | 3.92 | 3.07 | 2.68 | 2.45 | 2.29 | 2.18 | 2.09 | 2.02 | 1.96 | 1.91 |
| 400 | 3.86 | 3.02 | 2.63 | 2.39 | 2.24 | 2.12 | 2.03 | 1.96 | 1.90 | 1.85 |
| ∞ | 3.84 | 3.00 | 2.60 | 2.37 | 2.21 | 2.10 | 2.01 | 1.94 | 1.88 | 1.83 |

| $v_1$ | 11 | 12 | 15 | 20 | 24 | 30 | 40 | 60 | 120 | 400 |
|---|---|---|---|---|---|---|---|---|---|---|
| 5 | 4.70 | 4.68 | 4.62 | 4.56 | 4.53 | 4.50 | 4.46 | 4.43 | 4.40 | 4.38 |
| 6 | 4.03 | 4.00 | 3.94 | 3.87 | 3.84 | 3.81 | 3.77 | 3.74 | 3.70 | 3.68 |
| 7 | 3.60 | 3.57 | 3.51 | 3.44 | 3.41 | 3.38 | 3.34 | 3.31 | 3.27 | 3.24 |
| 8 | 3.31 | 3.28 | 3.22 | 3.15 | 3.12 | 3.08 | 3.04 | 3.00 | 2.97 | 2.94 |
| 9 | 3.10 | 3.07 | 3.01 | 2.94 | 2.90 | 2.86 | 2.83 | 2.79 | 2.75 | 2.72 |
| 10 | 2.94 | 2.91 | 2.85 | 2.77 | 2.74 | 2.70 | 2.66 | 2.62 | 2.58 | 2.55 |
| 11 | 2.82 | 2.79 | 2.72 | 2.65 | 2.61 | 2.57 | 2.53 | 2.49 | 2.45 | 2.42 |
| 12 | 2.72 | 2.69 | 2.62 | 2.54 | 2.51 | 2.47 | 2.43 | 2.38 | 2.34 | 2.31 |
| 13 | 2.63 | 2.60 | 2.53 | 2.46 | 2.42 | 2.38 | 2.34 | 2.30 | 2.25 | 2.22 |
| 14 | 2.57 | 2.53 | 2.46 | 2.39 | 2.35 | 2.31 | 2.27 | 2.22 | 2.18 | 2.15 |
| 15 | 2.51 | 2.48 | 2.40 | 2.33 | 2.29 | 2.25 | 2.20 | 2.16 | 2.11 | 2.08 |

*(Continued)*

285

|  | $v_2$ | | | | | | | | | |
| $v_1$ | 11 | 12 | 15 | 20 | 24 | 30 | 40 | 60 | 120 | 400 |
|---|---|---|---|---|---|---|---|---|---|---|
| 16 | 2.46 | 2.42 | 2.35 | 2.28 | 2.24 | 2.19 | 2.15 | 2.11 | 2.06 | 2.02 |
| 17 | 2.41 | 2.38 | 2.31 | 2.23 | 2.19 | 2.15 | 2.10 | 2.06 | 2.01 | 1.98 |
| 18 | 2.37 | 2.34 | 2.27 | 2.19 | 2.15 | 2.11 | 2.06 | 2.02 | 1.97 | 1.93 |
| 19 | 2.34 | 2.31 | 2.23 | 2.16 | 2.11 | 2.07 | 2.03 | 1.98 | 1.93 | 1.89 |
| 20 | 2.31 | 2.28 | 2.20 | 2.12 | 2.08 | 2.04 | 1.99 | 1.95 | 1.90 | 1.86 |
| 21 | 2.28 | 2.25 | 2.18 | 2.10 | 2.05 | 2.01 | 1.96 | 1.92 | 1.87 | 1.83 |
| 22 | 2.26 | 2.23 | 2.15 | 2.07 | 2.03 | 1.98 | 1.94 | 1.89 | 1.84 | 1.80 |
| 23 | 2.24 | 2.20 | 2.13 | 2.05 | 2.00 | 1.96 | 1.91 | 1.86 | 1.81 | 1.77 |
| 24 | 2.22 | 2.18 | 2.11 | 2.03 | 1.98 | 1.94 | 1.89 | 1.84 | 1.79 | 1.75 |
| 25 | 2.20 | 2.16 | 2.09 | 2.01 | 1.96 | 1.92 | 1.87 | 1.82 | 1.77 | 1.73 |
| 26 | 2.18 | 2.15 | 2.07 | 1.99 | 1.95 | 1.90 | 1.85 | 1.80 | 1.75 | 1.71 |
| 27 | 2.17 | 2.13 | 2.06 | 1.97 | 1.93 | 1.88 | 1.84 | 1.79 | 1.73 | 1.69 |
| 28 | 2.15 | 2.12 | 2.04 | 1.96 | 1.91 | 1.87 | 1.82 | 1.77 | 1.71 | 1.67 |
| 29 | 2.14 | 2.10 | 2.03 | 1.94 | 1.90 | 1.85 | 1.81 | 1.75 | 1.70 | 1.66 |
| 30 | 2.13 | 2.09 | 2.01 | 1.93 | 1.89 | 1.84 | 1.79 | 1.74 | 1.68 | 1.64 |
| 40 | 2.04 | 2.00 | 1.92 | 1.84 | 1.79 | 1.74 | 1.69 | 1.64 | 1.58 | 1.53 |
| 60 | 1.95 | 1.92 | 1.84 | 1.75 | 1.70 | 1.65 | 1.59 | 1.53 | 1.47 | 1.41 |
| 120 | 1.87 | 1.83 | 1.75 | 1.66 | 1.61 | 1.55 | 1.50 | 1.43 | 1.35 | 1.29 |
| 400 | 1.81 | 1.78 | 1.69 | 1.60 | 1.54 | 1.49 | 1.42 | 1.35 | 1.26 | 1.18 |
| ∞ | 1.79 | 1.75 | 1.67 | 1.57 | 1.52 | 1.46 | 1.39 | 1.32 | 1.22 | 1.00 |

Values of $f$ Such That $Pr(F \leq f) = .99$ Where $F$ has an
$F$ Distribution With $v_1$ and $v_2$ Degrees of Freedom

|  | $v_2$ | | | | | | | | | |
| $v_1$ | 1 | 2 | 3 | 4 | 5 | 6 | 7 | 8 | 9 | 10 |
|---|---|---|---|---|---|---|---|---|---|---|
| 5 | 16.26 | 13.27 | 12.06 | 11.39 | 10.97 | 10.67 | 10.46 | 10.29 | 10.16 | 10.05 |
| 6 | 13.75 | 10.92 | 9.78 | 9.15 | 8.75 | 8.47 | 8.26 | 8.10 | 7.98 | 7.87 |
| 7 | 12.25 | 9.55 | 8.45 | 7.85 | 7.46 | 7.19 | 6.99 | 6.84 | 6.72 | 6.62 |
| 8 | 11.26 | 8.65 | 7.59 | 7.01 | 6.63 | 6.37 | 6.18 | 6.03 | 5.91 | 5.81 |
| 9 | 10.56 | 8.02 | 6.99 | 6.42 | 6.06 | 5.80 | 5.61 | 5.47 | 5.35 | 5.26 |
| 10 | 10.04 | 7.56 | 6.55 | 5.99 | 5.64 | 5.39 | 5.20 | 5.06 | 4.94 | 4.85 |
| 11 | 9.65 | 7.21 | 6.22 | 5.67 | 5.32 | 5.07 | 4.89 | 4.74 | 4.63 | 4.54 |
| 12 | 9.33 | 6.93 | 5.95 | 5.41 | 5.06 | 4.82 | 4.64 | 4.50 | 4.39 | 4.30 |
| 13 | 9.07 | 6.70 | 5.74 | 5.21 | 4.86 | 4.62 | 4.44 | 4.30 | 4.19 | 4.10 |
| 14 | 8.86 | 6.51 | 5.56 | 5.04 | 4.69 | 4.46 | 4.28 | 4.14 | 4.03 | 3.94 |
| 15 | 8.68 | 6.36 | 5.42 | 4.89 | 4.56 | 4.32 | 4.14 | 4.00 | 3.89 | 3.80 |
| 16 | 8.53 | 6.23 | 5.29 | 4.77 | 4.44 | 4.20 | 4.03 | 3.89 | 3.78 | 3.69 |
| 17 | 8.40 | 6.11 | 5.18 | 4.67 | 4.34 | 4.10 | 3.93 | 3.79 | 3.68 | 3.59 |
| 18 | 8.29 | 6.01 | 5.09 | 4.58 | 4.25 | 4.01 | 3.84 | 3.71 | 3.60 | 3.51 |
| 19 | 8.18 | 5.93 | 5.01 | 4.50 | 4.17 | 3.94 | 3.77 | 3.63 | 3.52 | 3.43 |
| 20 | 8.10 | 5.85 | 4.94 | 4.43 | 4.10 | 3.87 | 3.70 | 3.56 | 3.46 | 3.37 |
| 21 | 8.02 | 5.78 | 4.87 | 4.37 | 4.04 | 3.81 | 3.64 | 3.51 | 3.40 | 3.31 |

*(Continued)*

|       |      |      |      |      | $v_2$ |      |      |      |      |      |
|-------|------|------|------|------|------|------|------|------|------|------|
| $v_1$ | 1    | 2    | 3    | 4    | 5    | 6    | 7    | 8    | 9    | 10   |
| 22    | 7.95 | 5.72 | 4.82 | 4.31 | 3.99 | 3.76 | 3.59 | 3.45 | 3.35 | 3.26 |
| 23    | 7.88 | 5.66 | 4.76 | 4.26 | 3.94 | 3.71 | 3.54 | 3.41 | 3.30 | 3.21 |
| 24    | 7.82 | 5.61 | 4.72 | 4.22 | 3.90 | 3.67 | 3.50 | 3.36 | 3.26 | 3.17 |
| 25    | 7.77 | 5.57 | 4.68 | 4.18 | 3.85 | 3.63 | 3.46 | 3.32 | 3.22 | 3.13 |
| 26    | 7.72 | 5.53 | 4.64 | 4.14 | 3.82 | 3.59 | 3.42 | 3.29 | 3.18 | 3.09 |
| 27    | 7.68 | 5.49 | 4.60 | 4.11 | 3.78 | 3.56 | 3.39 | 3.26 | 3.15 | 3.06 |
| 28    | 7.64 | 5.45 | 4.57 | 4.07 | 3.75 | 3.53 | 3.36 | 3.23 | 3.12 | 3.03 |
| 29    | 7.60 | 5.42 | 4.54 | 4.04 | 3.73 | 3.50 | 3.33 | 3.20 | 3.09 | 3.00 |
| 30    | 7.56 | 5.39 | 4.51 | 4.02 | 3.70 | 3.47 | 3.30 | 3.17 | 3.07 | 2.98 |
| 40    | 7.31 | 5.18 | 4.31 | 3.83 | 3.51 | 3.29 | 3.12 | 2.99 | 2.89 | 2.80 |
| 60    | 7.08 | 4.98 | 4.13 | 3.65 | 3.34 | 3.12 | 2.95 | 2.82 | 2.72 | 2.63 |
| 120   | 6.85 | 4.79 | 3.95 | 3.48 | 3.17 | 2.96 | 2.79 | 2.66 | 2.56 | 2.47 |
| 400   | 6.70 | 4.66 | 3.83 | 3.37 | 3.06 | 2.85 | 2.68 | 2.56 | 2.45 | 2.37 |
| ∞     | 6.63 | 4.61 | 3.78 | 3.32 | 3.02 | 2.80 | 2.64 | 2.51 | 2.41 | 2.32 |
| $v_1$ | 11   | 12   | 15   | 20   | 24   | 30   | 40   | 60   | 120  | 400  |
| 5     | 9.96 | 9.89 | 9.72 | 9.55 | 9.46 | 9.38 | 9.30 | 9.20 | 9.11 | 9.05 |
| 6     | 7.79 | 7.72 | 7.56 | 7.40 | 7.31 | 7.23 | 7.15 | 7.06 | 6.97 | 6.91 |
| 7     | 6.54 | 6.47 | 6.31 | 6.16 | 6.07 | 5.99 | 5.91 | 5.82 | 5.74 | 5.68 |
| 8     | 5.73 | 5.67 | 5.52 | 5.36 | 5.28 | 5.20 | 5.12 | 5.03 | 4.95 | 4.89 |
| 9     | 5.18 | 5.11 | 4.96 | 4.81 | 4.73 | 4.65 | 4.57 | 4.48 | 4.40 | 4.34 |
| 10    | 4.77 | 4.71 | 4.56 | 4.41 | 4.33 | 4.25 | 4.17 | 4.08 | 4.00 | 3.94 |
| 11    | 4.46 | 4.40 | 4.25 | 4.10 | 4.02 | 3.94 | 3.86 | 3.78 | 3.69 | 3.63 |
| 12    | 4.22 | 4.16 | 4.01 | 3.86 | 3.78 | 3.70 | 3.62 | 3.54 | 3.45 | 3.39 |
| 13    | 4.02 | 3.96 | 3.82 | 3.66 | 3.59 | 3.51 | 3.43 | 3.34 | 3.25 | 3.19 |
| 14    | 3.86 | 3.80 | 3.66 | 3.51 | 3.43 | 3.35 | 3.27 | 3.18 | 3.09 | 3.03 |
| 15    | 3.73 | 3.67 | 3.52 | 3.37 | 3.29 | 3.21 | 3.13 | 3.05 | 2.96 | 2.90 |
| 16    | 3.62 | 3.55 | 3.41 | 3.26 | 3.18 | 3.10 | 3.02 | 2.93 | 2.84 | 2.78 |
| 17    | 3.52 | 3.46 | 3.31 | 3.16 | 3.08 | 3.00 | 2.92 | 2.83 | 2.75 | 2.68 |
| 18    | 3.43 | 3.37 | 3.23 | 3.08 | 3.00 | 2.92 | 2.84 | 2.75 | 2.66 | 2.59 |
| 19    | 3.36 | 3.30 | 3.15 | 3.00 | 2.92 | 2.84 | 2.76 | 2.67 | 2.58 | 2.52 |
| 20    | 3.29 | 3.23 | 3.09 | 2.94 | 2.86 | 2.78 | 2.69 | 2.61 | 2.52 | 2.45 |
| 21    | 3.24 | 3.17 | 3.03 | 2.88 | 2.80 | 2.72 | 2.64 | 2.55 | 2.46 | 2.39 |
| 22    | 3.18 | 3.12 | 2.98 | 2.83 | 2.75 | 2.67 | 2.58 | 2.50 | 2.40 | 2.34 |
| 23    | 3.14 | 3.07 | 2.93 | 2.78 | 2.70 | 2.62 | 2.54 | 2.45 | 2.35 | 2.29 |
| 24    | 3.09 | 3.03 | 2.89 | 2.74 | 2.66 | 2.58 | 2.49 | 2.40 | 2.31 | 2.24 |
| 25    | 3.06 | 2.99 | 2.85 | 2.70 | 2.62 | 2.54 | 2.45 | 2.36 | 2.27 | 2.20 |
| 26    | 3.02 | 2.96 | 2.81 | 2.66 | 2.58 | 2.50 | 2.42 | 2.33 | 2.23 | 2.16 |
| 27    | 2.99 | 2.93 | 2.78 | 2.63 | 2.55 | 2.47 | 2.38 | 2.29 | 2.20 | 2.13 |
| 28    | 2.96 | 2.90 | 2.75 | 2.60 | 2.52 | 2.44 | 2.35 | 2.26 | 2.17 | 2.10 |
| 29    | 2.93 | 2.87 | 2.73 | 2.57 | 2.49 | 2.41 | 2.33 | 2.23 | 2.14 | 2.07 |
| 30    | 2.91 | 2.84 | 2.70 | 2.55 | 2.47 | 2.39 | 2.30 | 2.21 | 2.11 | 2.04 |
| 40    | 2.73 | 2.66 | 2.52 | 2.37 | 2.29 | 2.20 | 2.11 | 2.02 | 1.92 | 1.84 |
| 60    | 2.56 | 2.50 | 2.35 | 2.20 | 2.12 | 2.03 | 1.94 | 1.84 | 1.73 | 1.64 |
| 120   | 2.40 | 2.34 | 2.19 | 2.03 | 1.95 | 1.86 | 1.76 | 1.66 | 1.53 | 1.43 |
| 400   | 2.29 | 2.23 | 2.08 | 1.92 | 1.84 | 1.75 | 1.64 | 1.53 | 1.39 | 1.26 |
| ∞     | 2.24 | 2.18 | 2.04 | 1.88 | 1.79 | 1.70 | 1.59 | 1.47 | 1.32 | 1.00 |

*Note.* This table was computed with IMSL subroutine MDFI.

## TABLE A.4
### Value of $x$ Such That $P=\Pr(\chi_v^2 \le x)$

| $v$ | $.01$ | $.025$ | $.05$ | $.10$ | $.25$ | $.50$ |
|---|---|---|---|---|---|---|
| 2 | 0.0201 | 0.0507 | 0.1025 | 0.2107 | 0.5754 | 1.3861 |
| 3 | 0.1148 | 0.2158 | 0.3518 | 0.5843 | 1.2126 | 2.3663 |
| 4 | 0.2971 | 0.4839 | 0.7107 | 1.0636 | 1.9226 | 3.3570 |
| 5 | 0.5530 | 0.8310 | 1.1452 | 1.6101 | 2.6745 | 4.3518 |
| 6 | 0.8715 | 1.2364 | 1.6345 | 2.2036 | 3.4541 | 5.3485 |
| 7 | 1.2372 | 1.6876 | 2.1671 | 2.8330 | 4.2548 | 6.3462 |
| 8 | 1.6422 | 2.1789 | 2.7320 | 3.4893 | 5.0706 | 7.3445 |
| 9 | 2.0860 | 2.6987 | 3.3239 | 4.1677 | 5.8987 | 8.3431 |
| 10 | 2.5546 | 3.2439 | 3.9383 | 4.8643 | 6.7370 | 9.3421 |
| 11 | 3.0473 | 3.8145 | 4.5741 | 5.5764 | 7.5837 | 10.3412 |
| 12 | 3.5676 | 4.4018 | 5.2249 | 6.3035 | 8.4378 | 11.3405 |
| 13 | 4.1022 | 5.0056 | 5.8902 | 7.0410 | 9.2982 | 12.3399 |
| 14 | 4.6532 | 5.6241 | 6.5681 | 7.7888 | 10.1642 | 13.3393 |
| 15 | 5.2256 | 6.2601 | 7.2600 | 8.5457 | 11.0365 | 14.3389 |
| 16 | 5.8067 | 6.9047 | 7.9603 | 9.3108 | 11.9121 | 15.3385 |
| 17 | 6.4000 | 7.5600 | 8.6699 | 10.0833 | 12.7918 | 16.3385 |
| 18 | 7.0042 | 8.2251 | 9.3880 | 10.8644 | 13.6751 | 17.3366 |
| 19 | 7.6268 | 8.9038 | 10.1138 | 11.6502 | 14.5618 | 18.3377 |
| 20 | 8.2523 | 9.5871 | 10.8495 | 12.4417 | 15.4515 | 19.3375 |
| 21 | 8.8864 | 10.2781 | 11.5896 | 13.2384 | 16.3440 | 20.3373 |
| 22 | 9.5285 | 10.9761 | 12.3358 | 14.0400 | 17.2392 | 21.3371 |
| 23 | 10.1874 | 11.6854 | 13.0877 | 14.8462 | 18.1368 | 22.3370 |
| 24 | 10.8457 | 12.3971 | 13.8449 | 15.6565 | 19.0367 | 23.3368 |
| 25 | 11.5103 | 13.1146 | 14.6071 | 16.4708 | 19.9387 | 24.3367 |
| 26 | 12.1898 | 13.8375 | 15.3772 | 17.2910 | 20.8427 | 25.3366 |
| 27 | 12.8679 | 14.5654 | 16.1490 | 18.1129 | 21.7486 | 26.3365 |
| 28 | 13.5514 | 15.3036 | 16.9250 | 18.9380 | 22.6562 | 27.3364 |
| 29 | 14.2401 | 16.0418 | 17.7049 | 19.7663 | 23.5655 | 28.3363 |
| 30 | 14.9429 | 16.7843 | 18.4885 | 20.5975 | 24.4764 | 29.3362 |
| 40 | 22.1394 | 24.4229 | 26.5080 | 29.0555 | 33.6676 | 39.3370 |
| 50 | 29.6845 | 32.3485 | 34.7634 | 37.6933 | 42.9486 | 49.3363 |
| 60 | 37.4646 | 40.4739 | 43.1874 | 46.4633 | 52.2998 | 59.3358 |
| 70 | 45.4230 | 48.7504 | 51.7389 | 55.3331 | 61.7038 | 69.3354 |
| 80 | 53.5226 | 57.1466 | 60.3912 | 64.2818 | 71.1497 | 79.3352 |
| 90 | 61.7377 | 65.6405 | 69.1259 | 73.2949 | 80.6295 | 89.3350 |
| 100 | 70.0494 | 74.2162 | 77.9294 | 82.3618 | 90.1378 | 99.3348 |

*(Continued)*

| v | .75 | .9 | .95 | .975 | .99 | .995 |
|---|---|---|---|---|---|---|
| 2 | 2.7723 | 4.6035 | 5.9948 | 7.3790 | 9.2205 | 10.5895 |
| 3 | 4.1085 | 6.2525 | 7.8167 | 9.3563 | 11.3247 | 12.8192 |
| 4 | 5.3856 | 7.7815 | 9.4917 | 11.1502 | 13.2797 | 14.8242 |
| 5 | 6.6262 | 9.2375 | 11.0733 | 12.8383 | 15.0876 | 16.7617 |
| 6 | 7.8416 | 10.6464 | 12.5961 | 14.4589 | 16.8104 | 18.5495 |
| 7 | 9.0382 | 12.0197 | 14.0702 | 16.0203 | 18.4705 | 20.2700 |
| 8 | 10.2202 | 13.3629 | 15.5117 | 17.5458 | 20.0820 | 21.9379 |
| 9 | 11.3891 | 14.6855 | 16.9252 | 19.0315 | 21.6542 | 23.5634 |
| 10 | 12.5493 | 15.9897 | 18.3112 | 20.4954 | 23.1940 | 25.1537 |
| 11 | 13.7012 | 17.2782 | 19.6806 | 21.9295 | 24.7545 | 26.7142 |
| 12 | 14.8460 | 18.5510 | 21.0297 | 23.3493 | 26.2460 | 28.2489 |
| 13 | 15.9846 | 19.8140 | 22.3668 | 24.7455 | 27.7167 | 29.8779 |
| 14 | 17.1178 | 21.0667 | 23.6908 | 26.1316 | 29.1692 | 31.3761 |
| 15 | 18.2460 | 22.3103 | 24.9997 | 27.4982 | 30.6054 | 32.8566 |
| 16 | 19.3699 | 23.5456 | 26.3011 | 28.8578 | 32.0269 | 34.3211 |
| 17 | 20.4899 | 24.7710 | 27.5932 | 30.2007 | 33.4352 | 35.7711 |
| 18 | 21.6062 | 25.9917 | 28.8767 | 31.5385 | 34.8314 | 37.2079 |
| 19 | 22.7192 | 27.2063 | 30.1484 | 32.8673 | 36.2165 | 38.6326 |
| 20 | 23.8293 | 28.4151 | 31.4163 | 34.1813 | 37.5914 | 40.0461 |
| 21 | 24.9365 | 29.6187 | 32.6776 | 35.4931 | 38.9570 | 41.4494 |
| 22 | 26.0411 | 30.8175 | 33.9327 | 36.7918 | 40.3138 | 42.8430 |
| 23 | 27.1432 | 32.0117 | 35.1780 | 38.0890 | 41.6625 | 44.2278 |
| 24 | 28.2432 | 33.1987 | 36.4215 | 39.3800 | 43.0036 | 45.6042 |
| 25 | 29.3410 | 34.3844 | 37.6600 | 40.6590 | 44.3375 | 46.9728 |
| 26 | 30.4368 | 35.5664 | 38.8938 | 41.9380 | 45.6648 | 48.3340 |
| 27 | 31.5308 | 36.7448 | 40.1191 | 43.2062 | 46.9857 | 49.6883 |
| 28 | 32.6230 | 37.9199 | 41.3439 | 44.4746 | 48.3007 | 51.0361 |
| 29 | 33.7136 | 39.0919 | 42.5647 | 45.7383 | 49.6101 | 52.3777 |
| 30 | 34.8004 | 40.2610 | 43.7817 | 46.9920 | 50.9141 | 53.7134 |
| 40 | 45.6097 | 51.7963 | 55.7534 | 59.3447 | 63.7104 | 66.8024 |
| 50 | 56.3279 | 63.1594 | 67.5006 | 71.4232 | 76.1719 | 79.5229 |
| 60 | 66.9762 | 74.3900 | 79.0783 | 83.3007 | 88.3961 | 91.9820 |
| 70 | 77.5717 | 85.5206 | 90.5279 | 95.0262 | 100.4408 | 104.2431 |
| 80 | 88.1256 | 96.5723 | 101.8765 | 106.6315 | 112.3435 | 116.3475 |
| 90 | 98.6455 | 107.5595 | 113.1425 | 118.1388 | 124.1303 | 128.3240 |
| 100 | 109.1370 | 118.4928 | 124.3396 | 129.5640 | 135.8201 | 140.1933 |

*Note.* This table was computed with IMSL subroutine MDCHI.

# Appendix B

The Bentler–Weeks model is only one of several models that can be used in PROC CALIS to evaluate structural equation models. Of these, perhaps the most well-known structural equation modeling system is the Keesling–Wiley–Jöreskog LISREL (linear structural relations) model. The term *LISREL* is used both to refer to the model and to the program devoted to its implementation (Jöreskog & Sörbom, 1993). The LISREL model can best be understood by considering its two components: a structural model,

$$\eta = B\eta + \Gamma\xi + \zeta$$

and two measurement models,

$$y = \Lambda_y \eta + \varepsilon$$
$$x = \Lambda_x \xi + \delta$$

where $\eta$ and $\xi$ are vectors of latent variables; $y$ and $x$ are vectors of observed variables; $\varepsilon$ and $\delta$ are vectors of measurement errors; and $\zeta$ is a vector of structural errors. Of these, $\eta$ and $y$ are endogenous variables; $\xi$ and $x$ are exogenous variables; and as error variables, $\varepsilon$, $\delta$, and $\zeta$ are exogenous as well. All variables have zero means. LISREL makes the following five assumptions:

1. $\zeta$ is uncorrelated with $\xi$.

2. $\varepsilon$ is uncorrelated with $\eta$.

3. $\delta$ is uncorrelated with $\xi$.

4. $\zeta$ is uncorrelated with $\varepsilon$ and $\delta$.

5. $I - B$ is nonsingular.

Imposing these five assumptions leads to the following expression for the implied covariance structure ($\Sigma$) between $y$ and $x$:

$$\Sigma = (y,x)(y,x)' = \begin{bmatrix} yy' & yx' \\ xy' & xx' \end{bmatrix}$$

$$= \begin{bmatrix} \Lambda_y(I-B)^{-1}(\Gamma\Phi\Gamma + \Psi)(I-B)^{-1}\Lambda_y' + \Theta_\varepsilon & \Lambda_y(I-B)^{-1}\Gamma\Phi\Lambda_x' + \Theta_{\delta,\varepsilon} \\ \Lambda_x\Phi\Gamma'(I-B')^{-1}\Lambda_y' + \Theta_{\delta,\varepsilon} & \Lambda_x\Phi\Lambda_x' + \Theta_\delta \end{bmatrix}$$

The implied covariance structure between $y$ and $x$ is therefore a function of nine parameter matrices (in contrast to the Bentler–Weeks model, which uses four parameter matrices to describe the same covariance structure). With $p$ observed $y$ variables, $q$ observed $x$ variables, $m$ latent endogenous $\eta$ variables, and $n$ latent exogenous $\xi$ variables, the nine parameter matrices are:

$\Lambda_y$ (LAMBDA-$y$, LY) = a matrix of factor loadings between $y$ and $\eta$ ($p \times m$)

$\Lambda_x$ (LAMBDA-$x$, LX) = a matrix of factor loadings between $x$ and $\xi$ ($q \times n$)

$\Theta_\varepsilon$ (THETA-epsilon, TE) = a variance–covariance matrix among the measurement errors of $y$ ($p \times p$)

$\Theta_\delta$ (THETA-delta, TD) = a variance–covariance matrix among the measurement errors of $x$ ($q \times q$)

$\Theta_{\delta,\varepsilon}$ (THETA, TH) = matrix of covariances between the measurement errors of $x$ and the measurement errors of $y$ ($q \times p$)

B (BETA, BE) = a matrix of regression coefficients among the $\eta$ ($m \times m$)

$\Gamma$ (GAMMA, GA) = a matrix of regression coefficients between $\eta$ and $\xi$ ($m \times n$)

$\Phi$ (PHI, PH) = a variance–covariance matrix of the $\xi$ ($n \times n$)

$\Psi$ (PSI, PS) = a variance–covariance matrix of the structural errors ($\zeta$) of the $\eta$ ($m \times m$)

These nine matrices describe the variance–covariance structure among the $\eta$ in terms of the variances and covariances among the $\xi$, between $\xi$ and $\eta$, between $\xi$ and $x$, and between $\eta$ and $y$. Note then that there is no special 10th matrix describing the variances and covariances among the η; this matrix would be completely redundant (a linear function of at least four of the other nine matrices):

$$\eta\eta' = (I - B)^{-1}(\Gamma\Phi\Gamma' + \Psi)(I - B')^{-1}$$

The redundancy of this "10th matrix" underscores the primary purpose of structural equation modeling: to evaluate models describing the variance–covariance structure among the endogenous latent variables, or to explain why the η may be significantly intercorrelated.

We first illustrate the LISREL model by evaluating the confirmatory factor model of Fig. 8.4 using the LISREL program. Because this model is a confirmatory factor model, only the measurement model portion of the LISREL model is necessary. Further, it is completely arbitrary whether the $y$ or $x$ measurement model is used; we illustrate the use of both. In LISREL $x$ measurement model notation, the structural equations underlying the model of Fig. 8.4 are:

$$x_1 = \lambda_{11}^{(x)}\xi_1 + \delta_1$$
$$x_2 = \lambda_{21}^{(x)}\xi_1 + \delta_2$$
$$x_3 = \lambda_{31}^{(x)}\xi_1 + \delta_3$$
$$x_4 = \lambda_{42}^{(x)}\xi_2 + \delta_4$$
$$x_5 = \lambda_{52}^{(x)}\xi_2 + \delta_5$$
$$x_6 = \lambda_{62}^{(x)}\xi_2 + \delta_6$$
$$x_7 = \lambda_{73}^{(x)}\xi_3 + \delta_7$$
$$x_8 = \lambda_{83}^{(x)}\xi_3 + \delta_8$$
$$x_9 = \lambda_{93}^{(x)}\xi_3 + \delta_9$$

In matrix notation, these equations can be written as:

$$
\begin{bmatrix} x_1 \\ x_2 \\ x_3 \\ x_4 \\ x_5 \\ x_6 \\ x_7 \\ x_8 \\ x_9 \end{bmatrix}
=
\begin{bmatrix} \lambda_{11}^{(x)} \\ \lambda_{21}^{(x)} \\ \lambda_{31}^{(x)} \\ \lambda_{42}^{(x)} \\ \lambda_{52}^{(x)} \\ \lambda_{62}^{(x)} \\ \lambda_{73}^{(x)} \\ \lambda_{83}^{(x)} \\ \lambda_{93}^{(x)} \end{bmatrix}
\begin{bmatrix} \xi_1 \\ \xi_2 \\ \xi_3 \end{bmatrix}
+
\begin{bmatrix} \delta_1 \\ \delta_2 \\ \delta_3 \\ \delta_4 \\ \delta_5 \\ \delta_6 \\ \delta_7 \\ \delta_8 \\ \delta_9 \end{bmatrix}
$$

We wish to model the covariance structure among the nine observed $x$ variables. This covariance structure is obtained as:

$$
\Sigma = \begin{bmatrix} x_1 \\ x_2 \\ x_3 \\ x_4 \\ x_5 \\ x_6 \\ x_7 \\ x_8 \\ x_9 \end{bmatrix} [x_1 \; x_2 \; x_3 \; x_4 \; x_5 \; x_6 \; x_7 \; x_8 \; x_9]
$$

$$
= \left[ \begin{bmatrix} \lambda_{11}^{(x)} \\ \lambda_{21}^{(x)} \\ \lambda_{31}^{(x)} \\ \lambda_{42}^{(x)} \\ \lambda_{52}^{(x)} \\ \lambda_{62}^{(x)} \\ \lambda_{73}^{(x)} \\ \lambda_{83}^{(x)} \\ \lambda_{93}^{(x)} \end{bmatrix} \begin{bmatrix} \xi_1 \\ \xi_2 \\ \xi_3 \end{bmatrix} + \begin{bmatrix} \delta_1 \\ \delta_2 \\ \delta_3 \\ \delta_4 \\ \delta_5 \\ \delta_6 \\ \delta_7 \\ \delta_8 \\ \delta_9 \end{bmatrix} \right] \left[ \begin{bmatrix} \lambda_{11}^{(x)} \\ \lambda_{21}^{(x)} \\ \lambda_{31}^{(x)} \\ \lambda_{42}^{(x)} \\ \lambda_{52}^{(x)} \\ \lambda_{62}^{(x)} \\ \lambda_{73}^{(x)} \\ \lambda_{83}^{(x)} \\ \lambda_{93}^{(x)} \end{bmatrix} \begin{bmatrix} \xi_1 \\ \xi_2 \\ \xi_3 \end{bmatrix} + \begin{bmatrix} \delta_1 \\ \delta_2 \\ \delta_3 \\ \delta_4 \\ \delta_5 \\ \delta_6 \\ \delta_7 \\ \delta_8 \\ \delta_9 \end{bmatrix} \right]'
$$

$$
= \begin{bmatrix} \lambda_{11}^{(x)} & 0 & 0 \\ \lambda_{21}^{(x)} & 0 & 0 \\ \lambda_{31}^{(x)} & 0 & 0 \\ 0 & \lambda_{42}^{(x)} & 0 \\ 0 & \lambda_{52}^{(x)} & 0 \\ 0 & \lambda_{62}^{(x)} & 0 \\ 0 & 0 & \lambda_{73}^{(x)} \\ 0 & 0 & \lambda_{83}^{(x)} \\ 0 & 0 & \lambda_{93}^{(x)} \end{bmatrix} \begin{bmatrix} \phi_1 & 0 & 0 \\ 0 & \phi_2 & 0 \\ 0 & 0 & \phi_3 \end{bmatrix} \begin{bmatrix} \lambda_{11}^{(x)} & 0 & 0 \\ \lambda_{21}^{(x)} & 0 & 0 \\ \lambda_{31}^{(x)} & 0 & 0 \\ 0 & \lambda_{42}^{(x)} & 0 \\ 0 & \lambda_{52}^{(x)} & 0 \\ 0 & \lambda_{62}^{(x)} & 0 \\ 0 & 0 & \lambda_{73}^{(x)} \\ 0 & 0 & \lambda_{83}^{(x)} \\ 0 & 0 & \lambda_{93}^{(x)} \end{bmatrix}' + \begin{bmatrix} \theta_{\delta 1} & 0 & 0 & 0 & 0 & 0 & 0 & 0 & 0 \\ 0 & \theta_{\delta 2} & 0 & 0 & 0 & 0 & 0 & 0 & 0 \\ 0 & 0 & \theta_{\delta 3} & 0 & 0 & 0 & 0 & 0 & 0 \\ 0 & 0 & 0 & \theta_{\delta 4} & 0 & 0 & 0 & 0 & 0 \\ 0 & 0 & 0 & 0 & \theta_{\delta 5} & 0 & 0 & 0 & 0 \\ 0 & 0 & 0 & 0 & 0 & \theta_{\delta 6} & 0 & 0 & 0 \\ 0 & 0 & 0 & 0 & 0 & 0 & \theta_{\delta 7} & 0 & 0 \\ 0 & 0 & 0 & 0 & 0 & 0 & 0 & \theta_{\delta 8} & 0 \\ 0 & 0 & 0 & 0 & 0 & 0 & 0 & 0 & \theta_{\delta 9} \end{bmatrix}
$$

Alternatively, we could evaluate this same confirmatory factor model by assuming that $x_1$ through $x_9$ are $y$ variables $y_1$ through $y_9$ (i.e., we evaluate the model using LISREL's $y$ measurement model instead of its $x$ measurement model):

$$
y_1 = \lambda_{11}^{(y)}\eta_1 + \varepsilon_1
$$
$$
y_2 = \lambda_{21}^{(y)}\eta_1 + \varepsilon_2
$$
$$
y_3 = \lambda_{31}^{(y)}\eta_1 + \varepsilon_3
$$
$$
y_4 = \lambda_{42}^{(y)}\eta_2 + \varepsilon_4
$$

$$y_5 = \lambda_{52}^{(y)}\eta_2 + \varepsilon_5$$
$$y_6 = \lambda_{62}^{(y)}\eta_2 + \varepsilon_6$$
$$y_7 = \lambda_{73}^{(y)}\eta_3 + \varepsilon_7$$
$$y_8 = \lambda_{83}^{(y)}\eta_3 + \varepsilon_8$$
$$y_9 = \lambda_{93}^{(y)}\eta_3 + \varepsilon_9$$

$$
\begin{bmatrix} y_1 \\ y_2 \\ y_3 \\ y_4 \\ y_5 \\ y_6 \\ y_7 \\ y_8 \\ y_9 \end{bmatrix} =
\begin{bmatrix} \lambda_{11}^{(y)} \\ \lambda_{21}^{(y)} \\ \lambda_{31}^{(y)} \\ \lambda_{42}^{(y)} \\ \lambda_{52}^{(y)} \\ \lambda_{62}^{(y)} \\ \lambda_{73}^{(y)} \\ \lambda_{83}^{(y)} \\ \lambda_{93}^{(y)} \end{bmatrix}
\begin{bmatrix} \eta_1 \\ \eta_2 \\ \eta_3 \end{bmatrix} +
\begin{bmatrix} \varepsilon_1 \\ \varepsilon_2 \\ \varepsilon_3 \\ \varepsilon_4 \\ \varepsilon_5 \\ \varepsilon_6 \\ \varepsilon_7 \\ \varepsilon_8 \\ \varepsilon_9 \end{bmatrix}
$$

TABLE B.1

Confirmatory Factor Analysis Example Using LISREL $x$ Measurement Model

```
DA NI=9 NO=255 MA=CM
LA
SC1 SC2 SC3 HS1 HS2 HS3 OG1 OG2 OG3
KM
1.00
0.40 1.00
0.45 0.37 1.00
0.52 0.49 0.36 1.00
0.41 0.44 0.28 0.63 1.00
0.35 0.43 0.20 0.74 0.60 1.00
0.34 0.20 0.51 0.35 0.15 0.20 1.00
0.19 0.01 0.39 0.19 0.07 0.07 0.69 1.00
0.27 0.14 0.42 0.31 0.18 0.21 0.71 0.71 1.00
SD
1.41 1.02 0.91 2.11 1.85 1.78 2.45 2.05 1.92
MO NX=9 NK=3 PH=SY,FI LX=FU,FI TD=SY,FI
VA 1.00 PH(1,1) PH(2,2) PH(3,3)
FREE LX(1,1) LX(2,1) LX(3,1)
FREE LX(4,2) LX(5,2) LX(6,2)
FREE LX(7,3) LX(8,3) LX(9,3)
FREE TD(1,1) TD(2,2) TD(3,3) TD(4,4) TD(5,5) TD(6,6) TD(7,7) TD(8,8) TD(9,9)
OU ND=4
```

$$\Sigma = \begin{bmatrix} y_1 \\ y_2 \\ y_3 \\ y_4 \\ y_5 \\ y_6 \\ y_7 \\ y_8 \\ y_9 \end{bmatrix} [y_1\ y_2\ y_3\ y_4\ y_5\ y_6\ y_7\ y_8\ y_9]$$

$$= \begin{bmatrix} \lambda_{11}^{(y)} \\ \lambda_{21}^{(y)} \\ \lambda_{31}^{(y)} \\ \lambda_{42}^{(y)} \\ \lambda_{52}^{(y)} \\ \lambda_{62}^{(y)} \\ \lambda_{73}^{(y)} \\ \lambda_{83}^{(y)} \\ \lambda_{93}^{(y)} \end{bmatrix} \begin{bmatrix} \eta_1 \\ \eta_2 \\ \eta_3 \end{bmatrix} + \begin{bmatrix} \varepsilon_1 \\ \varepsilon_2 \\ \varepsilon_3 \\ \varepsilon_4 \\ \varepsilon_5 \\ \varepsilon_6 \\ \varepsilon_7 \\ \varepsilon_8 \\ \varepsilon_9 \end{bmatrix} \begin{bmatrix} \begin{bmatrix} \lambda_{11}^{(y)} \\ \lambda_{21}^{(y)} \\ \lambda_{31}^{(y)} \\ \lambda_{42}^{(y)} \\ \lambda_{52}^{(y)} \\ \lambda_{62}^{(y)} \\ \lambda_{73}^{(y)} \\ \lambda_{83}^{(y)} \\ \lambda_{93}^{(y)} \end{bmatrix} \begin{bmatrix} \eta_1 \\ \eta_2 \\ \eta_3 \end{bmatrix} + \begin{bmatrix} \varepsilon_1 \\ \varepsilon_2 \\ \varepsilon_3 \\ \varepsilon_4 \\ \varepsilon_5 \\ \varepsilon_6 \\ \varepsilon_7 \\ \varepsilon_8 \\ \varepsilon_9 \end{bmatrix} \end{bmatrix}'$$

$$= \begin{bmatrix} \lambda_{11}^{(y)} & 0 & 0 \\ \lambda_{21}^{(y)} & 0 & 0 \\ \lambda_{31}^{(y)} & 0 & 0 \\ 0 & \lambda_{42}^{(y)} & 0 \\ 0 & \lambda_{52}^{(y)} & 0 \\ 0 & \lambda_{62}^{(y)} & 0 \\ 0 & 0 & \lambda_{73}^{(y)} \\ 0 & 0 & \lambda_{83}^{(y)} \\ 0 & 0 & \lambda_{93}^{(y)} \end{bmatrix} \begin{bmatrix} \Psi_1 & 0 & 0 \\ 0 & \Psi_2 & 0 \\ 0 & 0 & \Psi_3 \end{bmatrix} \begin{bmatrix} \lambda_{11}^{(y)} & 0 & 0 \\ \lambda_{21}^{(y)} & 0 & 0 \\ \lambda_{31}^{(y)} & 0 & 0 \\ 0 & \lambda_{42}^{(y)} & 0 \\ 0 & \lambda_{52}^{(y)} & 0 \\ 0 & \lambda_{62}^{(y)} & 0 \\ 0 & 0 & \lambda_{73}^{(y)} \\ 0 & 0 & \lambda_{83}^{(y)} \\ 0 & 0 & \lambda_{93}^{(y)} \end{bmatrix}' + \begin{bmatrix} \theta_{\delta 1} & 0 & 0 & 0 & 0 & 0 & 0 & 0 & 0 \\ 0 & \theta_{\delta 2} & 0 & 0 & 0 & 0 & 0 & 0 & 0 \\ 0 & 0 & \theta_{\varepsilon 3} & 0 & 0 & 0 & 0 & 0 & 0 \\ 0 & 0 & 0 & \theta_{\varepsilon 4} & 0 & 0 & 0 & 0 & 0 \\ 0 & 0 & 0 & 0 & \theta_{\varepsilon 5} & 0 & 0 & 0 & 0 \\ 0 & 0 & 0 & 0 & 0 & \theta_{\varepsilon 6} & 0 & 0 & 0 \\ 0 & 0 & 0 & 0 & 0 & 0 & \theta_{\varepsilon 7} & 0 & 0 \\ 0 & 0 & 0 & 0 & 0 & 0 & 0 & \theta_{\varepsilon 8} & 0 \\ 0 & 0 & 0 & 0 & 0 & 0 & 0 & 0 & \theta_{\varepsilon 9} \end{bmatrix}$$

Therefore, if the confirmatory factor model uses the $x$ measurement model, the matrices $\Lambda_x$, $\Phi$, $\Theta_\delta$ are invoked; if the $y$ measurement model is used, the matrices $\Lambda_y$, $\Psi$, $\Theta_\varepsilon$ are invoked.

In order to communicate the model to the LISREL program, the user is required to describe the characteristics of each of the matrices to be invoked. Essentially, two characteristics of each matrix need to be described: (a) the dimensions of the matrix (i.e., the number of rows and columns), and (b) which elements of the matrix will be evaluated (i.e., which parameters will be solved). There are various ways in which the user can communicate this information to LISREL; the program in Table B.1 is one way of communicating the $x$ measurement model; the program in Table B.2 is one way of communicating the $y$ measurement model. The reader is encouraged to consult

TABLE B.2
Confirmatory Factor Analysis Example Using LISREL $y$ Measurement Model

```
DA NI=9 NO=255 MA=CM
LA
SC1 SC2 SC3 HS1 HS2 HS3 OG1 OG2 OG3
KM
1.00
0.40 1.00
0.45 0.37 1.00
0.52 0.49 0.36 1.00
0.41 0.44 0.28 0.63 1.00
0.35 0.43 0.20 0.74 0.60 1.00
0.34 0.20 0.51 0.35 0.15 0.20 1.00
0.19 0.01 0.39 0.19 0.07 0.07 0.69 1.00
0.27 0.14 0.42 0.31 0.18 0.21 0.71 0.71 1.00
SD
1.41 1.02 0.91 2.11 1.85 1.78 2.45 2.05 1.92
MO NY=9 NE=3 PS=SY,FI LY=FU,FI TE=SY,FI
VA 1.00 PS(1,1) PS(2,2) PS(3,3)
FREE LY(1,1) LY(2,1) LY(3,1)
FREE LY(4,2) LY(5,2) LY(6,2)
FREE LY(7,3) LY(8,3) LY(9,3)
FREE TE(1,1) TE(2,2) TE(3,3) TE(4,4) TE(5,5) TE(6,6) TE(7,7) TE(8,8) TE(9,9)
OU ND=4
```

the LISREL manual (Jöreskog & Sörbom, 1993) for more details and alternatives.

Focusing on the program in Table B.1, the MOdel line provides the initial specification of the matrices. There are 9 observed x variables (NX=9) and 3 latent exogenous variables (NK=3), implying that PH is 3 × 3, LX is 9 × 3, and TD is 9 × 9. PH has been declared as a symmetric (SY) matrix with all of its parameters fixed (FI) to zero; LX has been declared an asymmetric (FU) matrix with all of its parameters fixed (FI) to zero; and TD has been declared a symmetric matric (SY) with all of its parameters fixed (FI) to zero. Thus the initial specification, suggesting no parameters are to be evaluated, must be revised by freeing (releasing from zero) specific elements of PH, LX, and TD. The freeing of the parameters is done on the FREE line. In accordance with the covariance structure implied by the model, we allow $x_1$ through $x_3$ to load on $\xi_1$ (FREE LX(1,1) LX(2,1) LX(3,1)); $x_4$ through $x_6$ to load on $\xi_2$ (FREE LX(4,2) LX(5,2) (LX(6,2)); $x_7$ through $x_9$ to load on $\xi_3$ (FREE LX(7,3) LX(8,3) LX(9,3)). Note that the scales of $\xi_1$, $\xi_2$, $\xi_3$ have been set by assigning VAlues of ones to the diagonal of the PH matrix (VA 1.00 PH(1,1) PH(2,2) PH(3,3)). We also free the diagonal of the TD matrix in order to evaluate the measurement errors of the $x$ variables (FREE TD(1,1)–TD(9,9)). Table B.3 shows a portion of the LISREL output: the maximum likelihood

## TABLE B.3
### Confirmatory Factor Analysis Example Using LISREL
(Number of Iterations = 13), LISREL Estimates (Maximum Likelihood)

LAMBDA-X

|      | KSI 1    | KSI 2    | KSI 3    |
| ---- | -------- | -------- | -------- |
| SC1  | 0.9835   | - -      | - -      |
|      | (0.1100) |          |          |
|      | 8.9379   |          |          |
| SC2  | 0.5850   | - -      | - -      |
|      | (0.0749) |          |          |
|      | 7.8097   |          |          |
| SC3  | 0.5871   | - -      | - -      |
|      | (0.0692) |          |          |
|      | 8.4870   |          |          |
| HS1  | - -      | 1.8599   | - -      |
|      |          | (0.1159) |          |
|      |          | 16.0496  |          |
| HS2  | - -      | 1.3222   | - -      |
|      |          | (0.1068) |          |
|      |          | 12.3820  |          |
| HS3  | - -      | 1.4943   | - -      |
|      |          | (0.0991) |          |
|      |          | 15.0730  |          |
| OG1  | - -      | - -      | 2.0351   |
|      |          |          | (0.1329) |
|      |          |          | 15.3175  |
| OG2  | - -      | - -      | 1.7029   |
|      |          |          | (0.1112) |
|      |          |          | 15.3175  |
| OG3  | - -      | - -      | 1.6411   |
|      |          |          | (0.1031) |
|      |          |          | 15.9208  |

*(Continued)*

297

PHI

| KSI 1 | KSI 2 | KSI 3 |
|-------|-------|-------|
| - - - - - - | - - - - | - - - - |
| 1.0000 | 1.0000 | 1.0000 |

THETA-DELTA

| SC1 | SC2 | SC3 | HS1 | HS2 | HS3 |
|-----|-----|-----|-----|-----|-----|
| - - - - - - | - - - - | - - - - | - - - - | - - - - | - - - - |
| 1.0209 | 0.6982 | 0.4834 | 0.9928 | 1.6743 | 0.9354 |
| (0.1792) | (0.0827) | (0.0698) | (0.2128) | (0.1779) | (0.1501) |
| 5.6955 | 8.4467 | 6.9209 | 4.6659 | 9.4105 | 6.2333 |

THETA-DELTA

| OG1 | OG2 | OG3 |
|-----|-----|-----|
| - - - - - - | - - - - | - - - - |
| 1.8608 | 1.3028 | 0.9932 |
| (0.2515) | (0.1761) | (0.1516) |
| 7.3974 | 7.3974 | 6.5498 |

SQUARED MULTIPLE CORRELATIONS FOR X - VARIABLES

| SC1 | SC2 | SC3 | HS1 | HS2 | HS3 |
|-----|-----|-----|-----|-----|-----|
| - - - - - - | - - - - | - - - - | - - - - | - - - - | - - - - |
| 0.4865 | 0.3289 | 0.4163 | 0.7770 | 0.5108 | 0.7048 |

SQUARED MULTIPLE CORRELATIONS FOR X - VARIABLES

| OG1 | OG2 | OG3 |
|-----|-----|-----|
| - - - - - - | - - - - | - - - - |
| 0.6900 | 0.6900 | 0.7306 |

GOODNESS OF FIT STATISTICS

CHI-SQUARE WITH 27 DEGREES OF FREEDOM = 254.2636 (P = 0.0)
ESTIMATED NON-CENTRALITY PARAMETER (NCP) = 227.2636

MINIMUM FIT FUNCTION VALUE = 1.0010
POPULATION DISCREPANCY FUNCTION VALUE (F0) = 0.8947
ROOT MEAN SQUARE ERROR OF APPROXIMATION (RMSEA) = 0.1820
P-VALUE FOR TEST OF CLOSE FIT (RMSEA < 0.05) = .38900044D-06

parameter estimates and the chi-square goodness of fit index. As to be expected, the LISREL values are identical to the values produced by PROC CALIS (see Tables 8.8 and 8.11). If instead the model had been evaluated using LISREL's $y$ measurement model specification, the parameter estimates and goodness of fit value would also have been the same.

PROC CALIS allows the analyst to use the LISREL model indirectly through the RAM (recticular action model) of McArdle (1980) and McArdle and McDonald (1984). The RAM model is given by:

$$v = Av + u$$
$$v = (I - A)^{-1}u$$

where $v$ and $u$ are vectors of variables (observed, latent, or both) and $A$ is a matrix of coefficients. All variables have zero means.

The covariance structure implied by the RAM model is:

$$vv' = (I - A)^{-1}uu'(I - A)^{-1'}$$
$$vv' = (I - A)^{-1}P(I - A)^{-1'}$$

Further we define a matrix $J$ that selects out the observed variables from among the total number of latent and observed variables to model:

$$vv' = \Sigma = J(I - A)^{-1}P(I - A)^{-1'}J'$$

Matrix $A$ models asymmetric relations between variables, whereas matrix $P$ models symmetric relations between variables. Note an identical model can be represented either in LISREL's nine matrix system, EQS's four matrix system, or RAM's two matrix system. If we wish to conceptualize our model in LISREL notation while invoking the RAM specification in CALIS, we define:

$$A = \begin{bmatrix} 0 & 0 & \Lambda_y & 0 \\ 0 & 0 & 0 & \Lambda_x \\ 0 & 0 & B & \Gamma \\ 0 & 0 & 0 & 0 \end{bmatrix}$$

$$P = \begin{bmatrix} \Theta_\varepsilon & \Theta_{\varepsilon,\delta} & 0 & 0 \\ \Theta_{\varepsilon,\delta} & \Theta_\delta & 0 & 0 \\ 0 & 0 & \Psi & 0 \\ 0 & 0 & 0 & \Phi \end{bmatrix}$$

If there are $p$ $y$ variables, $q$ $x$ variables, $m$ $\eta$ variables, and $n$ $\xi$ variables, then $A$ is $(p + q + m + n) \times (p + q + m + n)$; $P$ is $(p + q + m + n) \times (p + q + m + n)$; and $J$ is $(m + n) \times (p + q + m + n)$.

In order to illustrate the LISREL model's specification in CALIS, we extend the previous confirmatory factor model by specifying the model simultaneously as an $x$ measurement model and as a $y$ measurement model:

$$
\begin{aligned}
y_1 &= \lambda_{11}^{(y)}\eta_1 + \varepsilon_1 & x_1 &= \lambda_{11}^{(x)}\xi_1 + \delta_1 \\
y_2 &= \lambda_{21}^{(y)}\eta_1 + \varepsilon_2 & x_2 &= \lambda_{21}^{(x)}\xi_1 + \delta_2 \\
y_3 &= \lambda_{31}^{(y)}\eta_1 + \varepsilon_3 & x_3 &= \lambda_{31}^{(x)}\xi_1 + \delta_3 \\
y_4 &= \lambda_{42}^{(y)}\eta_2 + \varepsilon_4 & x_4 &= \lambda_{42}^{(x)}\xi_2 + \delta_4 \\
y_5 &= \lambda_{52}^{(y)}\eta_2 + \varepsilon_5 & x_5 &= \lambda_{52}^{(x)}\xi_2 + \delta_5 \\
y_6 &= \lambda_{62}^{(y)}\eta_2 + \varepsilon_6 & x_6 &= \lambda_{62}^{(x)}\xi_2 + \delta_6 \\
y_7 &= \lambda_{73}^{(y)}\eta_3 + \varepsilon_7 & x_7 &= \lambda_{73}^{(x)}\xi_3 + \delta_7 \\
y_8 &= \lambda_{83}^{(y)}\eta_3 + \varepsilon_8 & x_8 &= \lambda_{83}^{(x)}\xi_3 + \delta_8 \\
y_9 &= \lambda_{93}^{(y)}\eta_3 + \varepsilon_9 & x_9 &= \lambda_{93}^{(x)}\xi_3 + \delta_9
\end{aligned}
$$

Then $A$, $P$, and $J$ are defined as:

$A =$

| | $y_1$ | $y_2$ | $y_3$ | $y_4$ | $y_5$ | $y_6$ | $y_7$ | $y_8$ | $y_9$ | $x_1$ | $x_2$ | $x_3$ | $x_4$ | $x_5$ | $x_6$ | $x_7$ | $x_8$ | $x_9$ | $\eta_1$ | $\eta_2$ | $\eta_3$ | $\xi_1$ | $\xi_2$ | $\xi_3$ |
|---|---|---|---|---|---|---|---|---|---|---|---|---|---|---|---|---|---|---|---|---|---|---|---|---|
| $y_1$ | 0 | 0 | 0 | 0 | 0 | 0 | 0 | 0 | 0 | 0 | 0 | 0 | 0 | 0 | 0 | 0 | 0 | 0 | $\lambda_{11}^{(y)}$ | 0 | 0 | 0 | 0 | 0 |
| $y_2$ | 0 | 0 | 0 | 0 | 0 | 0 | 0 | 0 | 0 | 0 | 0 | 0 | 0 | 0 | 0 | 0 | 0 | 0 | $\lambda_{21}^{(y)}$ | 0 | 0 | 0 | 0 | 0 |
| $y_3$ | 0 | 0 | 0 | 0 | 0 | 0 | 0 | 0 | 0 | 0 | 0 | 0 | 0 | 0 | 0 | 0 | 0 | 0 | $\lambda_{31}^{(y)}$ | 0 | 0 | 0 | 0 | 0 |
| $y_4$ | 0 | 0 | 0 | 0 | 0 | 0 | 0 | 0 | 0 | 0 | 0 | 0 | 0 | 0 | 0 | 0 | 0 | 0 | 0 | $\lambda_{42}^{(y)}$ | 0 | 0 | 0 | 0 |
| $y_5$ | 0 | 0 | 0 | 0 | 0 | 0 | 0 | 0 | 0 | 0 | 0 | 0 | 0 | 0 | 0 | 0 | 0 | 0 | 0 | $\lambda_{52}^{(y)}$ | 0 | 0 | 0 | 0 |
| $y_6$ | 0 | 0 | 0 | 0 | 0 | 0 | 0 | 0 | 0 | 0 | 0 | 0 | 0 | 0 | 0 | 0 | 0 | 0 | 0 | $\lambda_{62}^{(y)}$ | 0 | 0 | 0 | 0 |
| $y_7$ | 0 | 0 | 0 | 0 | 0 | 0 | 0 | 0 | 0 | 0 | 0 | 0 | 0 | 0 | 0 | 0 | 0 | 0 | 0 | 0 | $\lambda_{73}^{(y)}$ | 0 | 0 | 0 |
| $y_8$ | 0 | 0 | 0 | 0 | 0 | 0 | 0 | 0 | 0 | 0 | 0 | 0 | 0 | 0 | 0 | 0 | 0 | 0 | 0 | 0 | $\lambda_{83}^{(y)}$ | 0 | 0 | 0 |
| $y_9$ | 0 | 0 | 0 | 0 | 0 | 0 | 0 | 0 | 0 | 0 | 0 | 0 | 0 | 0 | 0 | 0 | 0 | 0 | 0 | 0 | $\lambda_{93}^{(y)}$ | 0 | 0 | 0 |
| $x_1$ | 0 | 0 | 0 | 0 | 0 | 0 | 0 | 0 | 0 | 0 | 0 | 0 | 0 | 0 | 0 | 0 | 0 | 0 | 0 | 0 | 0 | $\lambda_{11}^{(x)}$ | 0 | 0 |
| $x_2$ | 0 | 0 | 0 | 0 | 0 | 0 | 0 | 0 | 0 | 0 | 0 | 0 | 0 | 0 | 0 | 0 | 0 | 0 | 0 | 0 | 0 | $\lambda_{21}^{(x)}$ | 0 | 0 |
| $x_3$ | 0 | 0 | 0 | 0 | 0 | 0 | 0 | 0 | 0 | 0 | 0 | 0 | 0 | 0 | 0 | 0 | 0 | 0 | 0 | 0 | 0 | $\lambda_{31}^{(x)}$ | 0 | 0 |
| $x_4$ | 0 | 0 | 0 | 0 | 0 | 0 | 0 | 0 | 0 | 0 | 0 | 0 | 0 | 0 | 0 | 0 | 0 | 0 | 0 | 0 | 0 | 0 | $\lambda_{42}^{(x)}$ | 0 |
| $x_5$ | 0 | 0 | 0 | 0 | 0 | 0 | 0 | 0 | 0 | 0 | 0 | 0 | 0 | 0 | 0 | 0 | 0 | 0 | 0 | 0 | 0 | 0 | $\lambda_{52}^{(x)}$ | 0 |
| $x_6$ | 0 | 0 | 0 | 0 | 0 | 0 | 0 | 0 | 0 | 0 | 0 | 0 | 0 | 0 | 0 | 0 | 0 | 0 | 0 | 0 | 0 | 0 | $\lambda_{62}^{(x)}$ | 0 |
| $x_7$ | 0 | 0 | 0 | 0 | 0 | 0 | 0 | 0 | 0 | 0 | 0 | 0 | 0 | 0 | 0 | 0 | 0 | 0 | 0 | 0 | 0 | 0 | 0 | $\lambda_{73}^{(x)}$ |
| $x_8$ | 0 | 0 | 0 | 0 | 0 | 0 | 0 | 0 | 0 | 0 | 0 | 0 | 0 | 0 | 0 | 0 | 0 | 0 | 0 | 0 | 0 | 0 | 0 | $\lambda_{83}^{(x)}$ |
| $x_9$ | 0 | 0 | 0 | 0 | 0 | 0 | 0 | 0 | 0 | 0 | 0 | 0 | 0 | 0 | 0 | 0 | 0 | 0 | 0 | 0 | 0 | 0 | 0 | $\lambda_{93}^{(x)}$ |
| $\eta_1$ | 0 | 0 | 0 | 0 | 0 | 0 | 0 | 0 | 0 | 0 | 0 | 0 | 0 | 0 | 0 | 0 | 0 | 0 | 0 | 0 | 0 | 0 | 0 | 0 |
| $\eta_2$ | 0 | 0 | 0 | 0 | 0 | 0 | 0 | 0 | 0 | 0 | 0 | 0 | 0 | 0 | 0 | 0 | 0 | 0 | 0 | 0 | 0 | 0 | 0 | 0 |
| $\eta_3$ | 0 | 0 | 0 | 0 | 0 | 0 | 0 | 0 | 0 | 0 | 0 | 0 | 0 | 0 | 0 | 0 | 0 | 0 | 0 | 0 | 0 | 0 | 0 | 0 |
| $\xi_1$ | 0 | 0 | 0 | 0 | 0 | 0 | 0 | 0 | 0 | 0 | 0 | 0 | 0 | 0 | 0 | 0 | 0 | 0 | 0 | 0 | 0 | 0 | 0 | 0 |
| $\xi_2$ | 0 | 0 | 0 | 0 | 0 | 0 | 0 | 0 | 0 | 0 | 0 | 0 | 0 | 0 | 0 | 0 | 0 | 0 | 0 | 0 | 0 | 0 | 0 | 0 |
| $\xi_3$ | 0 | 0 | 0 | 0 | 0 | 0 | 0 | 0 | 0 | 0 | 0 | 0 | 0 | 0 | 0 | 0 | 0 | 0 | 0 | 0 | 0 | 0 | 0 | 0 |

$P =$

| | $y_1$ | $y_2$ | $y_3$ | $y_4$ | $y_5$ | $y_6$ | $y_7$ | $y_8$ | $y_9$ | $x_1$ | $x_2$ | $x_3$ | $x_4$ | $x_5$ | $x_6$ | $x_7$ | $x_8$ | $x_9$ | $\eta_1$ | $\eta_2$ | $\eta_3$ | $\xi_1$ | $\xi_2$ | $\xi_3$ |
|---|---|---|---|---|---|---|---|---|---|---|---|---|---|---|---|---|---|---|---|---|---|---|---|---|
| $y_1$ | $\theta_{\varepsilon 1}$ | 0 | 0 | 0 | 0 | 0 | 0 | 0 | 0 | 0 | 0 | 0 | 0 | 0 | 0 | 0 | 0 | 0 | 0 | 0 | 0 | 0 | 0 | 0 |
| $y_2$ | 0 | $\theta_{\varepsilon 2}$ | 0 | 0 | 0 | 0 | 0 | 0 | 0 | 0 | 0 | 0 | 0 | 0 | 0 | 0 | 0 | 0 | 0 | 0 | 0 | 0 | 0 | 0 |
| $y_3$ | 0 | 0 | $\theta_{\varepsilon 3}$ | 0 | 0 | 0 | 0 | 0 | 0 | 0 | 0 | 0 | 0 | 0 | 0 | 0 | 0 | 0 | 0 | 0 | 0 | 0 | 0 | 0 |
| $y_4$ | 0 | 0 | 0 | $\theta_{\varepsilon 4}$ | 0 | 0 | 0 | 0 | 0 | 0 | 0 | 0 | 0 | 0 | 0 | 0 | 0 | 0 | 0 | 0 | 0 | 0 | 0 | 0 |
| $y_5$ | 0 | 0 | 0 | 0 | $\theta_{\varepsilon 5}$ | 0 | 0 | 0 | 0 | 0 | 0 | 0 | 0 | 0 | 0 | 0 | 0 | 0 | 0 | 0 | 0 | 0 | 0 | 0 |
| $y_6$ | 0 | 0 | 0 | 0 | 0 | $\theta_{\varepsilon 6}$ | 0 | 0 | 0 | 0 | 0 | 0 | 0 | 0 | 0 | 0 | 0 | 0 | 0 | 0 | 0 | 0 | 0 | 0 |
| $y_7$ | 0 | 0 | 0 | 0 | 0 | 0 | $\theta_{\varepsilon 7}$ | 0 | 0 | 0 | 0 | 0 | 0 | 0 | 0 | 0 | 0 | 0 | 0 | 0 | 0 | 0 | 0 | 0 |
| $y_8$ | 0 | 0 | 0 | 0 | 0 | 0 | 0 | $\theta_{\varepsilon 8}$ | 0 | 0 | 0 | 0 | 0 | 0 | 0 | 0 | 0 | 0 | 0 | 0 | 0 | 0 | 0 | 0 |
| $y_9$ | 0 | 0 | 0 | 0 | 0 | 0 | 0 | 0 | $\theta_{\varepsilon 9}$ | 0 | 0 | 0 | 0 | 0 | 0 | 0 | 0 | 0 | 0 | 0 | 0 | 0 | 0 | 0 |
| $x_1$ | 0 | 0 | 0 | 0 | 0 | 0 | 0 | 0 | 0 | $\theta_{\delta 1}$ | 0 | 0 | 0 | 0 | 0 | 0 | 0 | 0 | 0 | 0 | 0 | 0 | 0 | 0 |
| $x_2$ | 0 | 0 | 0 | 0 | 0 | 0 | 0 | 0 | 0 | 0 | $\theta_{\delta 2}$ | 0 | 0 | 0 | 0 | 0 | 0 | 0 | 0 | 0 | 0 | 0 | 0 | 0 |
| $x_3$ | 0 | 0 | 0 | 0 | 0 | 0 | 0 | 0 | 0 | 0 | 0 | $\theta_{\delta 3}$ | 0 | 0 | 0 | 0 | 0 | 0 | 0 | 0 | 0 | 0 | 0 | 0 |
| $x_4$ | 0 | 0 | 0 | 0 | 0 | 0 | 0 | 0 | 0 | 0 | 0 | 0 | $\theta_{\delta 4}$ | 0 | 0 | 0 | 0 | 0 | 0 | 0 | 0 | 0 | 0 | 0 |
| $x_5$ | 0 | 0 | 0 | 0 | 0 | 0 | 0 | 0 | 0 | 0 | 0 | 0 | 0 | $\theta_{\delta 5}$ | 0 | 0 | 0 | 0 | 0 | 0 | 0 | 0 | 0 | 0 |
| $x_6$ | 0 | 0 | 0 | 0 | 0 | 0 | 0 | 0 | 0 | 0 | 0 | 0 | 0 | 0 | $\theta_{\delta 6}$ | 0 | 0 | 0 | 0 | 0 | 0 | 0 | 0 | 0 |
| $x_7$ | 0 | 0 | 0 | 0 | 0 | 0 | 0 | 0 | 0 | 0 | 0 | 0 | 0 | 0 | 0 | $\theta_{\delta 7}$ | 0 | 0 | 0 | 0 | 0 | 0 | 0 | 0 |
| $x_8$ | 0 | 0 | 0 | 0 | 0 | 0 | 0 | 0 | 0 | 0 | 0 | 0 | 0 | 0 | 0 | 0 | $\theta_{\delta 8}$ | 0 | 0 | 0 | 0 | 0 | 0 | 0 |
| $x_9$ | 0 | 0 | 0 | 0 | 0 | 0 | 0 | 0 | 0 | 0 | 0 | 0 | 0 | 0 | 0 | 0 | 0 | $\theta_{\delta 9}$ | 0 | 0 | 0 | 0 | 0 | 0 |
| $\eta_1$ | 0 | 0 | 0 | 0 | 0 | 0 | 0 | 0 | 0 | 0 | 0 | 0 | 0 | 0 | 0 | 0 | 0 | 0 | 1 | 0 | 0 | 0 | 0 | 0 |
| $\eta_2$ | 0 | 0 | 0 | 0 | 0 | 0 | 0 | 0 | 0 | 0 | 0 | 0 | 0 | 0 | 0 | 0 | 0 | 0 | 0 | 1 | 0 | 0 | 0 | 0 |
| $\eta_3$ | 0 | 0 | 0 | 0 | 0 | 0 | 0 | 0 | 0 | 0 | 0 | 0 | 0 | 0 | 0 | 0 | 0 | 0 | 0 | 0 | 1 | 0 | 0 | 0 |
| $\xi_1$ | 0 | 0 | 0 | 0 | 0 | 0 | 0 | 0 | 0 | 0 | 0 | 0 | 0 | 0 | 0 | 0 | 0 | 0 | 0 | 0 | 0 | 1 | 0 | 0 |
| $\xi_2$ | 0 | 0 | 0 | 0 | 0 | 0 | 0 | 0 | 0 | 0 | 0 | 0 | 0 | 0 | 0 | 0 | 0 | 0 | 0 | 0 | 0 | 0 | 1 | 0 |
| $\xi_3$ | 0 | 0 | 0 | 0 | 0 | 0 | 0 | 0 | 0 | 0 | 0 | 0 | 0 | 0 | 0 | 0 | 0 | 0 | 0 | 0 | 0 | 0 | 0 | 1 |

$J =$

| | $y_1$ | $y_2$ | $y_3$ | $y_4$ | $y_5$ | $y_6$ | $y_7$ | $y_8$ | $y_9$ | $x_1$ | $x_2$ | $x_3$ | $x_4$ | $x_5$ | $x_6$ | $x_7$ | $x_8$ | $x_9$ | $\eta_1$ | $\eta_2$ | $\eta_3$ | $\xi_1$ | $\xi_2$ | $\xi_3$ |
|---|---|---|---|---|---|---|---|---|---|---|---|---|---|---|---|---|---|---|---|---|---|---|---|---|
| $y_1$ | 1 | 0 | 0 | 0 | 0 | 0 | 0 | 0 | 0 | 0 | 0 | 0 | 0 | 0 | 0 | 0 | 0 | 0 | 0 | 0 | 0 | 0 | 0 | 0 |
| $y_2$ | 0 | 1 | 0 | 0 | 0 | 0 | 0 | 0 | 0 | 0 | 0 | 0 | 0 | 0 | 0 | 0 | 0 | 0 | 0 | 0 | 0 | 0 | 0 | 0 |
| $y_3$ | 0 | 0 | 1 | 0 | 0 | 0 | 0 | 0 | 0 | 0 | 0 | 0 | 0 | 0 | 0 | 0 | 0 | 0 | 0 | 0 | 0 | 0 | 0 | 0 |
| $y_4$ | 0 | 0 | 0 | 1 | 0 | 0 | 0 | 0 | 0 | 0 | 0 | 0 | 0 | 0 | 0 | 0 | 0 | 0 | 0 | 0 | 0 | 0 | 0 | 0 |
| $y_5$ | 0 | 0 | 0 | 0 | 1 | 0 | 0 | 0 | 0 | 0 | 0 | 0 | 0 | 0 | 0 | 0 | 0 | 0 | 0 | 0 | 0 | 0 | 0 | 0 |
| $y_6$ | 0 | 0 | 0 | 0 | 0 | 1 | 0 | 0 | 0 | 0 | 0 | 0 | 0 | 0 | 0 | 0 | 0 | 0 | 0 | 0 | 0 | 0 | 0 | 0 |
| $y_7$ | 0 | 0 | 0 | 0 | 0 | 0 | 1 | 0 | 0 | 0 | 0 | 0 | 0 | 0 | 0 | 0 | 0 | 0 | 0 | 0 | 0 | 0 | 0 | 0 |
| $y_8$ | 0 | 0 | 0 | 0 | 0 | 0 | 0 | 1 | 0 | 0 | 0 | 0 | 0 | 0 | 0 | 0 | 0 | 0 | 0 | 0 | 0 | 0 | 0 | 0 |
| $y_9$ | 0 | 0 | 0 | 0 | 0 | 0 | 0 | 0 | 1 | 0 | 0 | 0 | 0 | 0 | 0 | 0 | 0 | 0 | 0 | 0 | 0 | 0 | 0 | 0 |
| $x_1$ | 0 | 0 | 0 | 0 | 0 | 0 | 0 | 0 | 0 | 1 | 0 | 0 | 0 | 0 | 0 | 0 | 0 | 0 | 0 | 0 | 0 | 0 | 0 | 0 |
| $x_2$ | 0 | 0 | 0 | 0 | 0 | 0 | 0 | 0 | 0 | 0 | 1 | 0 | 0 | 0 | 0 | 0 | 0 | 0 | 0 | 0 | 0 | 0 | 0 | 0 |
| $x_3$ | 0 | 0 | 0 | 0 | 0 | 0 | 0 | 0 | 0 | 0 | 0 | 1 | 0 | 0 | 0 | 0 | 0 | 0 | 0 | 0 | 0 | 0 | 0 | 0 |
| $x_4$ | 0 | 0 | 0 | 0 | 0 | 0 | 0 | 0 | 0 | 0 | 0 | 0 | 1 | 0 | 0 | 0 | 0 | 0 | 0 | 0 | 0 | 0 | 0 | 0 |
| $x_5$ | 0 | 0 | 0 | 0 | 0 | 0 | 0 | 0 | 0 | 0 | 0 | 0 | 0 | 1 | 0 | 0 | 0 | 0 | 0 | 0 | 0 | 0 | 0 | 0 |
| $x_6$ | 0 | 0 | 0 | 0 | 0 | 0 | 0 | 0 | 0 | 0 | 0 | 0 | 0 | 0 | 1 | 0 | 0 | 0 | 0 | 0 | 0 | 0 | 0 | 0 |
| $x_7$ | 0 | 0 | 0 | 0 | 0 | 0 | 0 | 0 | 0 | 0 | 0 | 0 | 0 | 0 | 0 | 1 | 0 | 0 | 0 | 0 | 0 | 0 | 0 | 0 |
| $x_8$ | 0 | 0 | 0 | 0 | 0 | 0 | 0 | 0 | 0 | 0 | 0 | 0 | 0 | 0 | 0 | 0 | 1 | 0 | 0 | 0 | 0 | 0 | 0 | 0 |
| $x_9$ | 0 | 0 | 0 | 0 | 0 | 0 | 0 | 0 | 0 | 0 | 0 | 0 | 0 | 0 | 0 | 0 | 0 | 1 | 0 | 0 | 0 | 0 | 0 | 0 |

If we wish to model the confirmatory factor model using the $x$ measurement model alone, $A$, $P$, and $J$ become:

$$A = $$

| | $x_1$ | $x_2$ | $x_3$ | $x_4$ | $x_5$ | $x_6$ | $x_7$ | $x_8$ | $x_9$ | $\xi_1$ | $\xi_2$ | $\xi_3$ |
|---|---|---|---|---|---|---|---|---|---|---|---|---|
| $x_1$ | 0 | 0 | 0 | 0 | 0 | 0 | 0 | 0 | 0 | $\lambda_{11}^{(x)}$ | 0 | 0 |
| $x_2$ | 0 | 0 | 0 | 0 | 0 | 0 | 0 | 0 | 0 | $\lambda_{21}^{(x)}$ | 0 | 0 |
| $x_3$ | 0 | 0 | 0 | 0 | 0 | 0 | 0 | 0 | 0 | $\lambda_{31}^{(x)}$ | 0 | 0 |
| $x_4$ | 0 | 0 | 0 | 0 | 0 | 0 | 0 | 0 | 0 | 0 | $\lambda_{42}^{(x)}$ | 0 |
| $x_5$ | 0 | 0 | 0 | 0 | 0 | 0 | 0 | 0 | 0 | 0 | $\lambda_{52}^{(x)}$ | 0 |
| $x_6$ | 0 | 0 | 0 | 0 | 0 | 0 | 0 | 0 | 0 | 0 | $\lambda_{62}^{(x)}$ | 0 |
| $x_7$ | 0 | 0 | 0 | 0 | 0 | 0 | 0 | 0 | 0 | 0 | 0 | $\lambda_{73}^{(x)}$ |
| $x_8$ | 0 | 0 | 0 | 0 | 0 | 0 | 0 | 0 | 0 | 0 | 0 | $\lambda_{83}^{(x)}$ |
| $x_9$ | 0 | 0 | 0 | 0 | 0 | 0 | 0 | 0 | 0 | 0 | 0 | $\lambda_{93}^{(x)}$ |
| $\xi_1$ | 0 | 0 | 0 | 0 | 0 | 0 | 0 | 0 | 0 | 0 | 0 | 0 |
| $\xi_2$ | 0 | 0 | 0 | 0 | 0 | 0 | 0 | 0 | 0 | 0 | 0 | 0 |
| $\xi_3$ | 0 | 0 | 0 | 0 | 0 | 0 | 0 | 0 | 0 | 0 | 0 | 0 |

$$P = $$

| | $x_1$ | $x_2$ | $x_3$ | $x_4$ | $x_5$ | $x_6$ | $x_7$ | $x_8$ | $x_9$ | $\xi_1$ | $\xi_2$ | $\xi_3$ |
|---|---|---|---|---|---|---|---|---|---|---|---|---|
| $x_1$ | $\theta_{\delta1}$ | 0 | 0 | 0 | 0 | 0 | 0 | 0 | 0 | 0 | 0 | 0 |
| $x_2$ | 0 | $\theta_{\delta2}$ | 0 | 0 | 0 | 0 | 0 | 0 | 0 | 0 | 0 | 0 |
| $x_3$ | 0 | 0 | $\theta_{\delta3}$ | 0 | 0 | 0 | 0 | 0 | 0 | 0 | 0 | 0 |
| $x_4$ | 0 | 0 | 0 | $\theta_{\delta4}$ | 0 | 0 | 0 | 0 | 0 | 0 | 0 | 0 |
| $x_5$ | 0 | 0 | 0 | 0 | $\theta_{\delta5}$ | 0 | 0 | 0 | 0 | 0 | 0 | 0 |
| $x_6$ | 0 | 0 | 0 | 0 | 0 | $\theta_{\delta6}$ | 0 | 0 | 0 | 0 | 0 | 0 |
| $x_7$ | 0 | 0 | 0 | 0 | 0 | 0 | $\theta_{\delta7}$ | 0 | 0 | 0 | 0 | 0 |
| $x_8$ | 0 | 0 | 0 | 0 | 0 | 0 | 0 | $\theta_{\delta8}$ | 0 | 0 | 0 | 0 |
| $x_9$ | 0 | 0 | 0 | 0 | 0 | 0 | 0 | 0 | $\theta_{\delta9}$ | 0 | 0 | 0 |
| $\xi_1$ | 0 | 0 | 0 | 0 | 0 | 0 | 0 | 0 | 0 | 1 | 0 | 0 |
| $\xi_2$ | 0 | 0 | 0 | 0 | 0 | 0 | 0 | 0 | 0 | 0 | 1 | 0 |
| $\xi_3$ | 0 | 0 | 0 | 0 | 0 | 0 | 0 | 0 | 0 | 0 | 0 | 1 |

$$J = $$

| | $x_1$ | $x_2$ | $x_3$ | $x_4$ | $x_5$ | $x_6$ | $x_7$ | $x_8$ | $x_9$ | $\xi_1$ | $\xi_2$ | $\xi_3$ |
|---|---|---|---|---|---|---|---|---|---|---|---|---|
| $x_1$ | 1 | 0 | 0 | 0 | 0 | 0 | 0 | 0 | 0 | 0 | 0 | 0 |
| $x_2$ | 0 | 1 | 0 | 0 | 0 | 0 | 0 | 0 | 0 | 0 | 0 | 0 |
| $x_3$ | 0 | 0 | 1 | 0 | 0 | 0 | 0 | 0 | 0 | 0 | 0 | 0 |
| $x_4$ | 0 | 0 | 0 | 1 | 0 | 0 | 0 | 0 | 0 | 0 | 0 | 0 |
| $x_5$ | 0 | 0 | 0 | 0 | 1 | 0 | 0 | 0 | 0 | 0 | 0 | 0 |
| $x_6$ | 0 | 0 | 0 | 0 | 0 | 1 | 0 | 0 | 0 | 0 | 0 | 0 |
| $x_7$ | 0 | 0 | 0 | 0 | 0 | 0 | 1 | 0 | 0 | 0 | 0 | 0 |
| $x_8$ | 0 | 0 | 0 | 0 | 0 | 0 | 0 | 1 | 0 | 0 | 0 | 0 |
| $x_9$ | 0 | 0 | 0 | 0 | 0 | 0 | 0 | 0 | 1 | 0 | 0 | 0 |

Table B.4 illustrates the RAM specification for the LISREL model in CALIS and Tables B.5 and B.6 the output. The RAM specification in CALIS requires that each parameter in the model be identified by three numbers. The first number, which is either 1 or 2, signifies whether the parameter is in the $A$ or $P$ matrix, respectively. The second and third numbers represent the row and column location of the parameter. As an additional, optional specification, the parameter can be assigned a name or a constant value. For example in Table B.4, the specification

```
1  1 10   L1X1,
```

## TABLE B.4
### RAM Specification of LISREL Model in CALIS

```
OPTIONS NODATE;
DATA ONE(TYPE=CORR);
_TYPE_='CORR';
INPUT _TYPE_ $ _NAME_ $ V1-V9;
LABEL V1='SCI' V2='SC2' V3='SC3' V4='HS1' V5='HS2' V6='HS3'
      V7='OG1' V8='OG2' V9='OG3';
CARDS;
STD    .     1.41 1.02 0.91 2.11 1.85 1.78 2.45 2.05 1.92
CORR V1      1.00 . . . . . . . .
CORR V2      0.40 1.00 . . . . . . .
CORR V3      0.45 0.37 1.00 . . . . . .
CORR V4      0.52 0.49 0.36 1.00 . . . . .
CORR V5      0.41 0.44 0.28 0.63 1.00 . . . .
CORR V6      0.35 0.43 0.20 0.74 0.60 1.00 . . .
CORR V7      0.34 0.20 0.51 0.35 0.15 0.20 1.00 . .
CORR V8      0.19 0.01 0.39 0.19 0.07 0.07 0.69 1.00 .
CORR V9      0.27 0.14 0.42 0.31 0.18 0.21 0.71 0.71 1.00
;
RUN;
PROC CALIS DATA=ONE ALL COV MAXITER=1000 NOBS=255;
RAM

           1    1   10   L1X1,
           1    2   10   L1X2,
           1    3   10   L1X3,
           1    4   11   L2X4,
           1    5   11   L2X5,
           1    6   11   L2X6,
           1    7   12   L3X7,
           1    8   12   L3X8,
           1    9   12   L3X9,
           2    1    1   THE1,
           2    2    2   THE2,
           2    3    3   THE3,
           2    4    4   THE4,
           2    5    5   THE5,
           2    6    6   THE6,
           2    7    7   THE7,
           2    8    8   THE8,
           2    9    9   THE9,
           2   10   10   1,
           2   11   11   1,
           2   12   12   1;

RUN;
```

303

## TABLE B.5
### LISREL Model Output, Part I Using SAS CALIS

```
Fit criterion..................................................   1.0010
Goodness of Fit Index (GFI) ...................................   0.8277
GFI Adjusted for Degrees of Freedom (AGFI) ....................   0.7129
Root Mean Square Residual (RMR) ...............................   0.6706
Parsimonious GFI (Mulaik, 1989) .....'.........................   0.6208
Chi-square = 254.2636        df = 27          Prob>chi**2 = 0.0001
Null Model Chi-square:       df = 36                       1112.0091
RMSEA Estimate..........................  0.1820   90%C.I.[0.1620, 0.2028]
Probability of Close Fit.......................................   0.0000
ECVI Estimate...........................  1.1486   90%C.I.[0.9589, 1.3689]
Bentler's Comparative Fit Index................................   0.7888
Normal Theory Reweighted LS Chi-square.........................  237.8681
Akaike's Information Criterion.................................   200.2636
Bozdogan's (1987) CAIC.........................................   77.6495
Schwarz's Bayesian Criterion...................................  104.6495
McDonald's (1989) Centrality...................................   0.6404
Bentler & Bonett's (1980) Non-normed Index.....................   0.7184
Bentler & Bonett's (1980) NFI..................................   0.7713
James, Mulaik, & Brett (1982) Parsimonious NFI.................   0.5785
Z-Test of Wilson & Hilferty (1931) ............................  12.3451
Bollen (1986) Normed Index Rho1................................   0.6951
Bollen (1988) Non-normed Index Delta2..........................   0.7905
Hoelter's (1983) Critical N....................................        42
```

### Residual Matrix

|     | V1 | V2 | V3 | V4 | V5 | |
|-----|----|----|----|----|----|---|
| V1 | -0.000055 | 0.000004 | 0.000032 | 1.547052 | 1.069485 | SC1 |
| V2 | 0.000004 | -0.000000 | -0.000002 | 1.054578 | 0.830280 | SC2 |
| V3 | 0.000032 | -0.000002 | -0.000018 | 0.691236 | 0.471380 | SC3 |
| V4 | 1.547052 | 1.054578 | 0.691236 | -0.000390 | 0.000084 | HS1 |
| V5 | 1.069485 | 0.830280 | 0.471380 | 0.000084 | -0.000018 | HS2 |
| V6 | 0.878430 | 0.780708 | 0.323960 | 0.000343 | -0.000074 | HS3 |
| V7 | 1.174530 | 0.499800 | 1.137045 | 1.809325 | 0.679875 | OG1 |
| V8 | 0.549195 | 0.020910 | 0.727545 | 0.821845 | 0.265475 | OG2 |
| V9 | 0.730944 | 0.274176 | 0.733824 | 1.255872 | 0.639360 | OG3 |

|     | V6 | V7 | V8 | V9 | |
|-----|----|----|----|----|---|
| V1 | 0.878430 | 1.174530 | 0.549195 | 0.730944 | SC1 |
| V2 | 0.780708 | 0.499800 | 0.020910 | 0.274176 | SC2 |
| V3 | 0.323960 | 1.137045 | 0.727545 | 0.733824 | SC3 |
| V4 | 0.000343 | 1.809325 | 0.821845 | 1.255872 | HS1 |
| V5 | -0.000074 | 0.679875 | 0.265475 | 0.639360 | HS2 |
| V6 | -0.000302 | 0.872200 | 0.255430 | 0.717696 | HS3 |
| V7 | 0.872200 | -0.000070 | 0.000062 | -0.000002 | OG1 |
| V8 | 0.255430 | 0.000062 | -0.000055 | 0.000002 | OG2 |
| V9 | 0.717696 | -0.000002 | 0.000002 | -0.000000 | OG3 |

Average Absolute Residual = 0.4625

Average Off-diagonal Absolute Residual = 0.5781

| Term & Matrix | | Row & Column | | Parameter | Estimate | Standard Error |
|---|---|---|---|---|---|---|
| 1 | 2 | 1 | 10 | L1X1 | 0.983421 | 0.110035 |
| | | V1 | F1 | | | |
| 1 | 2 | 2 | 10 | L1X2 | 0.584974 | 0.074904 |
| | | V2 | F1 | | | |
| 1 | 2 | 3 | 10 | L1X3 | 0.587097 | 0.069180 |
| | | V3 | F1 | | | |
| 1 | 2 | 4 | 11 | L2X4 | 1.859733 | 0.115904 |
| | | V4 | F2 | | | |
| 1 | 2 | 5 | 11 | L2X5 | 1.322298 | 0.106789 |
| | | V5 | F2 | | | |
| 1 | 2 | 6 | 11 | L2X6 | 1.494273 | 0.099152 |
| | | V6 | F2 | | | |
| 1 | 2 | 7 | 12 | L3X7 | 2.035106 | 0.132865 |
| | | V7 | F3 | | | |
| 1 | 2 | 8 | 12 | L3X8 | 1.702842 | 0.111173 |
| | | V8 | F3 | | | |
| 1 | 2 | 9 | 12 | L3X9 | 1.641115 | 0.103079 |
| | | V9 | F3 | | | |
| 1 | 3 | 1 | 1 | THE1 | 1.021038 | 0.179253 |
| | | E1 | E1 | | | |
| 1 | 3 | 2 | 2 | THE2 | 0.698206 | 0.082665 |
| | | E2 | E2 | | | |
| 1 | 3 | 3 | 3 | THE3 | 0.483436 | 0.069850 |
| | | E3 | E3 | | | |
| 1 | 3 | 4 | 4 | THE4 | 0.993883 | 0.212851 |
| | | E4 | E4 | | | |
| 1 | 3 | 5 | 5 | THE5 | 1.674046 | 0.177925 |
| | | E5 | E5 | | | |
| 1 | 3 | 6 | 6 | THE6 | 0.935850 | 0.150125 |
| | | E6 | E6 | | | |
| 1 | 3 | 7 | 7 | THE7 | 1.860914 | 0.251556 |
| | | E7 | E7 | | | |
| 1 | 3 | 8 | 8 | THE8 | 1.302885 | 0.176121 |
| | | E8 | E8 | | | |
| 1 | 3 | 9 | 9 | THE9 | 0.993143 | 0.151641 |
| | | E9 | E9 | | | |
| 1 | 3 | 10 | 10 | . | 1.000000 | 0 |
| | | D1 | D1 | | | |
| 1 | 3 | 11 | 11 | . | 1.000000 | 0 |
| | | D2 | D2 | | | |
| 1 | 3 | 12 | 12 | . | 1.000000 | 0 |
| | | D3 | D3 | | | |

*(Continued)*

| Term & Matrix | | Row & Column | | Parameter | Estimate | Standard Error |
|---|---|---|---|---|---|---|
| 1 | 2 | 1 | 10 | L1X1 | 8.937 | |
| | | V1 | F1 | | | |
| 1 | 2 | 2 | 10 | L1X2 | 7.810 | |
| | | V2 | F1 | | | |
| 1 | 2 | 3 | 10 | L1X3 | 8.487 | |
| | | V3 | F1 | | | |
| 1 | 2 | 4 | 11 | L2X4 | 16.045 | |
| | | V4 | F2 | | | |
| 1 | 2 | 5 | 11 | L2X5 | 12.382 | |
| | | V5 | F2 | | | |
| 1 | 2 | 6 | 11 | L2X6 | 15.071 | |
| | | V6 | F2 | | | |
| 1 | 2 | 7 | 12 | L3X7 | 15.317 | |
| | | V7 | F3 | | | |
| 1 | 2 | 8 | 12 | L3X8 | 15.317 | |
| | | V8 | F3 | | | |
| 1 | 2 | 9 | 12 | L3X9 | 15.921 | |
| | | V9 | F3 | | | |
| | | | | | | |
| 1 | 3 | 1 | 1 | THE1 | 5.696 | |
| | | E1 | E1 | | | |
| 1 | 3 | 2 | 2 | THE2 | 8.446 | |
| | | E2 | E2 | | | |
| 1 | 3 | 3 | 3 | THE3 | 6.921 | |
| | | E3 | E3 | | | |
| 1 | 3 | 4 | 4 | THE4 | 4.669 | |
| | | E4 | E4 | | | |
| 1 | 3 | 5 | 5 | THE5 | 9.409 | |
| | | E5 | E5 | | | |
| 1 | 3 | 6 | 6 | THE6 | 6.234 | |
| | | E6 | E6 | | | |
| 1 | 3 | 7 | 7 | THE7 | 7.398 | |
| | | E7 | E7 | | | |
| 1 | 3 | 8 | 8 | THE8 | 7.398 | |
| | | E8 | E8 | | | |
| 1 | 3 | 9 | 9 | THE9 | 6.549 | |
| | | E9 | E9 | | | |
| 1 | 3 | 10 | 10 | . | 0.000 | |
| | | D1 | D1 | | | |
| 1 | 3 | 11 | 11 | . | 0.000 | |
| | | D2 | D2 | | | |
| 1 | 3 | 12 | 12 | . | 0.000 | |
| | | D3 | D3 | | | |

identifies a parameter in the $A$ matrix, located in row 1 and column 10, given the name of L1X1 (LAMBDA 1, X1). Referring to the $A$ matrix given earlier, we can see that this element does indeed correspond to the loading of $x_1$ on $\xi_1$. Similarly, the RAM specification for this parameter corresponds to the LISREL program specification for this same parameter in Table B.1: FREE LX(1,1). As another example, the specification

    2   10  10    1,

assigns a constant value of 1 to the variance of $\xi_1$ in the $P$ matrix. Referring to the $P$ matrix given earlier, this element corresponds to the 10th row and the 10th column. In the LISREL specification of Table B.1, the variance of $\xi_1$ is set to the constant value of 1 by VA 1.00 PH(1,1). It is not necessary to identify the elements of the $J$ matrix; CALIS does this automatically. The output for the RAM model given in Tables B.5 and B.6 shows results identical to the EQS output of Tables 8.8 and 8.11 and identical to the LISREL output of Table B.3.

# References

Akaike, H. (1987). Factor analysis and AIC. *Psychometrika, 52,* 317–332.

Allen, D. A. (1971). *The prediction of sum of squares as a criterion for selecting predictor variables* (Tech. Rep. No. 23). Lexington, KY: University of Kentucky, Department of Statistics.

Andrews, D. F. (1973). Car accidents—Environmental aspects. *International Statistical Review, 41,* 235–239.

Baker, R. (1986). *Test-retest reliability of the leadership effectiveness questionnaire.* Working paper, Anderson Graduate School of Management, University of California, Los Angeles.

Bekker, P. A., Merckens, A., & Wansbeek, T. J. (1994). *Identification, equivalent models, and computer algebra.* New York: Academic Press.

Bentler, P. M. (1983). Some contributions to efficient statistics for structural models: Specification and estimation of moment structures. *Psychometrika, 48,* 493–517.

Bentler, P. M. (1989). *EQS, structural equations, program manual.* Los Angeles: BMDP Statistical Software, Inc.

Bentler, P. M. (1990). Comparative fit indexes in structural equation models. *Psychological Bulletin, 107,* 238–246.

Bentler, P. M. (1995). *EQS structural equation program manual.* Encino, CA: Multivariate Software, Inc.

Bentler, P. M., & Bonett, D. G. (1980). Significance tests and goodness of fit in the analysis of covariance structures. *Psychological Bulletin, 88,* 588–606.

Bickel, P. J., & Doksum, K. A. (1977). *Mathematical statistics: Basic ideas and selected topics.* San Francisco: Holden-Day.

Bollen, K. A. (1986). Sample size and Bentler and Bonett's Nonnormed Fit Index. *Psychometrika, 51,* 375–377.

Bollen, K. A. (1988). A new incremental fit index for general structural equation models. *Sociological Methods and Research, 17,* 303–316.

Bollen, K. A. (1989). *Structural equations with latent variables.* New York: Wiley.

Bollen, K. A., & Lennox, R. (1991). Conventional wisdom on measurement: A structural equation perspective. *Psychological Bulletin, 110,* 305–314.

Bollen, K. A., & Long, J. S. (1993). *Testing structural equation models.* Newbury Park, CA: Sage.

Bozdogan, H. (1987). Model selection and Akaike's Information Criterion (AIC): The general theory and its analytical extensions. *Psychometrika, 52*, 345–370.

Bozdogan, H. (1990). On the information-based measure of covariance complexity and its application to the evaluation of multivariate linear models. *Communications in Statistics, Theory, and Methods, 19*, 221–278.

Browne, M. W., & Arminger, G. (1995). Specification and estimation of mean and covariance structure models. In G. Arminger, C. C. Clogg, & M. E. Sobel (Eds.), *Handbook of statistical modeling for the social and behavioral sciences* (pp. 185–250). New York: Plenum.

Browne, M. W., & Cudeck, R. (1992). Alternative ways of assessing model fit. In K. A. Bollen & J. S. Long (Eds.), *Testing structural equation models* (pp. 136–162). Newbury Park, CA: Sage.

Bruno, J. E., & Marcoulides, G. A. (1985). Equality of educational opportunity at racially isolated schools. *The Urban Review, 17*(3), 155–165.

Byrne, B. M. (1994). *Structural equation modeling with EQS and EQS/Windows: Basic concepts, applications, and programming.* Thousand Oaks, CA: Sage Publications.

Cattell, R. B. (1966). The scree test for the number of factors. *Multivariate Behavioral Research, 1*, 245–276.

Cohen, J., & Cohen P. (1983). *Applied multiple regression/correlation analysis for the behavioral sciences.* Hillsdale, NJ: Lawrence Erlbaum Associates.

Comrey, A. L. (1973). *A first course in factor analysis.* New York: Academic Press.

Comrey, A. L., & Lee, H. B. (1992). *A first course in factor analysis* (2nd ed.). Hillsdale, NJ: Lawrence Erlbaum Associates.

Cooley, W. W., & Lohnes, P. R. (1971). *Multivariate data analysis.* New York: John Wiley.

Cramer, E., & Nicewander, W. A. (1979). Some symmetric, invariant measures of multivariate association. *Psychometrika, 44*, 43–54.

Cudeck, R. (1989). Analysis of correlation matrices using covariance structure models. *Psychological Bulletin, 105*, 317–327.

Dielman, T. E. (1996). *Applied regression analysis for business and economics.* Belmont, CA: Wadsworth Publishing Company.

Dixon, W. J. (1990a). *BMDP statistical software manual* (Vol. 1). Berkeley: University of California Press.

Dixon, W. J. (1990b). *BMDP statistical software manual* (Vol. 2). Berkeley: University of California Press.

Fisher, F. M. (1966). *The identification problem in econometrics.* New York: McGraw-Hill.

Fisher, R. A. (1936). The use of multiple measurements in taxonomic problems. *Annals of Eugenics, 7*, 179–188.

Flamholtz, E. G. (1986). *How to make the transition from entrepreneurship to a professional managed firm.* San Francisco: Jossey-Bass.

Flury, B., & Riedwyl, S. H. (1988). *Multivariate statistics: A practical approach.* London: Chapman and Hall.

Ganster, D. C., Fusilier, M. R., & Mayes, B. T. (1986). Role of social support in the experience of stress at work. *Journal of Applied Psychology, 71*, 102–110.

Gibson, C., & Marcoulides, G. A. (1995). The cultural contingency approach to leadership: Examining a model across four countries. *Journal of Managerial Issues, 7*(2), 176–192.

Gnanadeskian, R. (1977). *Methods for statistical data analysis of multivariate observations.* New York: Wiley.

Gorsuch, R. L. (1983). *Factor analysis.* Hillsdale, NJ: Lawrence Erlbaum Associates.

Hand, D. J. (1982). *Kernel discriminant analysis.* New York: Research Studies Press.

Harman, H. H. (1976). *Modern factor analysis.* Chicago: University of Chicago Press.

Heck, R. H., Marcoulides, G. A., & Glasman, N. S. (1989). The application of causal modeling techniques to administrative decision making. *Educational Administration Quarterly, 25*(3), 253–267.

# 310

Heck, R. H., & Marcoulides, G. A. (1989). Examining the generalizability of administrative personnel allocation decision. *The Urban Review, 21*(1), 51–62.

Heise, D. R. (1975). *Causal analysis.* New York: Wiley.

Hendrickson, A. E., & White, P. O. (1964). Promax: A quick method for rotation to oblique simple structure. *British Journal of Mathematical and Statistical Psychology, 17,* 65–70.

Hershberger, S. L. (1994). The specification of equivalent models before the collection of data. In A. von Eye & C. C. Clogg (Eds.), *Latent variables analysis: Applications for developmental research* (pp. 68–108). Thousand Oaks, CA: Sage.

Hershberger, S. L. (1995, June). *A new goodness of fit index based on the number of equivalent models.* Paper presented at the meeting of the Psychometric Society, Minneapolis, MN.

Herzberg, P. A. (1969). The parameters of cross-validation [Monograph Suppl. 16]. *Psychometrika.*

Hoelter, J. W. (1983). The analysis of covariance structures: Goodness-of-fit indices. *Sociological Methods and Research, 11,* 325–344.

Holzinger, K. J. (1930). *Statistical resume of the Spearman two-factor theory.* Chicago: University of Chicago Press.

Holzinger, K. J., & Spearman, C. (1926). Note on the sampling error of tetrad differences. *British Journal of Educational Psychology, 20,* 91–97.

Hotelling, H. (1931). The generalization of Student's ratio. *Annals of Mathematical Statistics, 2,* 360–378.

Hotelling, H. (1933). Analysis of a complex of statistical variables into principal components. *Journal of Educational Psychology, 24,* 417–441; 498–520.

Huberty, C. J. (1989). Problems with stepwise methods: Better alternatives. In B. Thompson (Ed.), *Advances in social science methodology* (Vol. 1, pp. 43–70). Greenwich, CT: JAI Press.

Huberty, C. J. (1994). *Applied discriminant analysis.* New York: John Wiley.

Huberty, C. J., & Curry, A. R. (1978). Linear versus quadratic multivariate classification. *Multivariate Behavioral Research, 13,* 237–245.

Hull, J., Lehn, D., & Tedlie, J. (1991). A general approach to testing multifaceted personality constructs. *Journal of Personality and Social Psychology, 61,* 934–945.

Hurley, J. R., & Cattell, R. B. (1962). The Procrustes Program: Producing direct rotation to test a hypothesized factor structure. *Behavioral Science, 7,* 258–262.

James, L. R., Mulaik, S. A., & Brett, J. M. (1982). *Causal analysis.* Beverly Hills, CA: Sage.

Johnson, R. A., & Wichern, D. W. (1982). *Applied multivariate statistical analysis* (2nd ed.). Upper Saddle River, NJ: Prentice Hall.

Johnson, R. A., & Wichern, D. W. (1988). *Applied multivaritate statistical analysis.* Englewood Cliffs, NJ: Prentice Hall.

Jolliffe, I. T. (1972). Discarding variables in a principal component analysis, I: Artificial data. *Applied Statistics, 21,* 160–173.

Jöreskog, K. G. (1967). Some contributions to maximum likelihood factor analysis. *Psychometrika, 32,* 443–482.

Jöreskog, K. G. (1969). A general approach to confirmatory maximum likelihood factor analysis. *Psychometrika, 34,* 183–202.

Jöreskog, K. G. (1979). Author's addendum to: A general approach to confirmatory maximum likelihood factor analysis. In K. G. Jöreskog & D. Sörbom (Eds.), *Advances in factor analysis and structural equation models* (pp. 40–43). Cambridge, MA: Abt Books.

Jöreskog, K. G., & Sörbom, D. (1981). *LISREL V: Analysis of linear structural relationships by maximum likelihood, instrumental variables and least squares methods.* Mooresville, IN: Scientific Software, Inc.

Jöreskog, K. G., & Sörbom, D. (1989). *LISREL 7: A guide to the program and applications* (2nd ed.). Chicago: SPSS Inc.

Jöreskog, K. G., & Sörbom, D. (1993). *LISREL 8 user's reference guide.* Chicago: Scientific Software International, Inc.

Judge, T. A., & Watanabe, S. (1993). Another look at the job satisfaction–life satisfaction relationship. *Journal of Applied Psychology, 78,* 939–948.

Kaiser, H. F. (1958). The varimax criterion for analytic rotation in factor analysis. *Psychometrika*, *23*, 187–200.

Kaiser, H. F. (1960). The application of electronic computer to factor analysis. *Educational and Psychological Measurement*, *20*, 141–151.

Kelly, T. L. (1935). Essential traits of mental life. *Harvard Studies in Education*, *26*. Cambridge, MA: Harvard University Press.

Kenny, D. A. (1974). A test of a vanishing tetrad: The second canonical correlation equals zero. *Social Science Research*, *3*, 83–87.

Kenny, D. A. (1979). *Correlation and causality*. New York: Wiley.

Kirk, R. E. (1982). *Experimental design: Procedures for the behavioral sciences*. Monterey, CA: Brooks/Cole.

Lanchenbruch, P. A. (1975). *Discriminant analysis*. New York: Hafner Press.

Lawley, D. N., & Maxwell, A. E. (1963). *Factor analysis as a statistical method*. London: Butterworth.

Lawley, D. N., & Maxwell, A. E. (1971). *Factor analysis as a statistical method*. New York: American Elsevier.

Lee, S.-Y. (1985). Analysis of covariance and correlation structures. *Computational Statistics and Data Analysis*, *2*, 279–295.

Li, C.-C. (1975). *Path analysis: A primer*. Pacific Grove, CA: Boxwood Press.

Linden, M. (1977). A factor analytic study of Olympic decathlon data. *Research Quarterly*, *48*(3), 562–568.

Loehlin, J. C. (1992). *Latent variable models: An introduction to factor, path, and structural analysis* (2nd ed.). Hillsdale, NJ: Lawrence Erlbaum Associates.

Long, J. S. (1983). *Covariance structure models: An introduction to LISREL*. Beverly Hills, CA: Sage.

Lomax, R. G. (1993, April). A beginners guide to structural equation modeling. Paper presented at the Annual Meeting of the American Educational Research Association, Atlanta, Georgia.

MacDonald, P. D. M. (1975). Estimation of finite mixture distributions. In R. P. Gupta (Ed.), *Applied statistics* (pp. 231–245). Amsterdam: North-Holland.

Marcoulides, G. A. (1989a). Measuring computer anxiety: The Computer Anxiety Scale. *Educational and Psychological Measurement*, *37*(4), 733–739.

Marcoulides, G. A. (1989b). Using computer based programs to improve student performance. *Journal of Artificial Inteligence in Education*, *1*(2), 93–101.

Marcoulides, G. A. (1989c). Using statistical packages to perform factor analysis: A comparison of some estimation procedures. *MicroPsych Network*, *4*(2), 42–45.

Marcoulides, G. A. (1990). An alternative method for estimating variance components in generalizability theory. *Psychological Reports*, *66*(2), 379–386.

Marcoulides, G. A. (1995). Structural equation modeling. *Decision Line*, *26*(5), 23–26.

Marcoulides, G. A. (1996). Estimating variance components in generalizability theory: The covariance structure analysis approach. *Structural Equation Modeling*, *4*, 120–125.

Marcoulides, G. A., & Drezner, Z. (1993). A procedure for transforming points in multidimensional space to a two-dimensional representation. *Educational and Psychological Measurement*, *53*(4), 933–940.

Marcoulides, G. A., & Heck, R. H. (1993). Organizational culture and performance: Proposing and testing a model. *Organization Science*, *4*(2), 209–225.

Marcoulides, G. A., Mills, R. B., & Unterbrink, H. (1993). Improving pre-employment screening: Drug testing in the workplace. *Journal of Managerial Issues*, *5*(2), 290–302.

Marcoulides, G. A., & Papadopoulos, D. (1993). LISPATH: A program for generating structural equation path diagrams. *Educational and Psychological Measurement*, *53*(4), 268–271.

Marcoulides, G. A., & Schumacker, R. E. (1996). *Advanced structural equation modeling: Issues and techniques*. Hillsdale, NJ: Lawrence Erlbaum Associates.

McDonald, R. P. (1989). An index of goodness-of-fit based on noncentrality. *Journal of Classification, 6*, 97–103.

McArdle, J. J. (1980). Causal modeling applied to psychonomic systems simulation. *Behavior Research Methods and Instrumentation, 12*, 193–209.

McArdle, J. J., & McDonald, R. P. (1984). Some algebraic properties of the Reticular Action Model. *British Journal of Mathematical and Statistical Psychology, 37*, 59–72.

McCloy, R. A., Campbell, J. P., & Cudeck, R. (1994). A confirmatory test of a model of performance determinants. *Journal of Applied Psychology, 79*, 493–505.

McLachlan, G. J. (1992). *Discriminant analysis and statistical pattern recognition*. New York: Wiley.

McNemar, Q. (1951). The factors in factoring behavior. *Psychometrika, 16*, 353–359.

Morrison, D. F. (1976). *Multivariate statistical methods* (2nd ed.). San Francisco: McGraw Hill.

Morrison, D. F. (1990). *Multivariate statistical methods*. New York: McGraw-Hill.

Mosier, C. I. (1939). Determining a simple structure when loadings for certain tests are known. *Psychometrika, 4*, 149–162.

Mulaik, S. A. (1972). *The foundations of factor analysis*. New York: McGraw Hill.

Mulaik, S. A. (1988). Confirmatory factor analysis. In J. R. Nessrelroade & R. B. Cattell (Eds.), *Handbook of multivariate experimental psychology* (2nd ed., pp. 259–288). New York: Plenum.

Olson, C. L. (1974). Comparative robustness of six tests in multivariate analysis of variance. *Journal of the American Statistical Association, 69*, 894–908.

Olson, C. L. (1976). On choosing a test statistic in multivariate analysis of variance. *Psychological Bulletin, 83*(4), 579–586.

Pearson, K. (1901). The lines of closest fit to a system of points. *Philosophic Magazine, 2*, 559–572.

Pedhazur, E. J., & Schmelkin, L. P. (1991). *Measurement, design, and analysis: An integrated approach*. Hillsdale, NJ: Lawrence Erlbaum Associates.

Rao, C. R. (1952). *Advanced statistical methods in biometric research*. New York: Wiley.

Rao, C. R. (1965). *Linear statistical inference and its applications*. New York: Wiley.

Sadri, G., & Marcoulides, G. A. (1994). The dynamics of occupational stress: Proposing and testing a model. *Research and Practice in Human Resource Management, 2*(1), 1–19.

SAS Institute, Inc. (1979). *SAS user's guide, 1979 edition*. Raleigh, NC: Author.

SAS Institute, Inc. (1989a). *SAS/STAT user's guide, version 6* (4th ed., Vol. 1). Cary, NC: Author.

SAS Institute, Inc. (1989b). *SAS/STAT user's guide, version 6* (4th ed., Vol. 2). Cary, NC: Author.

Scalberg, E., & Doherty, B. (1983). *Validity and reliability testing of leadership questionnaire*. Working paper, Anderson Graduate School of Management, University of California, Los Angeles.

Schaafsma, W., & van Vark, G. N. (1979). Classification and discrimination problems with applications. Part IIa. *Statistica Neerlandica, 33*, 91–126.

Schatzoff, M. (1964). *Exact distributions of Wilks' likelihood ratio criterion and comparisons with competitive test*. Unpublished doctoral dissertation, Harvard University.

Schatzoff, M. (1966). Exact distributions of Wilks' likelihood ratio criterion. *Biometrika, 53*, 347–358.

Schumacker, R. E., & Lomax, R. G. (1996). *A beginner's guide to structural equation modeling*. Mahwah, NJ: Lawrence Erlbaum Associates.

Sclove, L. S. (1987). Application of model-selection criteria to some problems in multivariate analysis. *Psychometrika, 52*, 333–343.

Shavelson, R. J. (1988). *Statistical reasoning for the behavioral sciences*. Boston: Allyn and Bacon.

Sheehan, T. J., & Sanford, K. (1989, April). A structural model of medical student achievement. Paper presented at the Annual Meeting of the American Educational Research Association, San Francisco, California.

Spearman, C. (1904). General intelligence objectively determined and measured. *American Journal of Psychology, 15,* 201–293.

SPSS, Inc. (1990). *SPSS reference guide.* Chicago: Author.

Steiger, J. H. (1989). *Ez-Path: A supplementary module for SYSTAT and SYGRAPH.* Evanston, IL: SYSTAT.

Steiger, J. H. (1990). Structural model evaluation and modification: An interval estimation approach. *Multivariate Behavioral Research, 25,* 173–180.

Stewart, D., & Love, W. (1968). A general canonical correlation index. *Psychological Bulletin, 70,* 160–163.

Tatsuoka, M. M. (1971). *Multivariate analysis: Techniques for educational and psychological research.* New York: John Wiley.

Thurstone, L. L. (1935). *The vectors of mind.* Chicago: University of Chicago Press.

Thurstone, L. L. (1947). *Multiple factor analysis.* Chicago: University of Chicago Press.

Wilks, S. S. (1932). Certain generalizations in the analysis of variance. *Biometrika, 24,* 471–494.

Wilson, E. B., & Hilferty, M. M. (1931). The distribution of chi-square. *Proceedings of the National Academy of Sciences, USA, 17,* 694.

Wolfram, S. (1991). *Mathematica: A system for doing mathematics by computer* (2nd ed.). Reading, MA: Addison-Wesley.

Wright, S. (1921). Correlation and causation. *Journal of Agricultural Research, 20,* 557–585.

# Author Index

**315**

# Subject Index

**319**